Peter Brimblecombe

SCOPE 39

**Evolution of the Global
Biogeochemical Sulphur Cycle**

Scientific Committee on Problems of the Environment

SCOPE

Executive Committee, elected 10 June 1988

Officers

President: Professor F. di Castri, CEPE/CNRS, Centre L. Emberger, Route de Mende, BP 5051, 34033 Montpellier Cedex, France
Vice-President: Academician M. V. Ivanov, Institute of Microbiology, USSR Academy of Sciences, GSP-7 Prospekt 60 letija Oktjabrja 7,117811 Moscow, USSR.
Vice-President: Professor C. R. Krishna Murti, Scientific Commission for Continuing Studies on Effects of Bhopal Gas Leakage on Life Systems, Cabinet Secretariat, 2nd floor, Sardar Patel Bhavan, New Delhi 110001, India.
Treasurer: Doctor T. E. Lovejoy, Smithsonian Institution, Washington, DC 20560, USA.
Secretary-General: Professor J. W. B. Stewart, Saskatchewan Institute of Pedology, University of Saskatchewan, Saskatoon, S7N0W0 Saskatchewan, Canada.

Members

Professor M. O. Andreae (I.U.G.G. representative), Max-Planck-Institut für Chemie, Postfach 3060, D-6500 Mainz, FRG.
Professor M. A. Ayyad, Faculty of Science, Alexandria University, Moharram Bey, Alexandria, Egypt.
Professor R. Herrera (I.U.B.S. representative), Centro de Ecologia y Ciencias Ambientales (IVIC), Carretera Panamericana km. 11, Apartado 21827, Caracas, Venezuela.
Professor M. Kecskés, Department of Microbiology, University of Agricultural Sciences, Pater K. utca 1, 2103 Gödöllö, Hungary.
Professor R. O. Slatyer, School of Biological Sciences, Australian National University, P.O. Box 475, Canberra, ACT 2601, Australia.

SCOPE 39

Evolution of the Global Biogeochemical Sulphur Cycle

Edited by

Peter Brimblecombe
University of East Anglia, Norwich, UK

and

Alla Yu. Lein
USSR Academy of Sciences, Moscow, USSR

*Published on behalf of the
Scientific Committee on Problems of the Environment (SCOPE)
of the
International Council of Scientific Unions (ICSU)
by*
JOHN WILEY & SONS
Chichester · New York · Brisbane · Toronto · Singapore

Wiley Editorial Offices

John Wiley & Sons Ltd, Baffins Lane, Chichester,
West Sussex PO19 1UD, England

John Wiley & Sons Inc., 605 Third Avenue,
New York, NY 10158-0012, USA

Jacaranda Wiley Ltd, G.P.O. Box 859, Brisbane,
Queensland 4001, Australia

John Wiley & Sons (Canada) Ltd, 22 Worcester Road,
Rexdale, Ontario M9W 1L1, Canada

John Wiley & Sons (SEA) Pte Ltd, 37 Jalan Pemimpin #05-04,
Block B, Union Industrial Building, Singapore 2057

Copyright © 1989 by the Scientific Committee on the Problems of the Environment (SCOPE)

All rights reserved.

No part of this book may be reproduced by any means,
or transmitted, or translated into a machine language
without the written permission of the publisher.

Library of Congress Cataloging-in-Publication Data:

Evolution of the global biogeochemical sulphur cycle / edited by Peter
  Brimblecombe and Alla Yu. Lein.
      p. cm. — (SCOPE ; 39)
  "Published on behalf of the Scientific Committee on Problems of
the Environment (SCOPE) of the International Council of Scientific
Unions (ICSU)"
  Bibliography: p.
  Includes index.
  ISBN 0 471 92251 X
    1. Sulphur cycle.   I. Brimblecombe, Peter, 1949–   II. Lein,
Alla Yu.   III. International Council of Scientific Unions.
Scientific Committee on Problems of the Environment.   IV. Series:
SCOPE report ; 39.
QH344.E86   1989   89–5806
574.5′222—dc20   CIP

British Library Cataloguing in Publication Data:

Evolution of the global biogeochemical sulphur cycle.
  1. Sulphur cycle
  I. Brimblecombe, Peter, *1949–*   II. Lein, Alla Yu   III.
International Council of Scientific Unions
Scientific Committee on Problems of the Environment
IV. Series
574.19′218

ISBN 0 471 92251 X

Printed and bound in Great Britain by
Biddles Ltd, Guildford, Surrey

SCOPE 1: Global Environmental Monitoring 1971, 68 pp (out of print)

SCOPE 2: Man-Made Lakes as Modified Ecosystems, 1972, 76 pp (out of print)

SCOPE 3: Global Environmental Monitoring Systems (GEMS): Action Plan for Phase 1, 1973, 132 pp (out of print)

SCOPE 4: Environmental Sciences in Developing Countries, 1974, 72 pp (out of print)

Environment and Development, proceedings of SCOPE/UNEP Symposium on Environmental Sciences in Developing Countries, Nairobi, February 11–23, 1974, 418 pp (out of print)

SCOPE 5: Environmental Impact Assessment: Principles and Procedures, Second Edition, 1979, 208 pp

SCOPE 6: Environmental Pollutants: Selected Analytical Methods, 1975, 277 pp (out of print)

SCOPE 7: Nitrogen, Phosphorus and Sulphur: Global Cycles, 1975, 192 pp (out of print)

SCOPE 8: Risk Assessment of Environmental Hazard, 1978, 132 pp (out of print)

SCOPE 9: Simulation Modelling of Environmental Problems, 1978, 128 pp (out of print)

SCOPE 10: Environmental Issues, 1977, 242 pp (out of print)

SCOPE 11: Shelter Provision in Developing Countries, 1978, 112 pp (out of print)

SCOPE 12: Principles of Ecotoxicology, 1978, 372 pp (out of print)

SCOPE 13: The Global Carbon Cycle, 1979, 491 pp (out of print)

SCOPE 14: Saharan Dust: Mobilization, Transport, Deposition, 1979, 320 pp (out of print)

SCOPE 15: Environmental Risk Assessment, 1980, 176 pp

SCOPE 16: Carbon Cycle Modelling, 1981, 404 pp (out of print)

SCOPE 17: Some Perspectives of the Major Biogeochemical Cycles, 1981, 175 pp (out of print)

SCOPE 18: The Role of Fire in Northern Circumpolar Ecosystems, 1983, 344 pp

SCOPE 19: The Global Biogeochemical Sulphur Cycle, 1983, 495 pp

SCOPE 20: Methods for Assessing the Effects of Chemicals on Reproductive Functions, 1983, 568 pp

SCOPE 21: The Major Biogeochemical Cycles and Their Interactions, 1983, 554 pp (out of print)

SCOPE 22: Effects of Pollutants at the Ecosystem Level, 1984, 443 pp

SCOPE 23: The Role of Terrestrial Vegetation in the Global Carbon Cycle: Measurement by Remote Sensing, 1984, 272 pp

SCOPE 24: Noise Pollution, 1986, 472 pp

SCOPE 25: Appraisal of Tests to Predict the Environmental Behaviour of Chemicals, 1985, 400 pp

SCOPE 26: Methods for Estimating Risks of Chemical Injury: Human and Non-human Biota and Ecosystems, 1985, 712 pp

SCOPE 27: Climate Impact Assessment: Studies of the Interaction of Climate and Society, 1985, 650 pp

SCOPE 28: Environmental Consequences of Nuclear War
Volume I Physical and Atmospheric Effects, 1985, 342 pp
Volume II Ecological and Agricultural Effects, 1985, 562 pp

SCOPE 29: The Greenhouse Effect, Climatic Change, and Ecosystems, 1986, 574 pp

SCOPE 30: Methods for Assessing the Effects of Mixtures of Chemicals, 1987, 928 pp

SCOPE 31: Lead, Mercury, Cadmium and Arsenic in the Enviroment, 1987, 384 pp.

SCOPE 32: Land Transformation in Agriculture, 1987, 384 pp

SCOPE 33: Nitrogen Cycling in Coastal Marine Environments, 1988, 480 pp

SCOPE 34: Practitioner's Handbook on the Modelling of Dynamic Change in Ecosystems, 1988, 176 pp

SCOPE 35: Scales and Global Change: Spatial and Temporal Variability in Biospheric and Geospheric Processes, 1988, 376 pp

SCOPE 36: Acidification in Tropical Countries, 1988, 424 pp

SCOPE 37: Biological Invasions: a Global Perspective, 1989, 528 pp

SCOPE 38:   Ecotoxicology and Climate, 1989, 392 pp

SCOPE 39:   Evolution of the Global Biogeochemical Sulphur Cycle, 1989, 242 pp

Funds to meet SCOPE expenses are provided by contributions from SCOPE National Committees, an annual subvention from ICSU (and through ICSU, from UNESCO), an annual subvention from the French Ministère de l'Environment, contracts with UN Bodies, particularly UNEP, and grants from Foundations and industrial enterprises.

International Council of Scientific Unions (ICSU)
Scientific Committee on Problems of the Environment (SCOPE)

SCOPE is one of a number of committees established by a non-governmental group of scientific organizations, the International Council of Scientific Unions (ICSU). The membership of ICSU includes representatives from 74 National Academies of Science, 20 International Unions and 26 other bodies called Scientific Associates. To cover multidisciplinary activities which include the interests of several unions, ICSU has established 10 scientific committees, of which SCOPE is one. Currently, representatives of 35 member countries and 20 international scientific bodies participate in the work of SCOPE, which directs particular attention to the needs of developing countries. SCOPE was established in 1969 in response to the environmental concerns emerging at that time; ICSU recognized that many of these concerns required scientific inputs spanning several disciplines and ICSU Unions. SCOPE's first task was to prepare a report on Global Environmental Monitoring (SCOPE 1, 1971) for the UN Stockholm Conference on the Human Environment.

The mandate of SCOPE is to assemble, review, and assess the information available on man-made environmental changes and the effects of these changes on man; to assess and evaluate the methodologies of measurement of environmental parameters; to provide an intelligence service on current research; and by the recruitment of the best available scientific information and constructive thinking to establish itself as a corpus of informed advice for the benefit of centres of fundamental research and of organizations and agencies operationally engaged in studies of the environment.

SCOPE is governed by a General Assembly, which meets every three years. Between such meetings its activities are directed by the Executive Committee.

R.E. Munn
Editor-in-Chief
SCOPE Publications

Executive Secretary: V. Plocq

Secretariat: 51 Bld de Montmorency
75016 PARIS

Contents

Preface   xvii

Workshop Participants and Contributors to this Volume   xix

Members of the Scientific Advisory Committee on the SCOPE Project 'The Global Biogeochemical Sulphur Cycie'   xxv

List of Working Group Participants   xxvii

Introduction   xxix
A. Yu. Lein

Part I EVOLUTION OF THE SULPHUR CYCLE THROUGH GEOLOGICAL TIME

Chapter 1 Evolution of the Sulphur Cycle in the Precambrian   3
M. Schidlowski

    1.1 Introduction   3
    1.2 Outlines of the global sulphur cycle   5
    1.3 Evolution of the sulphur cycle   8
    1.4 Origin of the oceanic sulphate reservoir   8
    1.5 Antiquity of dissimilatory sulphate reduction   10
    1.6 Interaction of early oceanic crust and Archaean seawater: possible effects on the isotope geochemistry of marine sulphur   15

Chapter 2 Modelling the Natural Cycle of Sulphur Through Phanerozoic Time   21
W.T. Holser, J.B. Maynard and K.M. Cruikshank

		2.1 Introduction	21
		2.2 Geochemical background	22
		2.2.1 Overview	22
		2.2.2 Sulphate reduction	23
		2.2.3 Deposition of sulphate	24
		2.2.4 Weathering	25
		2.2.5 The question of steady state	25
		2.2.6 Is the exogenic sulphur cycle leaky?	25
		2.2.7 Reservoirs of sulphur and their changes through time	26
		2.2.8 Isotope geochemistry of sulphur	29
		2.2.9 A geochemical record of the sulphur cycle: the isotope age curve	31
		2.2.10 Interconnections of the exogenic sulphur cycle with those of carbon and other elements	32
		2.2.11 Relations of the sulphur cycle to other geological parameters	34
		2.2.12 Events in the sulphur cycle	36
		2.3 Models of the sulphur cycle through time	38
		2.3.1 A simple box model and its equations	38
		2.3.2 Assumptions and results of some previous models	39
		2.3.3 A new model using reservoir inventory data	43
		2.3.4 Comparison of isotope-based and inventory-based models	49
		2.4 Conclusions	52
Chapter 3	**Local and Global Aspects of the Sulphur Isotope Age Curve of Oceanic Sulphate**		**57**
	H. Nielsen		
		3.1 How 'oceanic' are the existing evaporite data?	57
		3.2 A new approach to the sulphur isotope age curve	63
Chapter 4	**Contribution of Endogenous Sulphur to the Global Biogeochemical Cycle in the Geological Past**		**65**
	A. Yu. Lein and M.V. Ivanov		
		4.1 Natural flux of endogenous sulphur in the recent sulphur cycle	65
		4.1.1 Sulphur flux during subaerial volcanism	65
		4.1.2 Flux of endogenous sulphur from submarine volcanism	66
		4.2 Contribution of endogenous sulphur to the global cycles in the geological past	67

Contents

        4.2.1 Distribution of mass of volcanic and sedimentary rocks during the Phanerozoic   67
        4.2.2 The variation of $\delta^{34}S$ in seawater, $\delta^{13}C$ of organic C, the isotopic composition of Sr and changes of the ocean level during the Phanerozoic   69
    4.3 Application of Holser's 'inventory' model in calculations of the endogenous sulphur flux during geological epochs   71
    4.4 Changes in the global sulphur cycle and cataclysms in the geological past   72

Part II   EVOLUTION OF THE SULPHUR CYCLE OVER RECENT MILLENNIA

Chapter 5   Human Influence on the Sulphur Cycle   77
P. Brimblecombe, C. Hammer, H. Rohde, A. Ryaboshapko and C.F. Boutron

    5.1 The cycle today   79
        5.1.1 Introduction   79
        5.1.2 Changes in industrial emission to the atmosphere   83
        5.1.3 Changes in the riverine flux   87
        5.1.4 Changes in aeolian contribution   89
        5.1.5 Changes in emission of reduced sulphur compounds   91
        5.1.6 Changes in chemical processes in the atmosphere   91
    5.2 Records of past changes — anthropogenic and natural   92
        5.2.1 Introduction   92
        5.2.2 Historical analyses   93
        5.2.3 Ice cores   94
        5.2.4 Lake sediments and soils   105
    5.3 Sulphur cycle prior to human activity   107
    5.4 Transport scales and modelling   108
    5.5 Environmental impact of changes in the sulphur cycle   111
    5.6 Future scenarios and directions   113
        5.6.1 Future perspectives   113
        5.6.2 Future work   114

Part III INTERACTIONS OF SULPHUR AND CARBON CYCLES IN SOME MODERN ECOSYSTEMS

Chapter 6 Interaction of Sulphur and Carbon Cycles in Marine Sediments   **125**
M.V. Ivanov, A.Yu. Lein, W.S. Reeburgh and G.W. Skyring

6.1 Quantitative relationships between sulphate reduction and carbon metabolism in marine sediments   125
   6.1.1 Introduction   125
   6.1.2 Organic substrates for the sulphate-reducing bacteria in marine sediments   125
   6.1.3 Methods and variability: primary productivity   127
   6.1.4 Methods and variability: sulphate reduction rates   134
   6.1.5 Quantitative relationships between organic carbon oxidation and sulphate reduction   136
   6.1.6 Temporal relationships between organic synthesis, preservation and sulphate reduction   138
   6.1.7 Concluding remarks   142
6.2 Interaction of the sulphur and carbon cycles in recent marine sediments   143
   6.2.1 Introduction   143
   6.2.2 Primary production and aerobic mineralization of organic matter in the oceans   143
   6.2.3 Estimation of organic matter consumption in anaerobic diagenetic processes   145
   6.2.4 Quantitative estimation of C_{org} consumption during anaerobic diagenesis   147
6.3 Coupling of the carbon and sulphur cycles through anaerobic methane oxidation   149
   6.3.1 Introduction   149
   6.3.2 Geochemical evidence for anaerobic methane oxidation   151
   6.3.3 Diagenetic models   152
   6.3.4 Methane oxidation rate measurements   153
   6.3.5 Stable carbon isotope distributions   153
   6.3.6 Importance of anaerobic methane oxidation in global methane and sulphur cycles   156
   6.3.7 Evidence for a link between anaerobic methane oxidation and sulphate reduction   157
   6.3.8 Possible mechanisms for anaerobic methane oxidation and future work   158

Contents

	6.4 Quantitative evaluation of biogenic methane generation and oxidation in oceanic sediments	159
Chapter 7	**Sulphur Emission and Transformations at Deep Sea Hydrothermal Vents** *H.W. Jannasch*	**181**
	7.1 Introduction	181
	7.2 Characteristics of deep sea vents	181
	7.3 Emissions of sulphur	183
	7.4 Transformations of sulphur	185
Chapter 8	**Interaction of Sulphur and Carbon Cycles in Microbial Mats** *Y. Cohen, V.M. Gorlenko and E.A. Bonch-Osmolovskaya*	**191**
	8.1 Photosynthesis in cyanobacterial mats and its relation to the sulphur cycle	191
	8.1.1 Oxygenic and anoxygenic photosynthesis	192
	8.1.2 The sulphide-rich environment and toxicity	193
	8.1.3 Anoxygenic photosynthesis in *Oscillatoria limnetica* and other cyanobacteria	193
	8.1.4 Sulphide–oxygen fluctuation in cyanobacterial mats	197
	8.1.5 Strategies of oxygenic and anoxygenic photosynthesis in mat-forming cyanobacterial isolates	198
	8.1.6 Coupling of primary production and sulphate reduction in cyanobacterial mats	201
	8.1.7 Fe^{2+}-dependent photosynthesis in benthic cyanobacteria	202
	8.1.8 Regulation of the sulphur cycle in recent cyanobacterial mats	203
	8.1.9 Precambrian cyanobacteria and stromatolites	204
	8.2 Microorganisms of the sulphur cycle and their activity in microbial mats of hot springs	204
	8.2.1 Introduction	204
	8.2.2 Types of hot springs	205
	8.2.3 Microorganisms of the sulphur cycle	207
	8.2.3.1 Cyanobacteria	207
	8.2.3.2 Anoxic phototrophs	208
	8.2.3.3 Non-photosynthetic aerobic bacteria of the sulphur cycle	212
	8.2.3.4 Anaerobic microorganisms of the sulphur cycle	213
	8.2.4 Extreme factors in the formation of microbial mats in hot sulphur springs	214

8.2.4.1 Temperature	214
8.2.4.2 Sulphide concentration	216
8.2.4.3 pH	217
8.2.4.4 Combined effect of salinity and temperature	217
8.2.5 Horizontal and vertical zonality in microbial associations of hot sulphur springs	218
8.2.6 Productivity of microbial associations in different types of springs	220
8.2.7 Anaerobic destruction of organic matter in thermophilic mats	222
8.2.8 Conclusion	229

Index **239**

Preface

The present volume represents the report on the second stage of activity of the SCOPE International Sulphur Unit set up in 1982 according to the decision of the Scientific Advisory Committee on the Biogeochemical Cycles under the chairmanship of Professor J.W.B. Stewart. Upon the completion of the first stage of work and publication of *The Global Biogeochemical Sulphur Cycle* (SCOPE 19, John Wiley and Sons Ltd, Chichester, 1983) it was decided to continue the work on the generalization of the global sulphur cycle data and to concentrate on the two problems relatively insufficiently touched upon in the above-mentioned book.

A conference which considered using stable isotopes in the assessment of natural and anthropogenic sulphur in the environment was held in September, 1983 at the Institute of Biochemistry and Physiology of Microorganisms in Pushchino (Moscow region), the International workshop of SCOPE.

In summer 1984 the Tallinn Botanical Garden of Estonia Academy of Sciences hosted the International workshop on 'Evolution of the Global Biogeochemical Sulphur Cycle'. Thirty scientists of different specialities representing universities and institutes of 13 countries of Europe, Asia, Northern America and Australia and 20 Soviet researchers (see 'Workshop Participants and Contributors to this Volume') took part in the work. The presentations of the participants of the Tallinn workshop and their comments in the working groups form the foundation of this volume.

According to the decision of the Scientific Advisory Committee on the Global Biogeochemical Sulphur Cycle project (see page xxv for a list of the members of this Committee), Dr A. Yu. Lein, Institute of Geochemistry and Analytic Chemistry, USSR Academy of Science, and Dr W. Krumbein, Oldenburg University, West Germany were appointed as editors of the SCOPE report 'Evolution of the Global Biogeochemical Sulphur Cycle'.

Later on at the meeting of the Scientific Advisory Committee on the project, which took place in November, 1985 in Moscow, Dr P. Brimblecombe, University of East Anglia, UK was appointed as the English version editor to replace Dr W. Krumbein. The same meeting confirmed the contents of the book and coordinated the group of authors of its separate chapters and parts. Finally, at the second meeting of the Scientific Consultative Committee (Moscow, November 1986) this volume was recommended for publication as an official SCOPE report.

On behalf of the members of the Scientific Advisory Committee on the SCOPE project 'Global Biogeochemical Sulphur Cycle' I would like to express my gratitude to all the participants of the Tallinn symposium and all the authors of the present volume for the amicable and fruitful work both at the meeting itself and after it. It is necessary to express deep appreciation to the staff of Tallinn Botanical Garden headed by Professor L. Martin and the staff of the Centre of International Projects, USSR UNEP Commission, who together with Tallinn colleagues made our stay in Estonia not only interesting, but also pleasant.

SCOPE Vice-President
Academician M.V. Ivanov

Workshop Participants and Contributors to this Volume

BAJBAKOV, S.N. — UNEP Committee, Moscow, USSR.

BAULD, J. — Baas-Beck Geobiological Laboratory BMR, PO Box 378, Canberra, ACT 2601, Australia.

BONCH-OSMOLOVSKAYA, E.A. — Institute for Microbiology, USSR Academy of Sciences, Prospect 60-letija Oktjabrja 7a, Moscow, 117811, USSR.

BONDAR, V.A. — Institute of Biochemistry and Physiology of Microorganisms, USSR Academy of Sciences, Pushchino, Moscow Region, 142292, USSR.

BOUTRON, C. — Laboratoire de Glaciologie, CNRS, 2 rue Très Cloitres, Grenoble, CEDEX 38031, France.

BRIMBLECOMBE, P. — School of Environmental Sciences, University of East Anglia, Norwich NR4 7TJ, UK.

COHEN, Y. — Steinits Marine Biology Laboratory, PO Box 469, Eilat 88103, Israel.

CRUIKSHANK, K.M. — Department of Earth and Atmospheric Science, Purdue University, West Lafayette, Indiana 47907, USA.

DEXTER-DYER, B. — Biology Department, Boston University, Boston, Massachusetts 02215, USA.

ELIASSEN, A. — Norwegian Meteorological Institute, Box 318, Blindern, Oslo-3, Norway.

Freney, J.	CSIRO, Division of Plant Industry, PO Box 1600, Canberra, ACT 2601, Australia.
Galtchenko, V.F.	Institute for Microbiology, USSR Academy of Sciences, Prospect 60-letiya Oktjabrja, 7a, Moscow, 117811, USSR.
Gorlenko, V.M.	Institute for Microbiology, USSR Academy of Sciences, Prospect 60-letiya Oktjabrja, 7a, Moscow, 117811, USSR.
Goodman, G.	The Beiger Institute, International Institute for Energy and Human Ecology, The Royal Swedish Academy of Sciences, Box 50005, Stockholm, Sweden.
Gorokhov, V.D.	Institute of Biochemistry and Physiology of Microorganisms, USSR Academy of Sciences, Pushchino, Moscow Region, 142292, USSR.
Grinenko, L.N.	Institute of Geochemistry and Analytical Chemistry, USSR Academy of Sciences, Kosigin Street, 19, Moscow, 117975, USSR.
Grinenko, V.A.	Institute of Geochemistry and Analytical Chemistry, USSR Academy of Sciences, Kosigin Street, 19, Moscow, 117975, USSR
Hammer, C.	Geophysical Isotope Laboratory, Haraldsgade 6, DK 2200, Copenhagen, Denmark.
Holser, W.	Department of Geology, University of Oregon, Eugene, Oregon 97403, USA.
Ivanov, M.V.	Institute for Microbiology, USSR Academy of Sciences, Prospect 60-letiya Oktjabrja, 7a, Moscow, 117811, USSR.
Jannasch, H.	Woods Hole Oceanographic Institution, Woods Hole, Massachusetts 02543, USA.

KONDRATIEV, K. YA. Institute of Limnology,
USSR Academy of Sciences,
Petrovskaja naberejnaja,
Leningrad, 197046, USSR.

KROUSE, R.N. Physics Department,
University of Calgary,
Calgary, Alberta T2N 1N4, Canada.

KRUMBEIN, W. Geomicrobiology Division,
University of Oldenburg, PO Box 2503,
D-2900, Oldenburg, West Germany.

LEIN, A. YU. Institute of Geochemistry and Analytical Chemistry,
USSR Academy of Sciences,
Kosigin Street, 19,
Moscow, 117975, USSR.

MARTIN, YU. L. Botanical Garden of Estonia SSR Academy of Sciences,
Klostrimetsa Street, 44,
Tallinn, USSR.

MAYNARD, J. Geology Department,
University of Cincinnati,
Cincinnati, Ohio 45221, USA.

NIELSEN, H. Institute of Geochemistry,
Central Isotope Laboratory,
Göttingen University,
D-3400 Göttingen, West Germany.

NRIAGU, J. Canadian Center for Inland Water,
PO Box 5050, Burlington, Ontario L7R LA6, Canada.

PACES, T. Geological Survey, Malostranske nam. 19,
11821 Praha 1, Czechoslovakia.

PLOCQ, V. SCOPE Secretariat,
51 Bld de Montmorency
F–75016 Paris, France.

PUNING, YA. M. Estonian Institute of Geochemistry
SSR Academy of Sciences,
Boulevard Estonia, 7,
Tallinn, USSR.

REEBURGH, W. Institute of Marine Science, University of Alaska,
Fairbanks, Alaska 99701, USA.

REES, E.	Geology Department, McMaster University, Hamilton, Ontario L8S 4L8, Canada.
RICHNOV, H.	GPI, Hamburg University, Bundestrasse 55, D-2000, Hamburg, West Germany.
RIVKINA, E.M.	Institute of Biochemistry and Physiology of Microorganisms, USSR Academy of Sciences, Pushchino, Moscow Region, 142292, USSR.
RODHE, H.	Department of Meteorology, University of Stockholm, Arrhenius Laboratory, S–10691, Stockholm, Sweden.
ROSANOV, A.G.	Institute of Oceanology, USSR Academy of Sciences, Krasikov Street, 23, Moscow, 117218, USSR.
RYABOSHAPKO, A.G.	Institute of Applied Geophysics, Glebovskaja Street, 20b, Moscow, USSR.
SAMARKIN, V.A.	Institute of Biochemistry and Physiology of Microorganisms, USSR Academy of Sciences, Pushchino, Moscow Region, 142292, USSR.
SCHIDLOWSKI, M.	Air Chemistry Department, Paleoatmosphere Research Group, Max-Planck-Institut, Postfach 3060, D-6500 Mainz, West Germany.
SKYRING, G.	Baas-Becking Geobiological Laboratory BMR, PO Box 378, Canberra, ACT 2601, Australia.
STAL, L.	Geomicrobiology Division, University of Oldenburg, PO Box 2503, D-2900 Oldenburg, West Germany.
STEWART, J.B.W.	Department of Soil Science, University of Saskatchewan, Saskatoon, Saskatchewan S7N 0W0, Canada.

SUBRAMANIAN, V.	School of Environmental Sciences, J. Nehru University, New Delhi 10067, India.
TRUDINGER, PH.	Baas-Becking Geobiological Laboratory, PO Box 378, Canberra, ACT 2601, Australia.
TRÜPER, H.	Institute of Microbiology, University of Bonn, D-5300 Bonn, West Germany.
VARHELYI, G.	Institute for Atmospheric Physics, PO Box 39, 1675 Budapest, Hungary.
WAINSTEIN, M.B.	Institute of Biochemistry and Physiology of Microorganisms, USSR Academy of Sciences, Pushchino, Moscow Region, 142292, USSR.
VOLKOV, I.I.	Institute of Oceanology, USSR Academy of Sciences, Krasikov Street, 23, Moscow, 117218, USSR.
ZAWARZIN, G.A.	Institute for Microbiology, USSR Academy of Sciences, Prospect 60-letjya Oktjabrja, 7a, Moscow, 117811, USSR.
ZYAKUN, A.M.	Institute of Biochemistry and Physiology of Microorganisms, USSR Academy of Sciences, Pushchino, Moscow Region, 142292, USSR.

Members of the Scientific Advisory Committee on the SCOPE Project 'The Global Biogeochemical Sulphur Cycle'

GORLENKO, V.M.	Institute for Microbiology, USSR Academy of Sciences, Prospect 60-letiya Oktjabrja, 7a, Moscow, 117811, USSR.
HOWARTH, K.W.	Section of Biology and Systematics, Cornell University, Ithaca, New York 14853–270, USA.
IVANOV, M.V.	Institute for Microbiology, USSR Academy of Sciences, Prospect 60-letija Oktjabrja, 7a, Moscow, 117811, USSR.
JAESCHKE, W.	Center of Environmental Protection, University of Frankfurt am Main, Frankfurt am Main, West Germany.
KROUSE, R.H.	Physics Department, University of Calgary, Calgary, Alberta T2N 1N4, Canada.
RYABOSHAPKO, A.G.	Institute of Applied Geophysics, Glebovskaja Street, 20b, Moscow, USSR.
STEWART, J.W.B.	Department of Soil Science, University of Saskatchewan, Saskatoon, Saskatchewan S7N 0W0, Canada.
TRUDINGER, PH.	Baas-Becking Geobiological Laboratory, PO Box 378, Canberra, ACT 2601, Australia.

List of Working Group Participants

I. Working group 'Sulphur cycle evolution on a geological time scale'.
Co-chairmen: A.Yu. Lein
M. Schidlowski

Reporter: W. Holser

Members: L.N. Grinenko
V.A. Grinenko
M.V. Ivanov
R.N. Krouse
J. Maynard
H. Nielsen
T. Paces
E. Rees
H. Richnov
E.M. Rivkina
V.A. Samarkin
P. Trudinger
A.M. Zyakun

II. Working group 'Preindustrial sulphur cycle'.
Co-chairmen: L. Martin
H. Rodhe

Reporters: A.G. Chertov
C. Hammer

Members: P. Brimblecombe
C. Boutron
A. Eliassen
J. Freney
K. Ya. Kondratiev
T. Paces
V. Plocq
Ya. M. Puning
A.G. Ryaboshapko

J.B.W. Stewart
V. Subramanian
G. Varhelyi

III. Working group 'Interactions between cycles of sulphur, nitrogen and oxygen in ecosystems enriched in sulphur'.

Co-chairmen: Y. Cohen
M.V. Ivanov

Reporters: W. Reeburgh
G. Skyring
G.A. Zawarzin

Members: J. Bauld
V.A. Bondar
B. Dexter-Dyer
J. Freney
V.F. Galtchenko
V.M. Gorlenko
H. Jannasch
W. Krumbein
A. Yu. Lein
J. Nriagu
A.G. Rosanov
L. Stal
J.B.W. Stewart
H. Trüper
I.I. Volkov
M.B. Wainstein
A.M. Zyakun

Introduction

ALLA YU. LEIN

In a decade or so, the human race will enter the 21st century. A major task facing the Earth's growing population will be to overcome the threat of anthropogenic change to the biosphere and human survival. In historical times the largest changes seem to have arisen from the development of agriculture, but now a primary focus of concern is acid rain; the result of perturbed flows of nitrogen and sulphur to the biosphere. The interest of scientists and politicians in the true health of our environment has led to scientific coordinating centres such as the SCOPE/UNEP International Sulphur Unit being established in the 1970s. The objectives of such units are to investigate the global biogeochemical cycles of essential elements.

The first monograph prepared by the sulphur unit was published as a SCOPE report in 1983 and titled *The Global Biogeochemical Sulphur Cycle*. This report was devoted to constructing a simulation model for the modern sulphur cycle. It also included an important body of experimental material on sulphur masses in the lithosphere in Phanerozoic times. A second monograph *Stable Isotopes in the Assessment of Natural and Anthropogenic Sulphur in the Environment* attempts, as indicated by the title, to separate the contributions of natural and anthropogenic sulphur to the global cycle using sulphur isotopic composition in various reservoirs, fluxes and ecosystems. The current volume, *Evolution of the Global Biogeochemical Sulphur Cycle* is the third in this series. The term evolution, deriving from the Latin *evolvere*, to unroll, embodies the notion of development. Evolutionary concepts have become important in the natural sciences since the 19th century. The past decade has witnessed an increased interest in evolutionary processes taking place on a global scale. Space research has eliminated all but the faintest hopes for the existence of life in our solar system apart from the Earth. The uniqueness of life on Earth and the need to assess its tolerance to changes imposed by human activity compels scientists to better understand the changes in the sulphur cycle throughout the entire length of the geological record.

In the modern biosphere substantial variations in sulphur concentration are found in similar ecosystems at different geographical locations. Consequently we also need to understand the variability and distribution of sulphur in the past. This calls for knowledge of the mechanisms which control the transformation of sulphur compounds on the Earth's surface. An overview of long-term geological and geochemical investigations of rocks from the Archaean to the Neogene, including the determination of sulphur concentrations and isotopic composition, has been the object of Part I of this book. Here authors have modelled parts of the global sulphur cycle that have been subject to major changes throughout the Earth's history. The large changes of the past are useful in assessing the ability of the biosphere to cope with the stress currently imposed as a result of human activities.

The evolution of the sulphur cycle began with high temperature volcanic emissions of reduced and sulphur gases to the atmosphere of the Archaean Earth. Chapter 1, written by M. Schidlowski, reviews the blurred record left in Precambrian rocks. These rocks have been attracting increasing attention, because they cover the period in which the Earth's earliest biosphere emerged. In terms of the sulphur cycle it is important to establish the appearance of sulphate-reducing bacteria, whose activities gave rise to an effective large-scale fractionation of sulphur isotopes during its exogenic cycling. Over the last 1.2 Ga the fractionation has acquired a clearly defined pattern. Some key questions arise about the Precambrian Earth: at what moment in Earth history was the sulphur flux split into reducing and oxidizing pathways and is the evolution of the sulphur cycle on the Precambrian Earth the result of microbial activity and is there a correspondence between ancient and modern microbial activity?

Schidlowski gives 2.75 Ga before the present (BP) as the date for the initiation of sulphate reduction based on the isotopic composition of sulphide in the Canadian Shield. However, the record has been so degraded by time that assigning a biological origin to this sulphur cannot be entirely certain. These sulphides may have arisen as the result of hydrothermal activity. Regardless of these arguments, though, the lack of apparently biogenic sulphur in most rocks of the Canadian Shield for the ensuing 0.5 Ga argues against widespread distribution of sulphate reduction at 2.75 Ga BP.

Chapter 1 also raises the question of the interactions between rocks of the mantle and the early ocean in the light of current understanding of the processes that occur at the basalt–seawater interface in the oceanic rift zones (see Chapter 7).

To some extent, Chapter 2 is a continuation of Chapter 1, considering the sulphur fluxes of the Phanerozoic when biological processes controlled the fractionation of sulphur isotopes. The authors of this chapter, Holser, Maynard and Cruikshank, propose a simple multi-box model of exogenic sulphur which derives from the Soviet work of Ronov and his colleagues on the volumes and composition of crustal rock and more recent publications of isotopic data. The model presented is novel in including erosion coefficients, updated reservoir sizes and the flux of exogenic sulphur.

Introduction xxxi

Variation in fluxes are assessed by comparing the changes in isotopic composition of oxidized and reduced sulphur on an isotope age curve. Correct interpretation of the results from the model and more specifically the application to the sulphur cycle requires good estimates for the input parameters, primarily the data concerned with evaporite isotopy in various regions. Quite often such data are not consistent with existing geological interpretations and Chapter 3, written by H. Nielsen, one of the most experienced researchers in the field of sulphur isotopes, examines a number of specific examples of departures from the isotope age curve which are difficult to interpret unequivocally.

Models of exogenic sulphur are usually taken as 'closed' systems and do not account for influx and efflux of sulphur to deep layers of the terrestrial crust and mantle. The model presented in Chapter 2 is no exception. The SCOPE 19 volume, *The Global Biogeochemical Sulphur Cycle*, provided a rather approximate estimate of the flux which withdraws sulphur from the exogenic cycle among marine sediments in the subduction zones of the modern biogeochemical sulphur cycle. This is comparatively small and may be neglected in models. It is not as easy to disregard volcanic and hydrothermal fluxes that supply sulphur to exogenic reservoirs. In Chapter 4, Lein and Ivanov attempt, for the first time, to introduce endogenic sulphur fluxes, modelling the evolution of the sulphur cycle in the Phanerozoic.

Overall Part I presents a picture of the sulphur cycle in the Precambrian and some dynamic models of the cycles from the Cambrian through to the Neogene. The quantitative nature of the models makes it possible to single out unusual periods in the history of the sulphur cycle which appear to be related to evolutionary changes in the biosphere.

The second part of the book is devoted to considering the evolution of the sulphur cycle in the Holocene, i.e. the last ten thousand years. The authors of this part (Chapter 5—Brimblecombe, Hammer, Rodhe, Ryaboshapko and Boutron) are interested in atmospheric aspects of the sulphur cycle and quite naturally concentrate on this component of the exogenic sulphur cycle. The bias is not accidental; the low mass of the atmosphere makes it very sensitive to change. As in *The Global Biogeochemical Sulphur Cycle*, the authors consider the basic fluxes of the global atmospheric balance of sulphur (Section 5.1) and update a few of the 1983 flux estimates. The contribution of anthropogenic emissions to the continental atmosphere is assessed as some 55% in the updated sulphur balance. In Section 5.2 the changes in sulphur fluxes through the Holocene are examined using ice cores from polar regions of the Earth remote from important centres of human population: e.g. glaciers of central Antarctica and the Greenland ice plateau. The background concentrations of sulphur in ice cores ranges from 6 to 35 ng (S) g^{-1}. Prior to the 20th century, variations in sulphur concentration were due to volcanic eruptions. Over the past hundred years the concentration of anthropogenic sulphur in the middle troposphere of the northern hemisphere has been as much as 2–4 times that of the natural background. No such clear increase is found in the Antarctic ice cores.

Section 5.3 attempts to reconstruct the natural sulphur flux of the Holocene, prior to human intervention. The authors infer that the continental and marine atmosphere were balanced with regard to sulphur fluxes in the pre-industrial period. Their scenario for the next 50 years does not anticipate a substantial increase in global anthropogenic emissions of sulphur, but suggests possible shifts in the geographical distribution of anthropologically induced sulphur stress. There may be increases in emissions from tropical and developing countries while industrialized countries of Europe and North America are expected to abate the present level of anthropogenic emissions through control practices.

Throughout the evolutionary history of the biosphere the sulphur cycle has been and remains closely linked with the cycles of other essential biogenic elements, most notably the carbon cycle. The only criteria available for assessing this interrelation through past geological epochs derive from studies of rock composition. In some instances this is illustrated by the correlation between C_{org} and sulphide-sulphur. Another approach uses a comparison of the isotope age curves of sulphate-sulphur and carbonate-carbon (as seen in Chapter 2). The inverse correlation that is observed between the curves for carbon and sulphur can be explained by the compensation that occurs during the reduction processes in the sulphur cycle (i.e. $SO_4^{2-} \rightarrow S^{2-}$) and the oxidative reactions in the carbon cycle (i.e. $C_{org} \rightarrow CO_2 \rightarrow CO_3^{2-}$).

The interrelation of the sulphur and carbon cycles may be clearly demonstrated in modern ecosystems, such as microbial mats of inland water bodies, marshes and marine sediments. Peculiar microbial communities were discovered in deep hydrothermal systems in 1977. Researchers are currently giving a great deal of attention to microbial mats in the volcanic regions of the Earth, as these fascinating ecosystems show how chemosynthetic and photosynthetic processes interact with the sulphur system. Such systems appear to be among the most ancient on the Earth, being present in the earliest biosphere. Thus studies of these special communities are of great interest for those concerned with the interactions between the sulphur and carbon cycles. The extensive experimental work on these systems is reviewed in Part III of the book.

The role of sulphur bacteria, especially the sulphur-reducing bacteria, in the C, N and P cycles has been emphasized during investigations of modern anaerobic marine systems; e.g. organic matter of sediments is oxidized to CO_2 during bacterial reduction of sulphate in marine interstitial water and re-enters the biosphere. Under some conditions this CO_2 is precipitated as isotopically light carbonate minerals—metabolites which are paragenic to pyrite sulphides in sediments. In 1975 Lein and Ivanov proposed a method of estimating C_{org} consumption during anaerobic biogeochemical processes in the early stages of marine diagenesis. Widespread use of labelled compounds (^{14}C and ^{35}S) over the past decade has allowed the C and S interactions in marine sediments to be assessed quantitatively.

In Chapter 6 (Section 6.1) Skyring reviews the data available on sulphate reduction and carbon metabolism in marine sediments, underlining the complicated dependence of quantitative estimates on radio-experimental conditions. He argues that the amount of carbon oxidized on reduction of sulphate is dependent on the specific organic acid in the substrate. This differs from the arguments of Lein and may arise from results obtained with pure cultures of sulphate-reducing bacteria under rather special conditions. Section 6.1 also attempts a mass balance of C_{org} in the global oceanic sediments using the available evidence on the rates of biogeochemical processes. It appears that about 15% of the organic carbon reaching the sea-bottom is mineralized under anaerobic conditions.

During anaerobic mineralization of C_{org}, anaerobic oxidation of methane apparently occurs in parallel to sulphate reduction. This explains the fact that the final parts of the chapter focus on methane production, a topic which might have seemed rather removed from the central theme of the book. In Section 6.3 Reeburgh discusses the evidence he has found for this process and provides a quantitative estimate of the portion of C_{org} consumed. In the concluding section of the chapter Ivanov and Lein examine the interrelation between sulphate reduction, methane production and methane oxidation.

It is only recently that techniques have become available to examine the fluxes and chemistry of sulphur compounds in deep sea hydrothermal vents. H. Jannasch summarizes some of the findings of this research in Chapter 7. As our knowledge of these particular sources increases we should be able to introduce them into models of the evolution of the sulphur cycle. It is probable that hydrothermal activity was particularly important in controlling ocean composition in the very earliest stages of the evolution of the Earth.

Chapter 8 examines the all-important bacterial mats which produce laminated sediments that are evident in sedimentary rocks of great antiquity. Thus it examines, in a different way, some of the areas of interest in the first chapter of the book. Yehuda Cohen describes the role sulphur plays as an electron donor in oxygenic and anoxygenic photosynthesis for these microorganisms which may display a remarkable tolerance to sulphide, which is normally toxic. V.M. Gorlenko and E.A. Bonch-Osmolovskaya discuss microbial mat communities in hot springs, paying attention to their importance in the sulphur cycle. They also examine the importance of environmental variables, such as salinity, temperature and pH in influencing these communities. Their work provides some estimates of productivity and the intensity of sulphate reduction in these communities. While the processes seen in these rather specialized ecosystems do not contribute significantly to the overall fluxes on the Earth of the present day, they do give important information about the biogeochemistry of sulphur on the ancient Earth.

The papers in this volume in no way complete the story of the evolution of the sulphur cycle: rather they map out areas that continue to yield valuable results. It is hoped that they will not only foster these research areas, but also help initiate investigations into topics that have been mentioned only in passing.

The editors would like to thank Jean Stokes for drawing the figures for this volume, and Julian Briggs for his work on the earlier manuscripts.

Part I

Evolution of the Sulphur Cycle through Geological Time

Evolution of the Global Biogeochemical Sulphur Cycle
Edited by P. Brimblecombe and A. Yu. Lein
© 1989 SCOPE Published by John Wiley & Sons Ltd

CHAPTER 1
Evolution of the Sulphur Cycle in the Precambrian

M. SCHIDLOWSKI

1.1 INTRODUCTION

The geochemical cycle of sulphur is one of those elemental cycles that appear to be basically controlled by life processes, as has been been confirmed by numerous investigators (for overviews see Goldhaber and Kaplan, 1974; Claypool *et al.*, 1980; Trudinger, 1979; Migdisov *et al.*, 1983; Berner, 1984). There is little doubt that this has been so over most of the Earth's history (Schidlowski, 1979, 1983a). As such, the sulphur cycle constitutes one of the most striking examples of the impact the biosphere has exerted on the chemistry of the Earth's crust. Continuous biological processing of sulphur in the exogenic reservoir is ultimately responsible for dividing its flux from the surface environment to the rock section of the cycle between an organic and an inorganic sink. As a result of such splitting, the sedimentary sulphur reservoir is typically bipartite, consisting of partial repositories of biogenic sulphide and inorganically precipitated sulphate (Figure 1.1).

As one of the key elements of life that accounts for some 0.5 to 1.5% (dry weight) of biological matter, sulphur is cycled through the biosphere in many ways, notably during synthesis and decomposition of proteins. However, the largest turnover rates in the biological sulphur cycle are not achieved in the synthesis of sulphur-containing cell constituents (by assimilatory reduction of inorganic sulphate), but in an energy-releasing dissimilatory process that couples the reduction of sulphate with the oxidation of organic substrates, i.e.

$$2CH_2O + SO_4^{2-} \rightarrow 2HCO_3^- + H_2S \qquad (1)$$

With SO_4^{2-} rather than free oxygen serving here as an electron acceptor, this process of *dissimilatory sulphate reduction* may be regarded as a form of anaerobic respiration (sulphate respiration). This process is utilized by several genera of strictly anaerobic sulphate-reducing bacteria (viz. *Desulfovibrio, Desulfomonas, Desulfotomaculum*). They are particularly

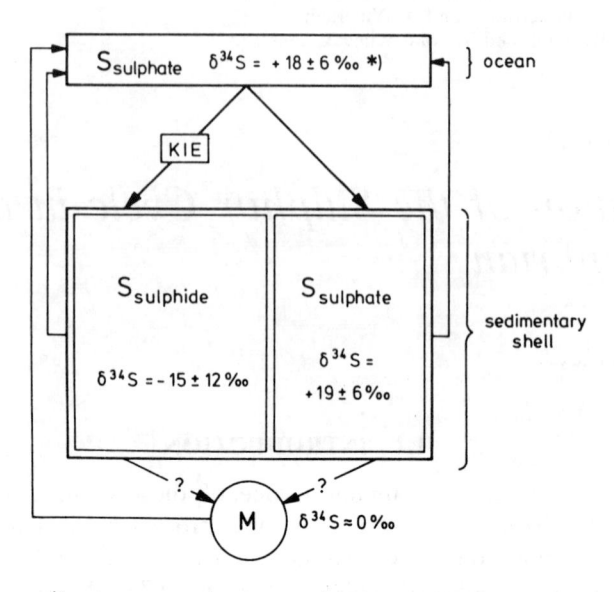

Figure 1.1 Box model of the global sulphur cycle showing partitioning of mantle (M)-derived sulphur between sulphide and sulphate [in sum, the reservoir triplet stores a total of 7.9×10^{21} g S (Migdisov et al., 1983)]. The kinetic isotope effect (KIE) inherent in bacterial sulphate reduction has been propagated from the exogenic exchange reservoir (hydrosphere, atmosphere) into the rock section of the cycle. This effect entails an average fractionation of 30–35‰ and is ultimately responsible for the isotopic disproportionation of terrestrial sulphur into a 'light' reduced and a 'heavy' oxidized phase (while bacteriogenic sulphide preferentially concentrates the light isotope (^{32}S), the non-biogenic phase (sulphate) displays a relative enrichment in heavy sulphur (^{34}S)). Note that sulphate evaporites stored in the sedimentary shell preserve the isotopic composition of their dissolved marine precursor with a minor shift of about +1‰. Empirically recorded fractionations between sulphide and sulphate from sedimentary rocks are often smaller than the range of ~ 35‰ chosen for this model; accordingly, some authors (Holser and Kaplan, 1966; Migdisov et al., 1983) prefer means between −12 and −9‰ for sedimentary sulphide

active in the anoxic regions just below the sediment–water interface, especially in near-coastal marine environments. The activities of these bacterial sulphate reducers are of paramount geological importance as a means of bringing about an important and uniquely biological low-temperature reduction of marine sulphate with a concomitant release of H_2S. Substantial portions of the H_2S generated in this process react with detrital iron minerals and thus end up as iron sulphide:

$$2FeOOH + 6CH_2O + 3SO_4^{2-} \rightarrow 6HCO_3^- + FeS + FeS_2 + 4H_2O \quad (2)$$

This sulphide is incorporated in newly formed sediments (preferentially in the form of pyrite) and consequently relegated to the rock section of the sulphur cycle.

Because sulphate can only be reduced by biological processes at environmental temperatures, the persistence of large quantities of reduced sulphur on the Earth's surface and finally the entire 'exogenous' sulphur cycle, depend on the activity of sulphate-reducing organisms. Annually they supply some 10^{13} to 10^{14} g sulphur to the reservoir of sedimentary sulphide (Berner, 1982; Ivanov, 1983) where it is apt to reside with a half-life of about 0.6 Ga. A still larger quantity of reduced sulphur compounds may escape to the surface waters and the atmosphere where they undergo rapid reoxidation in these oxygenated environments. The total turnover of sulphur in dissimilatory sulphate reduction is certainly much higher than in the assimilatory process that cycles sulphur through the Earth's standing biomass at a rate of $\sim 10^{14}$ g S per year. Moreover, as the sulphur content of the Earth's biomass (and notably of its protein fraction) is caught up in the rapid cycles of production and mineralization, sulphur is maintained at a steady state concentration of about one percent of the standing biomass, but does not feed major geochemical reservoirs. In contrast, the energy-yielding dissimilatory process has an impressive reservoir-generating capacity as is testified by the huge reservoir of sedimentary sulphide (Figure 1.1).

1.2 OUTLINES OF THE GLOBAL SULPHUR CYCLE

Figure 1.1 gives a simplified approach to the sulphur cycle. All the sulphur originally discharged from the Earth's mantle in the form of volatile compounds (mainly as SO_2 and H_2S) has now been condensed as sulphate and metal sulphides, namely (i) as dissolved sulphate in the ocean, (ii) as gypsum, anhydrite, etc., in evaporate beds and (iii) as pyrite in the sedimentary column. These three reservoirs account for 7.9×10^{21} g(S) (Migdisov et al., 1983) and therefore exceed by many orders of magnitude the sulphur content of the other reservoirs (standing biomass: $\sim 10^{16}$ g(S); atmospheric sulphate aerosol $\sim 10^{12}$ g(S); atmospheric SO_2: $\sim 10^{11}$ g(S); atmospheric H_2S: $\sim 10^{11}$ g(S). With the exception of evaporite deposition, all the transfers from one reservoir to another are (at least partly) mediated by life processes.

A typical reaction of H_2S oxidation to SO_4^{2-} in the course of the photosynthetic production of cell material follows the scheme:

$$H_2S + 2CO_2 + 2H_2O \xrightarrow{h\nu} 2CH_2O + 2H^+ + SO_4^{2-} \qquad (3)$$

Others are compiled in Chapter 2.

While the evaporite minerals, gypsum and anhydrite, are only slightly enriched in ^{34}S ($\cong 1.3‰$) with respect to the dissolved sulphate (Thode and

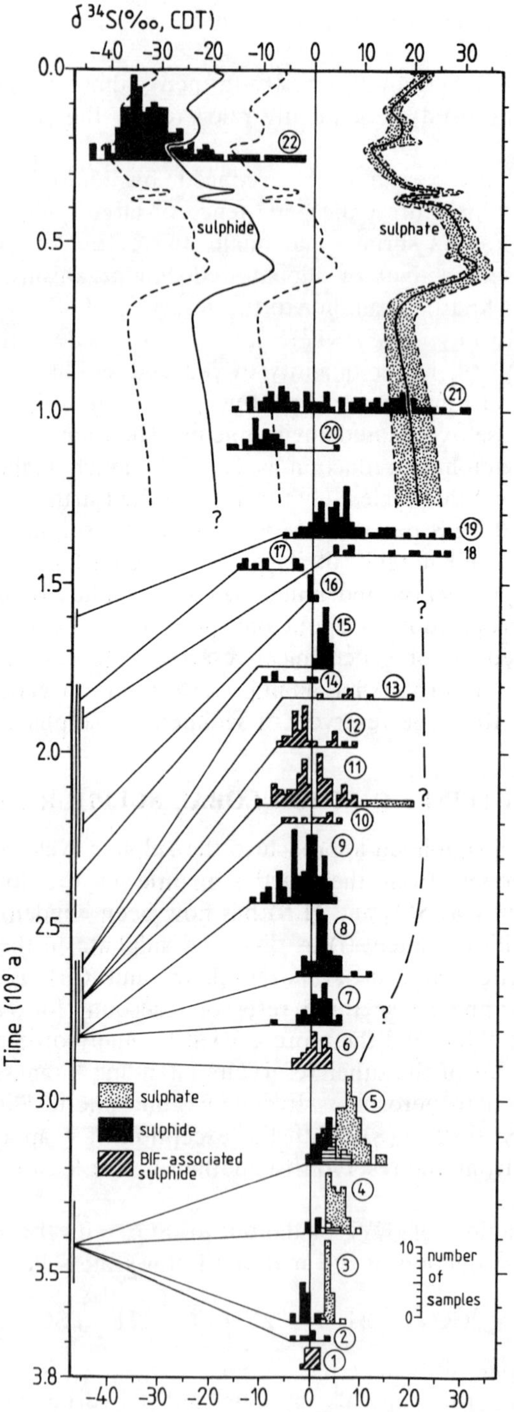
δ ³⁴S(‰, CDT)
-40 -30 -20 -10 0 10 20 30
Time (10⁹ a)
0.0 0.5 1.0 1.5 2.0 2.5 3.0 3.5 3.8
sulphide
sulphate
sulphate
sulphide
BIF-associated sulphide
number of samples
10
0
?
18

Monster, 1965), the bacteriogenic sulphides commonly display a distinct shift of $\delta^{34}S$ values towards negative readings compared with the parent sulphate source, the balance of heavy sulphur (^{34}S) accumulating in the residual seawater sulphate. The isotope effect (Figure 1.1) thus imposed on the flux of biogenic sulphide out of the surficial compartment is also responsible for the positive $\delta^{34}S$ values of seawater sulphate. Marine sulphate is rather uniform in its isotopic composition ($\delta^{34}S$ values between +20 and +21‰) in contemporary seawater (Rees et al., 1978) while the $\delta^{34}S$ values of bacteriogenic sulphide in marine sediments cover an extremely wide range. The variations in the marine $\delta^{34}S$ through geological history are discussed in more detail in Chapter 2. A reproduction of this 'age curve of marine sulphate' through the past 1.3 Ga is shown in Figure 1.2

During the Precambrian there is hardly any $CaSO_4$ left in the record and sulphur in the oldest 'evaporites' is represented solely by barites which are assumed to have replaced the original sulphate minerals. The record of $\delta^{34}S$ values of bacteriogenic sulphides inserted in Figure 1.2 goes further back in time but the information conveyed by these values is decidedly less specific due to the intrinsic scatter of bacteriogenic fractionation patterns (see, *inter alia*, Goldhaber and Kaplan, 1974; Schidlowski et al., 1983).

The cycle model shown in Figure 1.1 also considers the input of endogenic sulphur from volcanoes and from along thermally active portions of newly generated oceanic crust.

In the modern cycle where the influence of biogenic processes dominates, the contribution of these sources to the cycle is not discernible, but in the earliest period of the Earth's history, seawater chemistry and its isotopic record must have been largely mantle-dominated (Veizer et al., 1982). We

---

Figure 1.2 Isotope age functions of sedimentary sulphide and sulphate with special reference to the Precambrian record (from Schidlowski, 1983a, with minor additions; for references see also this paper). (1) Banded iron-formation from Isua, West Greenland. (2) Onverwacht Group, South Africa. (3) Warrawoona Group of Pilbara Block, Western Australia. (4) Fig Tree Group, South Africa. (5) Iengra Series of Aldan Shield, Siberia. (6) Banded iron-formation from Rhodesian schist belts, Zimbabwe. (7) Black shales from Yilgarn Block, Western Australia. (8) Deer Lake Greenstone Belt, Minnesota, USA. (9) Birch–Uchi Greenstone Belt, Canadian Shield. (10) Fortescue Group of Hamersley Basin, Australia. (11) and (12) Michipicoten and Woman River banded iron-formations, Canadian Shield. (13) Steep Rock Lake Series, Canadian Shield. (14) Ventersdorp Supergroup, South Africa. (15) Cahill Formation of Pine Creek geosyncline, Australia. (16) Frood Series of Sudbury District, Canadian Shield. (17) Black shales from Outokumpu, Finland. (18) Onwatin Slate of Sudbury Basin, Canadian Shield. (19) Sediments of McArthur Basin, Australia. (20) Adirondack sedimentary sulphides, Canadian Shield. (21) Nonesuch Shale, Canadian Shield. (22) Permian Kupferschiefer, central Europe. CDT = Canyon Diablo troilite furnishes the isotope ratio $^{32}S/^{34}S = 22.22$ that defines the zero permil point on the $\delta^{34}S$ scale ('CDT standard'); BIF = banded iron formation

shall come back to this in Section 1.6, and a more detailed treatment of this endogenic branch of the sulphur cycle will be given in Chapter 4 of this volume.

1.3 EVOLUTION OF THE SULPHUR CYCLE

Key biological processes in the global sulphur cycle are primarily (i) the oxidation of primordial (and recycled) sulphur compounds of lower oxidation states in the exogenic exchange reservoir (hydrosphere, atmosphere), and (ii) the dissimilatory reduction of oxidized sulphur (notably sulphate) by microbial sulphate reducers. If it were possible to find an indication of the onset of these processes in the geological record, this might provide us with important data on biological evolution.

The classical chemical methods fail to distinguish between a biologically mediated and an inorganic pedigree of the products of the principal sulphur-transforming processes. Geological and palaeontological observations may be helpful, but finally the distinction can only be achieved by decoding the isotopic signature notably of the sulphide minerals preserved in the record. This method utilizes the sizable isotope fractionation effects of bacterial processes. Hence, the isotopic approach plays a crucial role and provides the most promising tool for elucidating the history of the sulphur cycle. However, this approach must be used cautiously because inorganic processes in hydrothermal environments may result in quite large fractionation effects. For the basic principles of these processes see Ohmoto (1972) and Ohmoto and Rye (1979).

1.4 ORIGIN OF THE OCEANIC SULPHATE RESERVOIR

There is little doubt that the sulphate burden of the contemporary ocean (equivalent to 1.3×10^{21} g(S)) owes its origin to the oxidation of reduced sulphur compounds in the present weathering cycle which has probably been so over the last 2 Ga of geological history. Being particularly prone to preferential recycling, marine sulphate evaporites largely fade from the record prior to ~ 1. 3Ga ago, but rare older occurrences (Perry et al., 1971; Vinogradov et al., 1976; Heinrichs and Reimer, 1977; Lambert et al., 1978; Cameron, 1983) as well as ample indirect evidence (Button, 1976; Crick and Muir, 1980; Lowe and Knauth, 1977; Barley et al., 1979; and others) indicate that sulphate had been continuously present in the world oceans as from at least 3.5 Ga ago.

The sulphate in Archaean seas could have been readily produced by sulphur-oxidizing microbiota and does not necessarily imply contemporaneous atmospheric oxygen and oxidation weathering as some claim (e.g. Vinogradov et al., 1976). Microbial oxidation of reduced sulphur is linked to the electron supply for either photosynthetic or chemosynthetic activity by selected group of prokaryotes (Broda, 1975; Trüper, 1982). With bacterial (anoxygenic) photosynthesis requiring electron donors other than water, sulphur compounds at oxidation states lower than sulphate (H_2S, S^0, $S_2O_3^{2-}$) constitute convenient sources of reducing power for sulphur-oxidizing bacteria like the Chlorobiaceae and Chromatiaceae that have biosynthetic pathways such as

$$2H_2S + CO_2 \xrightarrow{h\nu} CH_2O + H_2O + 2S \qquad (4)$$

$$2S + 3CO_2 + 5H_2O \xrightarrow{h\nu} 3CH_2O + 4H^+ + 2SO_4^{2-} \qquad (5)$$

and

$$2Na_2S_2O_3 + 4CO_2 + 6H_2O \xrightarrow{h\nu} 4CH_2O + 4Na^+ + 4H^+ + 4SO_4^{2-} \qquad (6)$$

It is widely assumed that anoxygenic bacterial photosynthesis preceded the evolution of the oxygen-releasing variant of the photosynthetic process which latter had to overcome the formidable energy barrier posed by the H–O bonds of the water molecule by grafting an additional photosystem on the primary pathway. Accordingly, it is reasonable conjecture that the first photosynthetic oxidation equivalents released to the ancient environment were, in fact, elemental sulphur and oxidized sulphur compounds rather than molecular oxygen, with sulphate as a mild oxidant most probably accumulating in the Precambrian seas long before the buildup of appreciable oxygen levels in the atmosphere (Broda, 1975). Both direct (sedimentary barite, cf. Figure 1.2) and indirect evidence (pseudomorphs after primary gypsum and anhydrite) pertaining to the presence of sulphate evaporites in Archaean sediments are, therefore, consistent with the onset of a biologically mediated sulphur oxidation in the surficial exchange reservoir and the buildup of an oceanic sulphate burden as from at least 3.5 Ga ago (Lambert et al., 1978; Thorpe, 1979). Problems arising for this interpretation from an assumed near-saturation of the Archaean oceans with bivalent iron will be dealt with below.

Prior to the start of biological S-oxidation, the primordial sulphur compounds (H_2S, SO_2) degassed from the Earth's interior were apt to undergo a series of inorganic transformations, with sulphur dioxide dissolving as the sulphite ion (SO_3^{2-}) and the bulk of H_2S most probably reacting with dissolved metal ions (notably ferrous iron) in the oceans. As sulphite is known to undergo spontaneous disproportionation (Hayon et al., 1972), it is likely that some sulphate might have formed inorganically. An efficient way of raising the general oxidation state of exogenic sulphur in a prebiological world is photochemical oxidation with OH radical as oxidizing agent furnished by photodissociation of water vapour (Walker and Brimblecombe, 1985), e.g.

$$H_2O \xrightarrow{h\nu} H^. + OH^. \qquad (7)$$

However, the extent of the inorganic oxidation process would have been constrained by the photochemical OH source, some 10^{11} g a^{-1} (Walker, 1978), with probably little variation over the Earth's history. A definitive assessment of the impact of such a flux of OH radicals on the prebiological sulphur cycle has still to await the refinement of several geochemical background parameters and is, therefore, premature at this stage (notably, its implications for the buildup of the oldest oceanic sulphate reservoir). Since the rise of photosynthetic prokaryotes, inorganic sulphur oxidation was likely to be completely outrun by the biologically mediated processes.

1.5 ANTIQUITY OF DISSIMILATORY SULPHATE REDUCTION

A second crucial impact on the global sulphur cycle is exercised by bacterial (dissimilatory) sulphate reduction. As emphasized by Peck (1974), a combination of photosynthetic oxidation and dissimilatory reduction of sulphur would give rise to a closed biological sulphur cycle powered by radiant energy in which the dissimilatory process may be conceived as an adaptive reversal of a primary photosynthetic pathway relying on H$_2$S as electron source. An abstracted version of these cyclic transformations could be written as:

$$H_2S + 2HCO_3^- \underset{\text{dissimilation}}{\overset{\text{photosynthesis}}{\rightleftharpoons}} 2CH_2O + SO_4^{2-} \qquad (8)$$

Constituting a pivotal link in the global sulphur cycle, bacterial sulphate reduction primarily causes (i) a large-scale low-temperature reduction of oxidized sulphur at or near the Earth's surface that entails a sizable isotope shift, and (ii) a concomitant sulphur flux into the sedimentary sulphide reservoir. As such, the dissimilatory process is ultimately responsible for the global partitioning and isotopic disproportionation of terrestrial sulphur (Figure 1.1) between two major crustal reservoirs: sedimentary sulphide with a marked preponderance of isotopically light sulphur (^{32}S) and sedimentary sulphate with a relative enrichment in the heavy species (^{34}S).

The emergence of dissimilatory reduction of oxidized sulphur would have been a prerequisite for an extension of biological control to the lithospheric (crustal) section of the cycle and thus mark the onset of a fully fledged 'modern' sulphur cycle of the type depicted in Figure 1.1. This event and the underlying biochemical process would be coupled with a discrete step in prokaryotic evolution, namely, the rise of microbiota that made oxidized sulphur compounds serve their specific bioenergetic needs. Considering the geological age of prokaryotes, and accepting that photometabolism of H$_2$S has preceded that of H$_2$O (Broda, 1975; Trüper, 1982), this would *a priori* argue for a considerable antiquity of dissimilatory sulphate reduction.

Sulphate-reducing bacteria can be traced back in the record by their characteristic isotopic patterns preserved by pyrites deposited within their original sedimentary habitats. The prototype of a bacteriogenic sulphide pattern is exemplified by the frequency distribution of $\delta^{34}S$ values of the Permian Kupferschiefer (Figure 1.2), a fossil euxinic facies for which an impressive database is available (Marowsky, 1969). Typical features of such patterns are (i) a wholesale shift of $\delta^{34}S$ towards negative values as compared to the parent seawater sulphate, and (ii) a formidable scatter of the values, with an extended tailing of the frequency diagram towards the isotopic composition of the marine sulphate source (indicating that part of the fractionation had been achieved in a Rayleigh process operating in semiclosed compartments of the depositional basin).

Figure 1.2 compiles data which help us to trace back an 'index of biogenicity' in the sulphur isotope record. In the ideal case, isotopic inferences should be based on the analysis of coexisting (or at least coeval) sedimentary sulphide and sulphate, but in the Precambrian record gypsum and anhydrite are lacking due to their appreciable solubility. Sulphate occurrences older than 1.3 Ga are in the form of the much less soluble barite ($BaSO_4$); therefore an evaluation of the oldest record has to rely mainly on the isotope patterns of sedimentary sulphides.

Unfortunately, isotope data of Precambrian sulphides are still limited in number and scope which necessarily detracts from the confidence level of the oldest record. Figure 1.2 shows virtually the complete database currently available for $t > 1$ Ga, excluding only detrital and massive stratiform sulphides as well as those with ostensible hydrothermal overprints. The $\delta^{34}S$ patterns of the Proterozoic record clearly display a marked shift to the negative side of the δ-scale and distributional characteristics that suggest a bacteriogenic origin. These indicators progressively fade as we proceed beyond the Archaean–Proterozoic boundary. The oldest isotope patterns still meeting the essential criteria of biogenicity are those shown by Nos. 8, 11 and 12 of Figure 1.2, all of which seem to be bracketed by 2.6 Ga $< t <$ 2.8 Ga. The earlier record—and notably the oldest sedimentary sulphides from the Isua metasedimentary series of West Greenland—definitely lack biogenic features. If we were to rely on isotopic data as the principal evidence, the time interval 2.6–2.8 Ga would, accordingly, define the time of emergence of sulphate respirers.

A proper evaluation of the presumably oldest bacteriogenic isotope patterns should also address the question whether the distributional characteristics observed could have been obtained through inorganic (magmatic and hydrothermal) processes capable of isotopically differentiating a 'primitive' isotope mixture with an overall $\delta^{34}S$ value close to 0‰. In principle, such differentiation may readily occur in high-temperature systems when sulphur is partitioned among several S-bearing minerals as this is apt

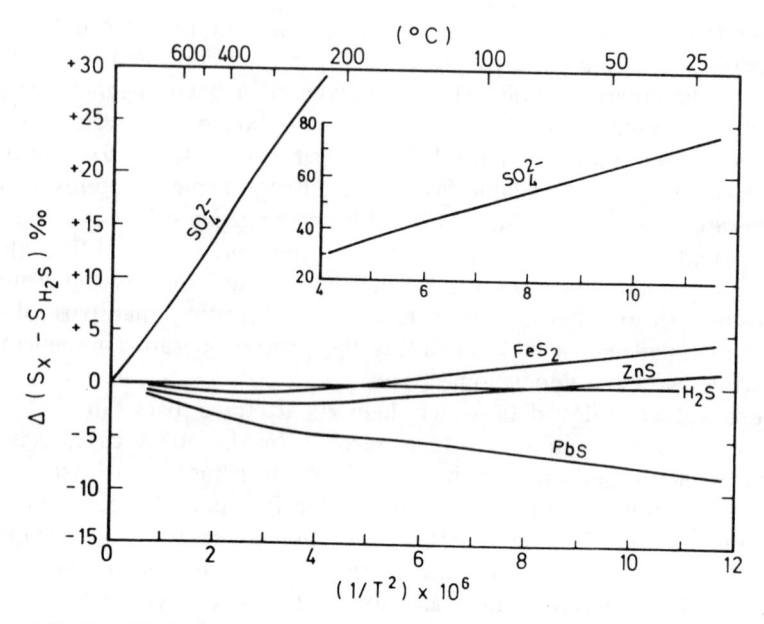

Figure 1.3 Equilibrium fractionation of sulphur isotopes between geochemically important sulphur-bearing phases (x) and hydrogen sulphide as a function of temperature [$\Delta(S_x - S_{H_2S}) = \delta^{34}S_x - \delta^{34}_{H_2S}$]. Note that isotopic differences among sulphides are generally moderate while there is very strong fractionation between sulphide and sulphate. After Ohmoto (1972)

to bring equilibrium chemistry into play. However, equilibrium fractionations among sulphide phases are relatively small (Figure 1.3) and would bring about only minor isotope effects. The only way to markedly differentiate a 'primitive' sulphur broth by purely inorganic means would be by introducing into such a system an oxidized phase which, by virtue of an inherent strength of the S–O bond, would preferentially sequester heavy sulphur (^{34}S) (note the pronounced difference in δ^{34}S between SO_4^{2-} and various sulphide phases in Figure 1.3). For instance, an increase in the relative proportion of sulphate in a hydrothermal system with an overall δ^{34}S = 0‰ would displace the δ^{34}S values of coexisting sulphides progressively towards negative readings, ultimately mimicking a bacteriogenic distribution pattern (Figure 1.4). However, as may be further inferred from Figure 1.4, the resulting δ^{34}S values would be predominantly negative. It is difficult to produce isotopically heavy sulphide from a source with δ^{34}S = 0‰ in high-temperature systems.

Thus some of the late Archaean sulphide patterns can be best interpreted in terms of bacterial reduction of an isotopically heavy pre-existing oxidized sulphur phase (Figure 1.2). (Additionally, relevant field evidence largely precludes a magmatogenic source of these sulphides.) Both direct and

Evolution of the Sulphur Cycle in the Precambrian 13

indirect evidence of sedimentary sulphate date back as far as 3.5 Ga, with the barite prevailing in the oldest record often demonstrably deriving from primary gypsum and anhydrite. This suggests that the sulphate content of the Archaean seas was high enough for calcium sulphates to be precipitated in suitable sedimentary environments and, by implication, indicates a remarkable degree of conservatism of seawater chemistry through time. Bacterial sulphate reduction is most probably responsible for some ~ 2.7 Ga old isotope patterns (Nos. 8, 11, 12; Figure 1.2), which strongly resemble bacteriogenic analogues from younger geological formations. This interpretation is strengthened by new data for the Michipicoten iron-formation (Thode and Goodwn, 1983) which brings out its bacteriogenicity more strongly (see No. 11 in Figure 1.2).

All older sedimentary sulphide occurrences thus far investigated lack unequivocal bacteriogenic features, i.e. they show relatively narrow isotope spreads around 0‰ and overall fractionations between −3 and −5 relative to coeval sulphate (barite). With the buildup of a marine sulphate reservoir at 3.5 Ga ago or earlier, the obvious time lag in the appearance of the first isotope patterns attributable to bacterial sulphate reduction is puzzling. Considering that prokaryotes had evolved very early on the ancient Earth,

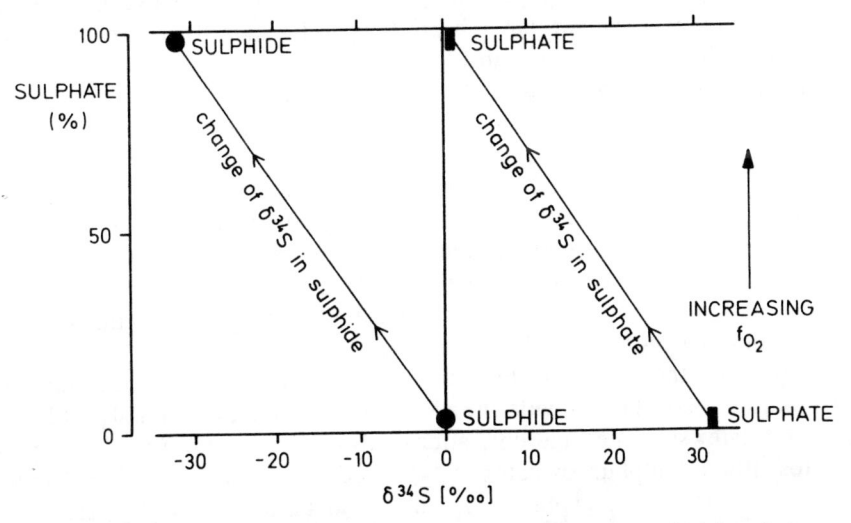

Figure 1.4 Effect of the introduction of an oxidized phase (sulphate) into a hydrothermal system at 200°C containing reduced sulphur with $\delta^{34}S = 0$‰. At this temperature, the equilibrium fractionation Δ ($^{34}SO_4^{2-} - {}^{34}S_{H_2S}$) = 32‰ (cf. Figure 1.5). As oxygen fugacity rises, $\delta^{34}S_{H_2S}$ falls because 'heavy' sulphur (^{34}S) is preferentially incorporated into SO_4^{2-}. Near-complete conversion of sulphide to sulphate under closed system conditions would impart a $\delta^{34}S$ close to −32‰ to the residual sulphide. The values for sulphide can never exceed those of the initial isotopic mixture in positive direction

the question may be asked whether the earliest biogenic isotope patterns could perhaps have been lost in the oldest rocks. If this is so then the first appearance of biological isotope patterns in the record some 2.7 Ga ago might merely indicate *a minimum age* for the emergence of bacterial sulphate reducers. Two counter indications are: firstly the two key deposits, i.e. Michipicoten and Woman River, are thought to have involved hydrothermal activity (Cameron, 1982). Secondly, the apparent lack of isotopic variation during the 500 million years following the Michipicoten–Woman River is regarded by Skyring and Donnelly (1982) as a strong argument against the 'biogenic' interpretation.

In view of the weakness of the isotopic information in the early Archaean, further attempts to trace the activity of bacterial sulphate reducers back in time will have to explore other lines of evidence. A promising approach might be a systematic survey of the organic carbon/sulphur ratios in pyrite-bearing rocks, which apparently closely correlate with the control of sulphate reduction by organic matter (Berner, 1984). Another independent check could be based on the presence or lack of isotopic equilibrium among different phases of sedimentary sulphide assemblages. If there is a bacteriogenic pedigree of the parent H_2S and ensuing low-temperature sulphide formation, then partitioning of sulphur isotopes between different sulphide minerals cannot have been thermodynamically controlled, and fractions will vary from those in Figure 1.3

In view of the scantiness of the Precambrian sulphur isotope record, Skyring and Donnelly (1982) have argued that *bona fide* bacteriogenic patterns probably do not appear prior to about 2 Ga ago, dismissing the biogenic features of several Archaean sulphide occurrences as being due to microbial reduction of sulphite which is invoked by the authors as a possible precursor process to sulphate reduction. While the assumption of an evolutionary sequence SO_3^{2-} reduction → SO_4^{2-}-reduction rests on a sound bioenergetic rationale, it cannot be correlated with specific features of the older sulphur isotope record.

Moreover, the proposal by these authors that prokaryotic sulphur metabolism had started with the dissimilatory process (initially utilizing sulphite, later sulphate) and only afterwards had evolved in the direction of photosynthetic sulphur oxidation is decidedly at variance with currently accepted concepts for evolution of sulphur metabolism (see Trüper, 1982, for an updated review). Moreover with a fully developed biological carbon cycle in operation since 3.5 Ga, if not 3.8 Ga, ago (Schidlowski, 1983b), any early dissimilatory process had most probably relied on organic substrates generated in a preceding step of autotrophic carbon fixation (which, for energetic reasons, must have extensively utilized H_2S as an electron source in its less evolved stage). Though the widely assumed near-saturation of large parts of the ancient ocean with ferrous iron may have imposed limits on the availability of H_2S in Archaean times, certain steady state levels were most probably sustained in selected compartments of the environment where this compound would have lent itself to biological utilization. Energetic considerations leave little doubt that photosynthetic utilization of H_2S must have preceded the evolution of the water-splitting process.

1.6 INTERACTION OF EARLY OCEANIC CRUST AND ARCHAEAN SEAWATER: POSSIBLE EFFECTS ON THE ISOTOPE GEOCHEMISTRY OF MARINE SULPHUR

As pointed out above, a conspicuous feature of the oldest isotope record is the disappearance of bacteriogenic sulphide patterns beyond 2.8 Ga ago (Figure 1.2). This has been taken as a tentative lower age limit for the emergence of dissimilatory sulphate reduction (Monster et al., 1979; Schidlowski, 1979). However, the existence of a marine sulphate reservoir as early as 3.5 Ga ago is indisputable. This would imply that, in principle, the stage had been set for the advent of sulphate respirers some 0.7 Ga before the earliest documentation of their isotopic signatures in the record. This time lag seems enigmatic as recent progress in the field of Precambrian palaeobiology indicates impressive levels of prokaryotic differentiation almost at the start of the sedimentary record (cf. Schopf and co-workers, 1983). Hence it is legitimate to ask whether the isotopic fingerprints of dissimilatory sulphate reduction could have been obliterated in the oldest rocks by processes that are as yet insufficiently understood. This cannot be metamorphism since bacteriogenic isotope patterns have been repeatedly shown not to undergo complete homogenization even in high-grade terranes (Buddington et al., 1969; Ohmoto and Rye, 1979; and others).

As the disappearance of biogenic signatures from the record is roughly correlated with the Archaean–Proterozoic boundary, this might point towards possible peculiarities of Archaean seawater chemistry. With vast areas of newly formed oceanic crust and a substantially higher heat-flow, basalt–seawater interaction should have been at a maximum during Archaean times ($t > 2.5$ Ga). As a result of correspondingly enhanced seawater circulation through ocean floor hydrothermal systems, the chemistry of the early ocean may have been largely mantle-dominated (Veizer et al., 1982).

Our understanding of the isotopic systematics of sulphur in basalt–seawater systems is still limited (for promising approaches see, inter alia, Shanks et al., 1981); accordingly, potential repercussions on the oldest isotope record of intense mantle buffering in the Archaean oceans are as yet difficult to assess. It seems, however, well established that inorganic reduction of seawater sulphate in submarine hydrothermal systems proceeds rapidly at temperatures $> 250°C$ and is complete as long as the quantity of ferrous iron available in the host rock stoichiometrically exceeds the influx of external sulphate. Moreover, wherever sulphate and sulphide coexist as dissolved species, isotopic equilibrium is likely to be attained for temperatures > 250°C. Hence, isotopic re-equilibration of exogenic sulphur by these processes on a global scale would establish a range of equilibrium fractionations between H_2S and SO_4^{2-} with temperature as the principal determinant (Figure 1.5). The isotopic composition of sulphide minerals formed from mixtures of primary (magmatogenic) and ocean-derived sulphur would, furthermore, strongly depend on the SO_4^{2-} content of the interacting seawater as well as the water/rock mass ratio of the specific reaction, i.e. ultimately on the intensity of the flux of marine sulphate through the hydrothermal system (Figure 1.6). Since anhydrite precipitation in the thermal aureoles of such systems has been shown to drastically curb the influx of SO_4^{2-}, the sulphate content of the primary seawater may, however, not be the most decisive parameter in these processes.

The quantitative impact of basalt–seawater interaction on the sulphur isotope geochemistry of the present ocean is still under debate. Thus assessment of the potential importance of mantle-buffering for the Archaean

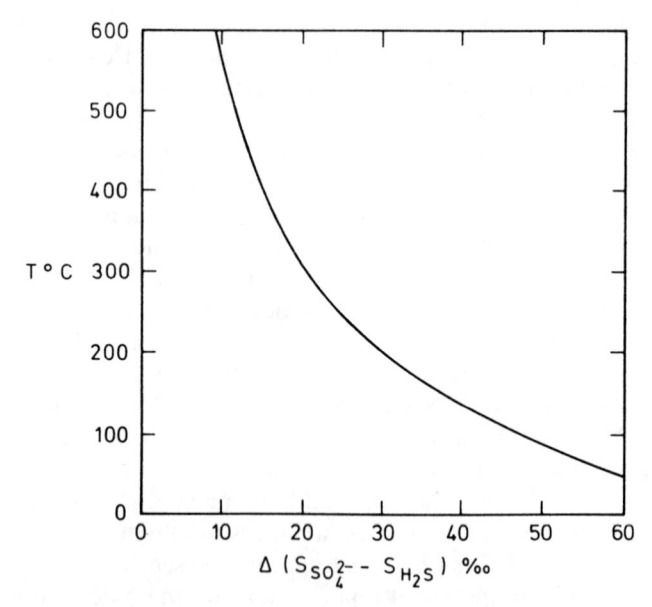

Figure 1.5 Equilibrium fractionation of sulphur isotopes between SO_4^{2-} and H_2S over the temperature range relevant for basalt–seawater interactions from experimental and theoretical data (adapted from Shanks *et al.*, 1981). If exogenic (marine) sulphate were completely processed in submarine hydrothermal systems, the temperature of the hydrothermal process would determine the isotopic difference between sulphide and sulphate

Evolution of the Sulphur Cycle in the Precambrian

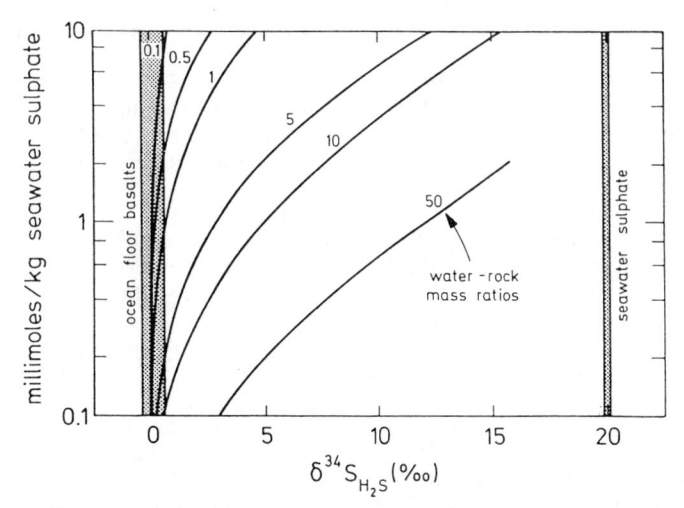

Figure 1.6 $\delta^{34}S$ values of hydrogen sulphide generated by mixing a magmatogenic sulphur source ($\delta^{34}S \cong 0‰$) with increasing proportions of sulphate-derived marine sulphur ($\delta^{34}S = 20‰$), assuming quantitative reduction of the sulphate (from experimental and theoretical data). Note that the isotopic composition of the resulting H$_2$S 'hybrid' is constrained by the $\delta^{34}S$ of the two sulphur sources. After Shanks *et al.* (1981)

sulphur cycle is certainly premature. However, as the relevant processes become clearer, their possible significance for the interpretation of the oldest sulphur isotope record should be periodically re-examined.

REFERENCES

Barley, M. E., Dunlop, J. S. R., Glover, J. E., and Groves, D. I. (1979). Sedimentary evidence for an Archaean shallow-water-volcanic-sedimentary facies, eastern Pilbara Block, Western Australia. *Earth Planet. Sci. Lett.*, **43**, 74–84.

Berner, R. A. (1982). Burial of organic carbon and pyrite sulfur in the modern ocean: Its geochemical and environmental significance. *Am. J. Sci.* **282**, 451–73.

Berner, R. A. (1984). Sedimentary pyrite formation: An update. *Geochim. Cosmochim. Acta*, **48**, 605–15.

Broda, E. (1975) *The Evolution of the Bioenergetic Processes*, Pergamon, Oxford, 220 pp.

Buddington, A. F., Jensen, M. L., and Mauger, R. L. (1969). Sulfur isotopes and origin of northwest Adirondack sulfide deposits. *Geol. Soc. Am. Mem.*, **115**, 423–51.

Button, A. (1976). Halite casts in the Umkondo System, south-eastern Rhodesia. *Trans. Geol. Soc. S. Afr.*, **79**, 177–8.

Cameron, E. M. (1982). Sulphate and sulphate reduction in early Precambrian oceans. *Nature*, **296**, 145–8.

Cameron, E. M. (1983). Evidence from early Proterozoic anhydrite for sulphur isotopic partitioning in Precambrian oceans. *Nature*, **304**, 54–6.

Claypool, G. E., Holser, W. T., Kaplan, I. R., Sakai, H., and Zak, I. (1980). The age curves of sulfur and oxygen isotopes in marine sulfate and their mutual interpretation. *Chem. Geol.,* **28**, 199–260.

Crick, I. H., and Muir, M. D. (1980). Evaporites and Uranium mineralization in the Pine Creek geosyncline. In: Ferguson, J., and Goleby, A. B. (Eds.) *Uranium in the Pine Creek Geosyncline*, International Atomic Energy Agency, Vienna, pp. 30–3.

Goldhaber, M. B., and Kaplan, I. R. (1974). The sulfur cycle. In: Goldberg, E. D. (Ed.) *The Sea*, Vol. **5**, Wiley, New York, pp. 569–655.

Hayon, E., Treinin, A., and Wilf, J. (1972). Electronic spectra, photochemistry, and autoxidation mechanism of the sulfite–bisulfite–pyrosulfite systems. The SO_2, SO_3, SO_4, and SO_5 radicals. *J. Am. Chem. Soc.,* **94**, 47–57.

Heinrichs, T. K., and Reimer, T. O. (1977). A sedimentary barite deposit from the Archaean Fig Tree Group of the Barberton Mountain Land (South Africa). *Econ. Geol.,* **72**, 1426–41.

Holser, W. T., and Kaplan, I. R. (1966). Isotope geochemistry of sedimentary sulfate. *Chem. Geol.,* **1**, 93–135.

Ivanov, M. V. (1983). Major fluxes of the global biogeochemical cycle of sulphur. In: Ivanov, M. V., and Freney, J. R. (Eds.) *The Global Biogeochemical Sulphur Cycle*, Wiley, New York, pp. 449–63.

Lambert, I. B., Donnelly, T. H., Dunlop, J. S. R., and Groves, D. I. (1978). Stable isotopic compositions of Early Archaean sulphate deposits of probable evaporitic and volcanogenic origins. *Nature,* **276**, 808–11.

Lowe, D. R., and Knauth, L. P. (1977). Sedimentology of the Onverwacht Group (3.4 billion years), Transvaal, South Africa, and its bearing on the characteristics and evolution of the early Earth. *J. Geol.,* **85**, 699–723.

Marowsky, G. (1969). Schwefel-, Kohlenstoff- und Sauerstoff-Isotopenuntersuchungen am Kupferschiefer als Beitrag zur genetischen Deutung. *Contr. Mineral. Petrol.,* **22**, 290–334.

Migdisov, A. A., Ronov, A. B., and Grinenko, V. A. (1983). The sulphur cycle in the lithosphere. I. Reservoirs. In: Ivanov, M. V., and Freney, J. R. (Eds.) *The Global Biogeochemical Sulphur Cycle*, Wiley, New York, 25–95.

Monster, J., Appel, P. W. U., Thode, H. G., Schidlowski, M., Carmichael, C. M., and Bridgwater, D. (1979). Sulfur isotope studies in Early Archaean sediments from Isua, West Greenland: implications for the antiquity of bacteria sulfate reduction. *Geochim. Cosmochim. Acta,* **43**, 405–13.

Ohmoto, H. (1972). Systematics of sulfur and carbon isotopes in hydrothermal ore deposits. *Econ. Geol.,* **67**, 551–78.

Ohmoto, H., and Rye, R. O. (1979). Isotopes of sulfur and carbon. In: Barnes, H. L. (Ed.) *Geochemistry of Hydrothermal Ore Deposits*, 2nd edn, Wiley, New York, pp. 509–67.

Peck, H. D. (1974). The evolutionary significance of inorganic sulfur metabolism. In: Carlile, M. J., and Skehel, J. J. (Eds.) *Symp. Soc. Gen. Microbiol.,* **24**, 241–62.

Perry, E. C., Monster, J., and Reimer, T. (1971). Sulfur isotopes in Swaziland System barites and the evolution of the Earth's atmosphere. *Science,* **171**, 1015–16.

Rees, C. E., Jenkins, W. E., and Monster, J. (1978). The sulphur isotopic composition of ocean water sulphate. *Geochim. Cosmochim. Acta,* **42**, 377–81.

Schidlowski, M. (1979). Antiquity and evolutionary status of bacterial sulfate reduction: sulfur isotope evidence. *Origins of Life,* **9**, 299–311.

Schidlowski, M. (1983a). Biologically mediated isotope fractionations: biochemistry, geochemical significance, and preservation in the Earth's oldest sediments. In: Ponnamperuma (Ed.) *Cosmochemistry and the Origin of Life*, Reidel, Dordrecht, pp. 277–322.

Schidlowski, M. (1983b). Evolution of photoautotrophy and early atmospheric oxygen levels. In: Nagy, B., Weber, R., Guerrero, J. C., and Schidlowski, M. (Eds.) *Developments and Interactions of the Precambrian Atmosphere, Lithosphere and Biosphere (Developments in Precambrian Geology 7)*, Elsevier, Amsterdam, pp. 211–27.

Schidlowski, M., Hayes, J. M., and Kaplan, I. R. (1983). Isotopic inferences of ancient biochemistries: Carbon, sulfur, hydrogen and nitrogen. In: Schopf, J. W. (Ed.) *Earth's Earliest Biosphere: Its Origin and Evolution*, Princeton University Press, Princeton NJ, pp. 149–86.

Schopf, J. W. (Ed.) (1983). *Earth's Earliest Biosphere: Its Origin and Evolution*, Princeton University Press, Princeton, NJ, 540 pp.

Shanks, W. C., Bischoff, J. L., and Rosenbauer, R. J. (1981). Seawater sulfate reduction and sulfur isotope fractionation in basaltic systems: interaction of seawater with fayalite and magnetite at 200–350°C. *Geochim. Cosmochim. Acta*, **45**, 1977–95.

Skyring, G. W., and Donnelly, T. H. (1982). Precambrian sulfur isotopes and a possible role for sulfite in the evolution of biological sulfate reduction. *Precambrian Res.*, **17**, 41–61.

Thode, H. G., and Goodwin, A. M. (1983). Further sulfur and carbon isotope studies of Late Archaean iron-formations of the Canadian Shield and the rise of sulfate-reducing bacteria. In: Nagy, B., Weber, R., Guerrero, J. C., and Schidlowski, M. (Eds.), *Developments and Interactions of the Precambrian Atmosphere, Lithosphere and Biosphere (Developments in Precambrian Geology 7)*, Elsevier, Amsterdam, pp. 229–48.

Thode, H. G., and Monster, J. (1965). Sulfur isotope geochemistry of petroleum, evaporites, and ancient seas. *Am. Assoc. Petr. Geol. Mem.*, **4**, pp. 367–77.

Thorpe, R. I. (1979). A sedimentary barite deposit from the Archaean Fig Tree Group of the Barberton Mountain Land (South Africa)—A Discussion. *Econ. Geol.*, **74**, 700–2,

Trudinger, P. A. (1979). The biological sulfur cycle. In: Swaine, D. J. and Trudinger, P. A. (Eds.) *Biogeochemical Cycling of Mineral-Forming Elements*, Elsevier, Amsterdam, pp. 293–313.

Trüper, H. G. (1982). Microbial processes in the sulfur cycle through time. In: Holland, H. D., and Schidlowski, M. (Eds.) *Mineral Deposits and the Evolution of the Biosphere*, Springer, Berlin, pp. 5–30.

Veizer, J., Compston, W., Hoefs, J., and Nielsen, H. (1982). Mantle buffering of the early oceans. *Naturwiss.*, **69**, 173–80.

Vinogradov, V. I., Reimer, T. O., Leites, A. M., and Smelov, S. B. (1976). The oldest sulfates in Archaean formations of the South African and Aldan Shields and the evolution of the Earth's oxygenic atmosphere. *Lithol. Min. Res*, **11**, 407–20.

Walker, J. C. G. (1978). Oxygen and hydrogen in the primitive atmosphere. *Pure Appl. Geophys.*, **116**, pp. 222–31.

Walker, J. C. G., and Brimblecombe, P. (1985). Iron and sulfur in the pre-biologic ocean. *Precambrian Res.*, **28**, 205–22.

Evolution of the Global Biogeochemical Sulphur Cycle
Edited by P. Brimblecombe and A. Yu. Lein
© 1989 SCOPE Published by John Wiley & Sons Ltd

CHAPTER 2
Modelling the Natural Cycle of Sulphur Through Phanerozoic Time*

W. T. HOLSER, J. B. MAYNARD AND K. M. CRUIKSHANK

2.1 INTRODUCTION

A geochemical model attempts to reduce the complexities of a natural system to the simplest terms that are operative through geological time, by integrating and abstracting its important features as aggregate reservoirs and the fluxes between them. A model of a geochemical cycle approximates a set of reservoirs and fluxes as a closed system, taking advantage of the constraints thus afforded by closed balance-equations. Models of exogenic geochemical cycles emphasize the nearly closed nature of the collective regime of the Earth's surface—the atmosphere, hydrosphere, biosphere and sedimentary rocks. This paper reviews the development and status of modelling the geochemical cycle of sulphur through Phanerozoic time up to the advent of Man, with primary attention to its important exogenic elements.

An important application of such modelling is to serve as a baseline for evaluating the impact of Man's activities on both the state and the dynamics of the sulphur system. Beyond that, however, modelling of the system should contribute to an understanding of its internal mechanisms, especially feedback loops that may reinforce or dampen fluctuations of the system. The models provide a system in which ideas about those mechanisms may be simulated and tested. A significant byproduct of geochemical modelling has been the recognition and estimation of fluxes or reservoirs that were unknown or inaccessible. Eriksson (1960) used this approach in the first estimate of fluxes of H_2S to the atmosphere. Such models have also been important in understanding the relations between the sulphur cycle and cycles of other major elements, such as carbon, oxygen and silicon, and in deducing the mechanisms by which these cycles are coupled. A final goal is greater understanding of the external influences on the cycle, given by, for example, correlation of variation in the cycle with changes in external conditions.

---

* Funded by US National Science Foundation Grant EAR 8115985 to the University of Oregon.

To begin with, some background data and concepts are extracted from more comprehensive recent treatments. The simplest box model of the sulphur cycle is set up, and then the previous models of the sulphur cycle are abstracted in terms of inputs, assumptions, outputs, and additional reservoirs or fluxes. This review focuses on the sulphur cycle itself, while recognizing its connections with the cycles of carbon and oxygen that have been of most interest in geochemical history.

In a following section some of the models are tested for sensitivity to variation of inputs and assumptions, and outputs are compared to each other and to geological observations.

2.2 GEOCHEMICAL BACKGROUND

2.2.1 Overview

The geochemistry of sulphur in the present exogenic cycle has been treated extensively in recent reviews (Nielsen, 1978; Ivanov and Freney, 1983; Holser *et al.*, 1988). These works include detailed discussion of reservoirs with estimates of their present sizes, and description of transfer mechanisms with estimates of consequent fluxes between reservoirs. Other background papers in this volume also provide new critical evaluations of certain features of the modern cycle. It seems unnecessary to repeat these data here—we arbitrarily assume, as a takeoff point for changes in past geological time, the balanced present-day cycle of Lein (1983).

Figure 2.1 reduces the cycle to its essentials of three reservoirs (reduced sulphide of sediments, oxidized sulphate of evaporites and other sediments, and seawater sulphate), for the purpose of following changes in the geological past. The fluxes are reduced to four fluxes into and out of seawater. At steady state the net fluxes between each pair would be zero.

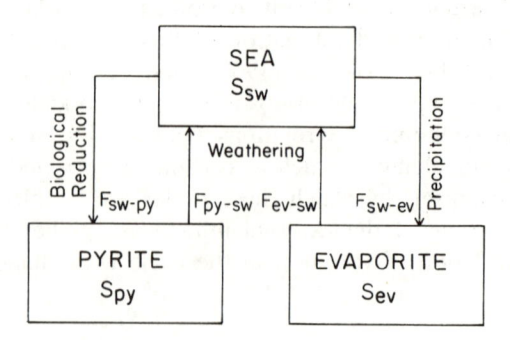

Figure 2.1 Simplified box model of the sulphur cycle

2.2.2 Sulphur reduction

A central feature of Figure 2.1 is the marine reservoir of sulphate, whose fate is either to be reduced and deposited as pyrite in marine sediments (principally on the continental margins), or, at specific times and places, to be crystallized as sulphate in evaporite deposits. The reduction process is dominated by the activity of anaerobic sulphur-reducing bacteria, which also require organic carbon as an energy source. The overall reaction can be schematically represented as

$$(SO_4)^{2-} + 2CH_2O \rightarrow S^{2-} + 2CO_2 + 2H_2O \qquad (1)$$

where the CH_2O represents any (degradable) organic carbon, and the S^{2-} represents any (completely reduced) sulphide. The present ocean has only a few anaerobic locales in which bacterial sulphate reduction is effective in the water column, such as the Black Sea. Immediately under the seawater–sediment interface, however, anaerobic conditions are widespread in both shelf and abyssal regimes. Differing deep-sea circulation at certain times and places in past geological periods probably increased the extent of the anaerobic regime in oceanic sediments (e.g. Arthur, 1979), although it is not clear to what extent such reducing conditions also extended above the sediment surface into the overlying water column. Reduced sulphide is significant in the long-term geochemical cycle of sulphur only after being fixed by ferric iron that was reduced and dissolved during the depositional process. The overall reaction may be written

$$8(SO_4)^{2-} + 2Fe_2O_3 + 8H_2O + 15C_{org} \rightarrow 4FeS_2 + 15CO_2 + 16(OH)^- \qquad (2)$$

Of the requirements for sulphur reduction, supplies of both sulphate and iron compounds are usually in excess in normal marine clastic sediments. In non-marine conditions, the production of FeS or FeS_2 may be limited by the supply of sulphate (Berner and Raiswell, 1983). In some marine carbonate facies the fixation of sulphide may be limited by the low concentration of iron; lack of iron may also be the reason for the predominance of native sulphur over pyrite in the bacterial reduction of sulphate in salt-dome cap rock and other evaporite rocks.

In normal marine clastic sedimentation, the parameter that seems to limit sulphate reduction is the supply of organic carbon in a form usable by the reducing bacteria (Goldhaber and Kaplan, 1974). In any case, the geological evidence seems clear that, despite the fact that sulphur reduction is limited by usable organic carbon, there is always a residue of organic carbon remaining in the sediment and in the rock that for some reason could not be used. Surprisingly, pyritic sulphur and residual carbon are in a relatively constant proportion of about 0.12 (in moles) in Holocene sediments

(Goldhaber and Kaplan, 1974). Ancient shales show a similar or somewhat higher S/C ratio that may be related to euxinic conditions (Williams, 1978; Leventhal, 1983). Such a constant ratio could result under two conditions (although not necessarily ones that are always easily satisfied in the environment): (i) the fraction of usable carbon was constant and (ii) the fraction of produced H_2S that was fixed as pyrite was constant. The relation was derived by Sweeney and Kaplan (Goldhaber and Kaplan, 1974, p. 600), but is more clearly developed by Williams (1978, p. 1039). Goldhaber and Kaplan (1974, p. 602) assumed that the fraction of H_2S lost (not used for pyrite fixation, but reoxidized) in the marine environment was 'relatively small', but more recent field experiments by Jorgensen et al. (1977) on tidal flats show a 90% loss.

Most of the deposition of sulphide occurs in the biologically productive shallows of the continental margins, but the rate is imperfectly known. Berner (1972) summarized available data on present sediments and guessed 8.7 Tg (S) a^{-1}. Holser et al. (1988) estimated an average sulphide deposition for previous geological periods at 17.6 Tg (S) a^{-1} by comparing Ronov's (Ronov et al., 1974) sulphur contents and sedimentation rates (for platform and geosynclinal sediments). The flux in past geological times would have been dependent on the prevalence of these biologically productive locales.

2.2.3 Deposition of sulphate

At the present time sulphate is not deposited in significant amounts. In past geological times sulphate evaporites occurred in a variety of palaeogeographic situations, including: prograding sabkha shorelines, lagoons, and deep basins of cratonic, mediterranean or rift-valley origin, where the brine may have varied from hundreds of metres deep to sub-sea level desiccation. All these conditions are very unusual, and are initiated by specific tectonic accidents. Once precipitation begins, it proceeds rapidly. Rates are typically about 700g (S) $m^{-2}a^{-1}$ (Holser, 1979, p. 266) which is three orders of magnitude faster than sulphide precipitation in muds. Available basins are quickly filled, especially if halite precipitation is also involved. But the special conditions necessary for evaporite concentration generally do not last long (hundreds of thousands of years) before tectonic conditions again change shorelines, barriers and basinal circulation to either fully marine or non-marine conditions. While evaporite deposition *is* going on, a large fraction of the sulphate in the affected seawater is precipitated (90% by the beginning of the halite facies: Holser, 1979, p. 245). Consequently, removal of sulphate from the ocean has been highly episodic (Zharkov, 1981; Holser et al., 1980).

2.2.4 Weathering

The present weathering flux of sulphur to the sea may be estimated from the flow and composition of the world's rivers and subtracting the sulphur added through atmospheric fallout over the continents from the burning of fossil fuels, etc. It can also be estimated from the rate of transport of sediments to the sea (Holeman, 1968), multiplied by the average sulphur content and mass of rock types being weathered.

2.2.5 The question of steady state

An important aspect of the geochemical cycle that figures prominently in calculations is the question as to whether the cycle is in a dynamic steady state, and especially whether the ocean reservoir and its fluxes are relatively constant or not. This question can be addressed separately in terms of constancy of reservoir mass and reservoir isotope ratio. Berner (1972) calculated that the ocean is presently grossly out of balance with respect to sulphate. Its concentration is increasing because more sulphur is transferred to the oceans in river water than is deposited as sulphide (and sulphate). The new central estimates by Holser *et al.* (1988) agree with this conclusion, but within the wide limits of precision of the data they found it possible to balance the system.

Some authors have simply assumed a constant ocean composition, estimating either weathering or sedimentation, and equating the two. Many modellers have assumed, not only that the ocean composition is presently in a steady state, but that it has remained so throughout most of geological time. Holland (1972) has reasoned that the composition of the ocean has remained constant within a factor of two or three during the Phanerozoic, because both the mineralogy of marine evaporites, and the sequence of their crystallization ($CaCO_3$, $CaSO_4$, NaCl; no sodium sulphates or carbonates) have not noticeably changed during that time. Within those limits, however, variations are not only possible, but required by the gross irregularity of the chemical composition of sediments through time (a theme that we shall develop below). Indeed, feedback mechanisms that act to maintain the chemical contents, even within Holland's limits, are difficult to pin down.

2.2.6 Is the exogenic sulphur cycle leaky?

Is it feasible to limit our model calculations of the history of the sulphur cycle to the simple three reservoirs of Figure 2.1, or does a significant amount of sulphur leak in or out of the oceanic or continental crust? A recent evaluation of the whole cycle (Holser *et al.*, 1988) suggests that the simple approximation holds reasonably well, but not all geochemists would agree with that conclusion. In a quest for ideal steady state balance of marine sulphur, some authors have sought an additional sink of sulphur in mid-ocean ridge (MOR) activity and plate subduction. Circulation of seawater in a sub-sea floor hydrothermal system may deliver sulphur to the oceanic crust by precipitation of anhydrite (resulting from its low solubility at high temperature) or it may precipitate sulphide by reduction with the ferrous iron of silicates. Alternatively seawater may leach sulphide from the intruded basalts. Perhaps each of these processes is active at some times or places, but for the sulphur cycle the question is whether the total flux of sulphur in either direction is significant. Specifically, Edmond *et al.* (1979) measured a sharply decreasing level of sulphate in approaching the Galapagos deep-sea hydrothermal vents, which indicated complete removal of sulphate from the circulating seawater at 380°C, and he extrapolated this result to infer a flux of 120 ± 20 Tg (S) a^{-1} into the MOR; this value was strengthened by McDuff and Edmund (1982) on the East Pacific Rise. Andrews (1979), Holser *et al.* (1979, p. 10) and Wolery and Sleep (1988) have argued some points against massive transfer of sulphur from the sea to MORs, e.g. major seawater circulation for even a few million years should have reduced the sulphate concentration of seawater to the very low solubility of anhydrite at high temperature, or if the sulphate underwent reduction then all the FeO in several kilometres of crust should have been oxidized. On the sea floor itself comparison of chemical analyses of altered basalt with quenched basaltic glass indicates that between 20 and 80% of the sulphur content of the erupted basalt was *added* to seawater. This represents an appreciable component of the seawater balance. Sulphide deposited on the *deep* ocean floor by sulphur-reducing bacteria represents an equally small flux, because of the small amount of organic carbon available there. Some undetermined fraction of this sedimentary sulphide is returned to the exogenic cycle by metamorphism and volcanism in subduction zones.

These subcycles are minor perturbations to the main exogenic cycle of sulphur, and most investigators have ignored them in modelling the exogenic sulphur cycle through time.

2.2.7 Reservoirs of sulphur and their changes through time

In line with the above discussions, we simplify the sulphur cycle to three main reservoirs: sediments (mainly shales) containing reduced (sulphide) sulphur, sediments (mainly evaporites) containing oxidized (sulphate) sulphur, and the world ocean (also oxidized); see Figure 2.1. At any point in time, each reservoir of sulphur represents an inventory of all sulphur in the defined reservoir. The most useful characteristics to be inventoried are the mass and the isotope ratio of sulphur. Other characteristics or subdivisions of these simplified reservoirs may also be useful: age, rock type, tectonic milieu. In the simple dynamic operation of the model the whole of a reservoir is regarded as equally accessible to a flux, if not actually homogeneously mixed. This may not always be true.

Table 2.1 lists the masses and isotope ratios of these reservoirs for the present time, as estimated by previous authors. The key figure, the fraction

Table 2.1. Balance of sulphur reservoirs in the exogenic cycle

Author	Mass (10^6 Tg) sedimentary S Reduced	Oxidized[a]	Reduced Total[b]	Mean δ^{34}S sediments (‰(CDT)*) Reduced	Oxidized[a]	Total[b]
Holser and Kaplan (1966)	2690	5190	0.29	−12.0	+17.0	+8.9
Li (1972)	7410	4780	0.55	−12.0	+17.0	+1.3
Garrels and Perry (1974)	9430	6410	0.55	−12.0	+17.0	+1.3
Schidlowski et al. (1977)	5900	6350	0.44	−16.0	+19.0	+3.8
Neilsen (1978)[c] (A)	5710	5290	0.46	−15.2	+17.0	+2.3
(B)	5710	1310	0.69	−15.2	+17.0	−4.7
Garrels and Lerman (1981)	5710	3460	0.55	−16.0	+19.0	0.0
Berner and Raiswell (1983)	8010	8010	0.42	—	—	—
Migdisov et al. (1983, Table 2.13)[d] (A)	3960	2630	0.50	−9.3	+8.7	+1.4
(B)	4970	1380	0.65	−9.3	+8.7	−1.1
Holser et al. (1988)[e] (A)	4910	4870	0.44	−7.9	8.4[f]	+2.5[f]
(B)	5770	4040	0.52	−7.4	+12.1[f]	+2.9[f]
Garrels and Lerman (1984)	6410	6410	0.45	−16.0	+19.0	+3.2
Present work	4970	2470	0.57	−7.0	+19.4	+4.6

[a] Not including seawater.
[b] Including about 1280×10^6 Tg sulphate sulphur in seawater with δ^{34}S = 20‰.
[c] (A) assumes the evaporite mass deduced by Holser and Kaplan (1966), whereas (B) assumes a smaller evaporite mass suggested by Otto Braitsch (K. H. Wedepohl, personal communication).
[d] (A) is data as measured by Ronov (1980, etc.), whereas (B) is corrected for primary sulphide oxidized to sulphate (Migdisov et al., 1983, p. 83).
[e] (A) takes Ronov's (1980) analyses for sulphate and sulphide at face value, except evaporite sulphate modified by Holser; (B) assumes that all sulphur in geosynclinal clastic sediments is sulphide (a correction similar to that made by Migdisov et al. (1983)—footnote ([d]) above).
[f] Recalculated from data on isotopes by Ronov et al. (1974) and on volumes by Ronov (1980) and Southam and Hay (1981)—see Holser et al. (1988).
* CDT = with respect to Canyon Diablo Troilite.

Table 2.2. Present inventory of sulphate sulphur in rock[a], 10^6 Tg

Period (1)	Elapsed time (Ma) (2)	Duration (Ma) (3)	Dispersed sulphate[b] Platforms (4)	Geosynclines (5)	Shelf/Slope/ Oceanic (6)	Total dispersed (7)	Evaporites[c] (8)	Total sulphate[d] (9)	δ^{34}S (VI)[e] (‰) (10)
Neogene	2	21	15	6	9	30	400	430	+20.6
Palaeogene	23	42	29	6	15	50	14	64	17.7
Cretaceous	65	75	77	17	40	134	190	324	16.5
Jurassic	140	55	46	14	25	84	520	604	16.9
Triassic	195	35	34	11	19	65	55	120	16.1
Permian	230	50	26	11	15	52	320	372	11.6
Carboniferous	280	65	26	15	18	59	30	89	15.9
Devonian	345	50	34	22	23	79	105	184	23.5
Silurian	395	40	11	7	7	25	6	31	24.6
Ordovician	435	65	18	10	12	40	2	42	22.8
Cambrian	500	70	37	12	21	70	150	220	29.5
Total Phanerozoic			353	133	204	690	1780	2470	
Eocambrian	570	130	8	3	5	16	100	116	28.8

[a] Sulphate sulphur in present oceans: 1280×10^6 Tg (S), at ^{34}S (VI) = +20.0‰.
[b] Aggregate analyses of dispersed sulphate (corrected for oxidized pyrite), from Migdisov et al. (1983, Table 2.13), apportioned to periods according to masses of rock in each period (Ronov, 1980, Table 13, column 16).
[c] Evaporite volumes for Permian–Neogene from Holser et al. (1980); for Cambrian–Carboniferous from Zharkov (1981, 1984).
[d] Total dispersed and evaporite sulphate.
[e] Mean ^{34}S of marine evaporite sulphate from Lindh (1983); Saltzman et al. (1982).

of reduced sulphur, varies from 0.29 to 0.69. Some of the differences are simply a measure of the inherent uncertainties of estimation, but other differences reflect real differences in definition and method. The classical method of estimating geochemical reservoirs is to take some list of rock types and determine their relative importance by geochemical balance of their key major constituents against an average igneous rock. Then independently published analyses (e.g. sulphur content) of rocks given the same type name are averaged to get a mean concentration. Ronov and co-workers used a more direct approach, made possible by the existence of inventories of the amount and composition of various rock types (Ronov, 1980; Migdisov et al., 1983). The reservoir estimates, based on Ronov's data, are indicated in the footnotes of Table 2.1.

If the inventory approach of Ronov is used, rather than a geochemical balance, then in the course of estimating the sulphate and sulphide reservoirs at the present time, one can also assign an inventory of sulphate and sulphide to each geological period. In Tables 2.2 and 2.3 we have recalculated the data, newly compiled by Migdisov et al. (1983), to give estimates of the present surviving masses of the two reservoirs for each geological period. Table 2.2 shows our own estimates of the mass of sulphate surviving (mainly as anhydrite) in giant evaporites (Holser et al., 1980; Zharkov, 1981); alternatively one could take the compilation of Migdisov et al. (1983, Table 2.17; note that they converted volumes as gypsum rather than anhydrite). The amounts of sulphates (Table 2.2, column 7) dispersed in other rocks (not including sulphate from surface oxidation of pyrite) were calculated from the data of Migdisov et al. (1983). The 'corrected' mass of dispersed sulphate from their Table 2.13 was divided among the geological periods according to the rock masses on the continent (from their Table 2.15). Our Table 2.3 similarly takes Migdisov's data for sulphide in platform rocks by period (their Table 2.8 corrected for oxidation by their Table 2.13), and for sulphide in geosynclinal rocks by era (their Table 2.12) divided among periods in proportion to their time.

These estimates will be used as a check on the operation of dynamic models.

2.2.8 Isotope geochemistry of sulphur

Fractionation of the stable isotopes of sulphur has played a central role in geochemical calculations. The isotope relations in the present and ancient sulphur cycles are discussed in some detail in other SCOPE Reports by Grinenko and Ivanov (1983) and Krouse and Grinenko, and only certain points relative to geochemical modelling will be reviewed here.

Departures of the isotope ratio in parts per thousand from a standard (Canyon Diablo Troilite) are designated by the '$\delta^{34}S$' notation. Dissimilatory

Table 2.3. Present inventory of sulphide sulphur in rock, 10^6 Tg

Period (1)	Platforms[a] (2)	Geosyn- clines[b] (3)	Shelf/ Slope/ Oceanic[c] (4)	Total sulphide[d] (5)	Total sulphur inventory[e] (6)	Sulphide[f] total (in rock) (7)	$\delta^{34}S(-II)$[g] (‰) (8)	$\Delta^{34}S$[h] (‰) (9)
Neogene	92	123	49	264	694	0.38	−9.6	30.2
Palaeogene	170	135	70	375	439	0.85		27.3
Cretaceous	293	359	150	802	1126	0.71	−18.7	35.2
Jurassic	291	266	128	685	1289	0.53	−6.0	22.9
Triassic	104	235	78	417	537	0.78	−5.8	21.9
Permian	166	10	86	462	834	0.55	−4.6	16.2
Carboniferous	114	293	93	500	589	0.85	−7.4	23.3
Devonian	80	412	113	605	789	0.77	−0.8	24.3
Silurian	60	60	27	147	178	0.83	−12.3	36.9
Ordovician	131	90	51	272	314	0.87	−6.0	28.8
Cambrian	224	123	80	427	647	0.66	−2.6	32.1
Total Phanerozoic	1734	2312	928	4974	7444	0.67		
Eocambrian	39	53	21	113	229	0.49	+2.4	26.4

[a] Migdisov et al. (1983, Table 2.8, corrected for oxidized pyrite according to aggregate of Table 2.13).
[b] Geosynclinal sulphide for each era (Migdisov et al., Table 2.12) apportioned among periods according to rock masses (Ronov, 1980, Table 13, column 16), corrected for oxidized pyrite according to geosynclines averaged over Phanerozoic (Migdisov et al., 1980, Table 2.13).
[c] Corrected Phanerozoic totals (Migdisov et al., 1983, Table 2.13) distributed among periods same as all continental rocks (columns 2 + 3).
[d] Columns 2 + 3 + 4.
[e] Table 2.2, column 9, plus Table 2.3, column 5.
[f] Columns 5/6.
[g] Grinenko et al. (1983), Table 2.16).
[h] $\Delta^{34}S = \delta^{34}S(VI) − \delta^{34}S(−II)$: Table 2.2, column 10, less Table 2.3, column 8.

sulphur metabolism during reduction by sulphur bacteria generates substantial isotope fractionation, particularly in the intermediate step of sulphite formation. The sulphide produced is light by $\triangle = \delta S (VI) - \delta S (-II) = +45$ to $+60‰$ (where VI and $-II$ refer to sulphur as sulphate and sulphide), depending on the overall rate of reduction, and on which of several steps is rate limiting. Natural environments may also involve local depletion of sulphate supply, depending on whether the system was open or closed to sulphate diffusion. Differences between the isotope ratios of pyrite and the seawater sulphate from which it was formed may vary from nearly $+60‰$ to nearly $0‰$, and many are near $+40‰$ (see discussion in Claypool et al., 1980). These conditions vary radically within the sediment, so that the isotope ratios in the sulphide of the resulting sedimentary rocks are extremely variable in place and time (Schwarcz and Burnie, 1973; Migdisov et al., 1983). Consequently, it is difficult to determine mean $\delta^{34}S$ for reduced sediments. Most modellers have been forced to assume that $\triangle_{sw-py}$ remained constant regardless of age, facies or rate of sedimentation of the rocks, but this assumption may be substantially in error, as we will show in a later section of this review.

In contrast to reduction, the process of crystallization of sulphate (as gypsum or anhydrite) results in only a small fractionation of about $-1.3‰$. Usually this fractionation can be ignored, and the $\delta^{34}S$ value of a marine evaporite rock approximates that of the surface seawater brine from which it crystallized. It has also been found by experience in most (but not all) cases that the seawater brine in a marine evaporite basin is so dominated by the inflowing marine sulphate that the deposited sulphate has the same $\delta^{34}S$ as the seawater (see discussion in Claypool et al., 1980).

However, Chapter 3 compiles some case studies where the assumption of an oceanic evaporite would lead to quite unreasonable conclusions.

2.2.9 A geochemical record of the sulphur cycle: the isotope age curve

Despite this problem marine sulphate evaporite rocks do generally hold a record of past changes in the fluxes and reservoirs of the sulphur cycle. As discussed above, the level of $\delta^{34}S$ (VI) in marine evaporite rocks usually closely mimics sulphur isotopes in the brine filling the evaporite basin, which in turn is mainly governed by that of the surface of the world ocean. The record of $\delta^{34}S$ (VI) in evaporite rocks (the isotope age curve) consequently corresponds to shifts in the oxidation/reduction system of sulphur through time. Several representations of the isotope shifts are shown in Fig. 2.2, where it is clear, no matter whose compilation is used, that isotope shifts in both directions have been large and irregular. At the risk of oversimplification, the record might be characterized in this way: lower values of $\delta^{34}S$, as in Permian time, indicate a pervasive oxidation of sulphide

32   *Evolution of the Global Biogeochemical Sulphur Cycle*

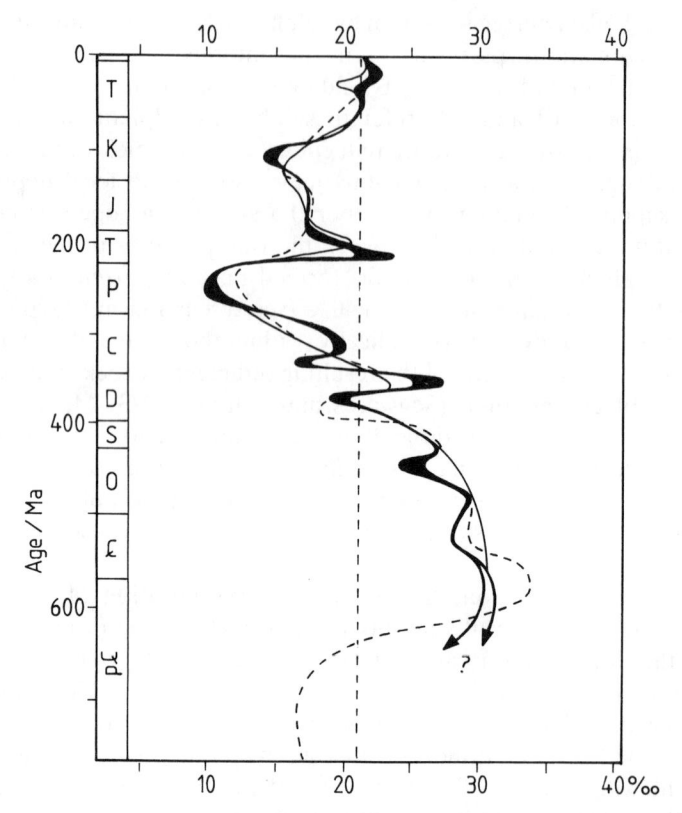

Figure 2.2 Isotope age curve for $\delta^{34}S$ (VI) of sulphur in evaporite rocks in equilibrium with surface seawater (by H. Nielsen, based on data to mid-1984). Curves previously published by Nielsen (1978) (———), and Claypool *et al.* (1980) (– – – –)

to sulphate, while high values of $\delta^{34}S$, as in early Palaeozoic time, represent massive reduction of sulphate to sulphide. The differences are substantial. Consequently the sulphur isotope age curve has been a starting point for all historical modelling of the sulphur cycle. Pilot has discussed the sulphur isotope age curve in another SCOPE Report (Krouse and Grinenko).

2.2.10 Interconnections of the exogenic sulphur cycle with those of carbon and other elements

The exogenic geochemical cycle of sulphur is clearly linked to that of oxygen, which led Holland (1973) and many later modellers to try to calculate putative changes in composition of the ancient atmosphere. Holland pointed out that any tendency to atmospheric changes owing to reduction (or oxidation) in the sulphur cycle may well be compensated by opposite oxidation (or reduction) changes in the carbon cycle, as exemplified by the box model of Fig. 2.3. Holland recognized that such compensation might be demonstrated by an inverse relation of $\delta^{34}S$ (VI) to $\delta^{13}C_{carb}$ (where C_{carb} is carbon as carbonate) through time, i.e. oxidation/reduction of sulphur compensated by reduction/oxidation of carbon. Garrels and Perry (1974) made this prediction more explicit. However, neither Holland nor Garrels and Perry (also Schidlowski *et al.*, 1977) could detect a clear variation in $\delta^{13}C$ in the data available to them. Veizer *et al.* (1980) were able to demonstrate by statistical treatment of several thousand analyses of $\delta^{13}C_{carb}$ that at least in the long-term the values were inversely related to $\delta^{34}S$ (VI). This relation was refined by Lindh (1983) (see also Saltzman *et al.*, 1982) as shown in Fig. 2.4, where several real deviations from the relationship are also evident (Holser, 1984).

More sophisticated models of the exogenic chemical cycles have further involved the isotope ratios of oxygen in sulphate (Holser *et al.*, 1979; Claypool *et al.*, 1980), the oxidation/reduction of iron (Schidlowski *et al.*,

HOLOCENE STEADY STATE MODEL OF GLOBAL CARBON-SULPHUR CYCLE

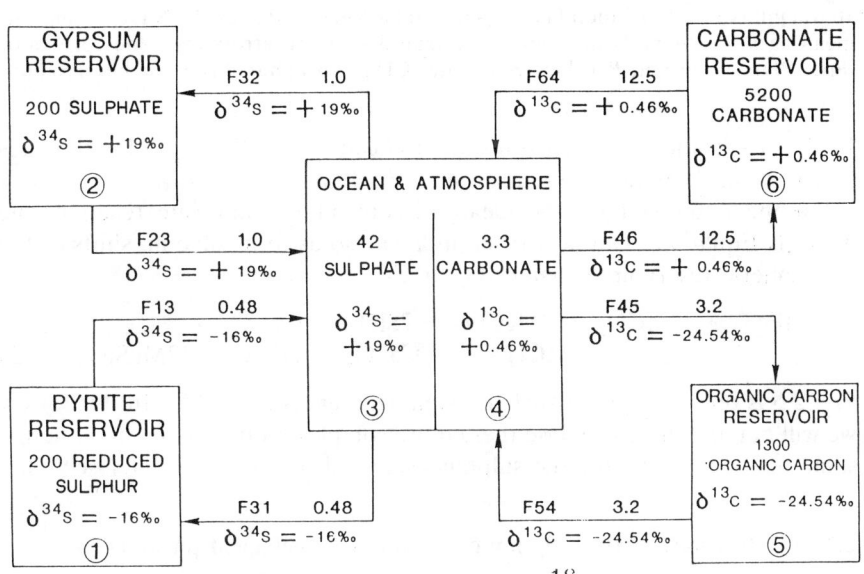

RESERVOIRS IN UNITS OF 10^{18} MOLES
FLUXES IN UNITS OF 10^{18} MOLES/MILLION YEARS

Figure 2.3 Model of the pre-anthropogenic coupled sulphur–carbon exogenic system, after Garrels and Lerman (1984). The model is in steady state with respect to masses, fluxes and isotope values

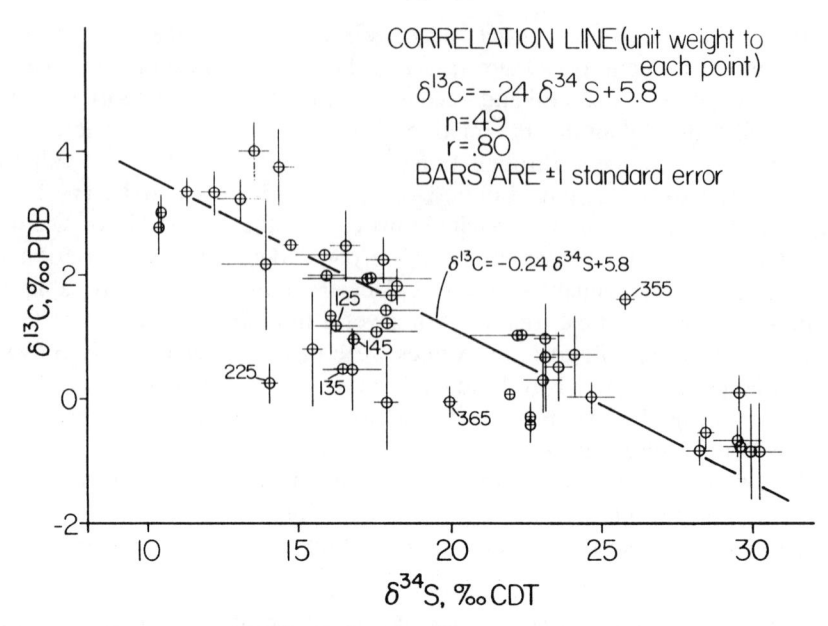

Figure 2.4 Correlation of $\delta^{34}S$ (VI) with $\delta^{13}C_{carb}$ for published data aggregated at 10 Ma intervals, after Lindh (1983) (see also Saltzman et al., 1982). Numbers indicate mean ages for intervals that plot more than 3 standard errors from the correlation line. PDB = carbon, Pee Dec Bellemite; CDT = sulphur, Caon Diablo Troilite)

1977), or silication (metamorphism)–desilication (weathering) of carbonates (Garrels and Lerman, 1981; Lasaga et al., 1985).

For the latter system, the idealized controlling composite reaction, the 'Garrels Equation', can be written in a way so as to involve no shifts of the atmospheric reservoirs of *either* O_2 or CO_2:

$$4FeS_2 + 8CaCO_3 + 7MgCO_3 + 7SiO_2 + 15H_2O \rightarrow$$
$$15CH_2O + 8CaSO_4 + 2Fe_2O_3 + 7MgSiO_3 \quad (3)$$

Fig. 2.5 is one example of such a system (Lasaga et al., 1985). In this review we will not attempt to analyse these more complex models in detail, although some of their results for the sulphur cycle will be cited in a later section.

2.2.11 Relations of the sulphur cycle to other geological parameters

Extensive variations in the sulphur or related chemical cycles through geological time, raise questions concerning the external forcing mechanisms responsible for these changes. Correlative variations in isotope ratios of marine strontium may suggest a proximate cause of some sulphur isotope variations. Although strontium basically owes its isotope ratio, $^{87}Sr/^{86}Sr$, to

Modelling the Sulphur Cycle Through Phanerozoic Time　　　　　　　　35

```
                3.0  CO₂                              2.90  CO₂
                Metamorphism                          Metamorphism
                        3.8  O.              3.80  CO₂
                        Weathering           C-Burial
        ┌──────┐        3.8  O₂    ┌──────┐           ┌──────┐
        │  O₂  │◄───────C° Burial──│ CO₂  │           │ Org C│
        │  38  │                   │0.055 │◄──────────│ 1250 │
        └──────┘        3.96  CO₂  └──────┘  3.80 CO₂ └──────┘
                        Weathering           Weathering
                        0.51  SO₄
        ┌──────┐        Reduction                7.84  CO₂
        │Pyrite│                                 Weathering
        │ 250  │        0.20  SO₄       17.38
        └──────┘        Volcanic-Seawater
                                                 18.4  HCO₃
                        0.20  SO₄
                        Metamorphism
                        2.10  Mg               18.4   Ca
                        Weathering             Precipitation
        ┌──────┐        2.10  Ca    ┌──────┐   16.14  HCO₃   ┌──────┐
        │Dolo- │                    │Oceans│   Weathering    │Calcite│
        │mite  │        0.12  SO₄   │Mg Ca │   8.30  Ca      │ 3000 │
        │1000  │                    │SO₄   │                 │      │
        └──────┘        8.16 HCO₃   │HCO₃  │   0.23  SO₄     └──────┘
                                    │75.6 14│
                                    │40  2.8│
                                    └──────┘
                                     │││││
                                     5.2  Mg
                                     Volcanic-Seawater
                                     5.2  Ca
                                     11.48 HCO₃
                                     3.10 Mg Weathering
                                     2.80 Ca
                                                 1.02  Ca
                                                 Weathering      ┌──────┐
                                                 1.02  Ca        │ CaSO₄│
                                                 Precipitation   │ 250  │
                                                                 └──────┘
                        1.5  Ca
                        Metamorphism  ┌─────────────┐  2.90  Ca
                        1.5  Mg       │Ca-Mg Silicates│ Metamorphism
                                      └─────────────┘
        0.16  SO₄
        Weathering
                        0.23  SO₄
                        Weathering
                                                 11.48  CO₂
                                                 Weathering
```

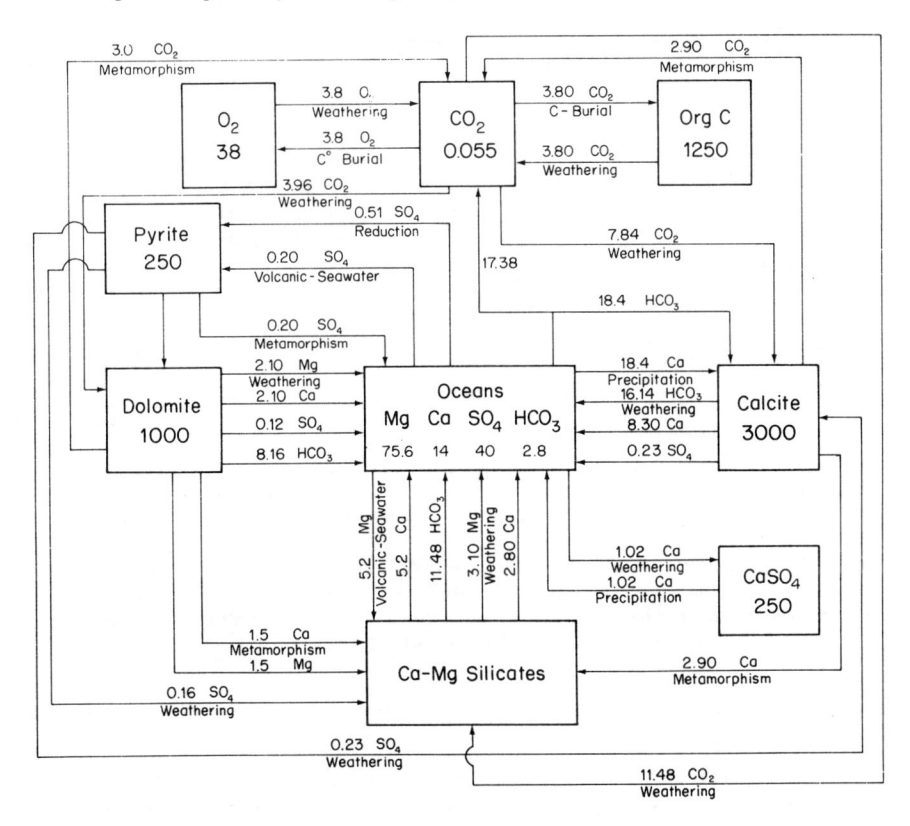

Figure 2.5 Enlarged model of present day cycles of sulphur, carbon and silica, after Lasaga *et al.* (1985). Reservoirs are in 10^{18} mols and fluxes in 10^{18} mol Ma^{-1}

radioactive decay of ^{87}Rb, the ratio in marine rocks can be attributed mainly to the erosion of old granitic rocks from shield areas, which provide ^{87}Sr/^{86}Sr that is high (~0.720), and interaction with young basaltic rocks such as at mid-ocean ridges, which provide ^{87}Sr/^{86}Sr that is low (~0.704). The isotope age curve for marine strontium, as determined in both carbonates (Burke *et al.*, 1982) and biogenic apatite (Kovach, 1980) is loosely correlated with the sulphur curve (Hoefs, 1981; Holser, 1984). World sea-level (Vail and Mitchum, 1979), which is thought to be driven mainly by variations in activity at mid-ocean ridges, also correlates with the sulphur curve (Hoefs, 1981), but less significantly and in the 'wrong' direction (Holser, 1984). Mackenzie and Pigott (1981) proposed a tectonic model to account for these correlations, and their model has been debated in Holser *et al.* (1988). A more quantitative, kinetic relation of these chemical cycles (particularly that of carbon) to geological variables (weathering, volcanic-seawater reaction, magmatism, and metamorphism, as functions of land area and seafloor

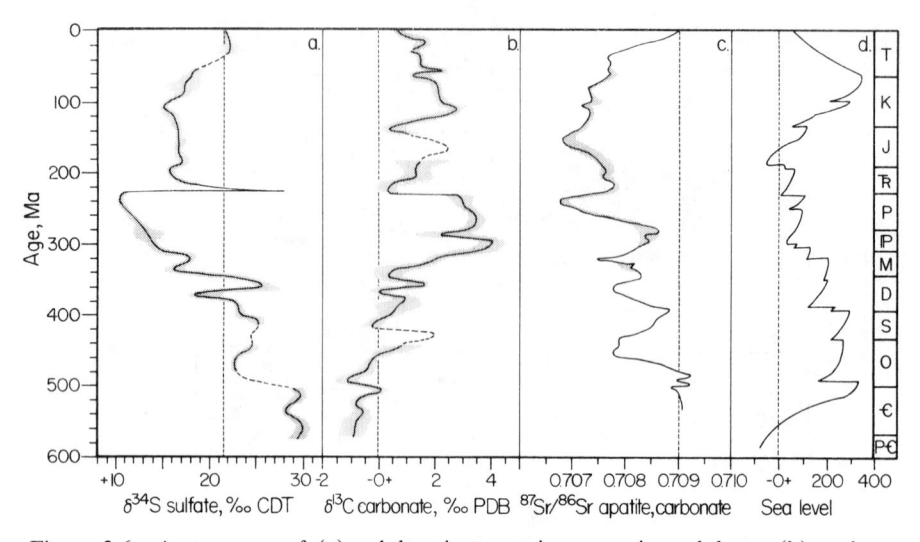

36   *Evolution of the Global Biogeochemical Sulphur Cycle*

Figure 2.6 Age curves of (a) sulphur isotopes in evaporite sulphate; (b) carbon isotopes in carbonate; (c) strontium isotopes in carbonate and in fossil apatite; and (d) sea level, after Holser (1984), who gives sources. Shading gives range of uncertainty; dashed lines indicate lack of data

spreading rate) was recently developed by Berner *et al.* (1983) and refined by Lasaga *et al.* (1985).

Figure 2.6 (Holser, 1984) presents an overall view of the isotope age curves of C, S and Sr, and the curve of eustatic sea-level changes. Several relations discussed in previous paragraphs are illustrated in this figure:

(a) The large swings of $\delta^{34}S$ through time are roughly balanced by smaller opposite swings of $\delta^{13}C$.
(b) This antithesis is imperfect even in its major aspects.
(c) The slow shifts of $\delta^{34}S$, $\delta^{13}C$ and $^{87}Sr/^{86}Sr$ are punctuated by short-term, uncorrelated events, which will be discussed in the following section.

2.2.12 Events in the sulphur cycle

Both in the papers mentioned above and in the detailed discussions in following sections, quantitative modelling of the sulphur cycle has necessarily been based on the long-term trends of the isotope age curve (or the means for whole geological periods). However, at particular geological times these trends are interrupted by short-term events for which modelling has been qualitative at best. Holser (1977) pointed out several such events, in the late Proterozoic, in the late Devonian, and in the early Triassic. The early

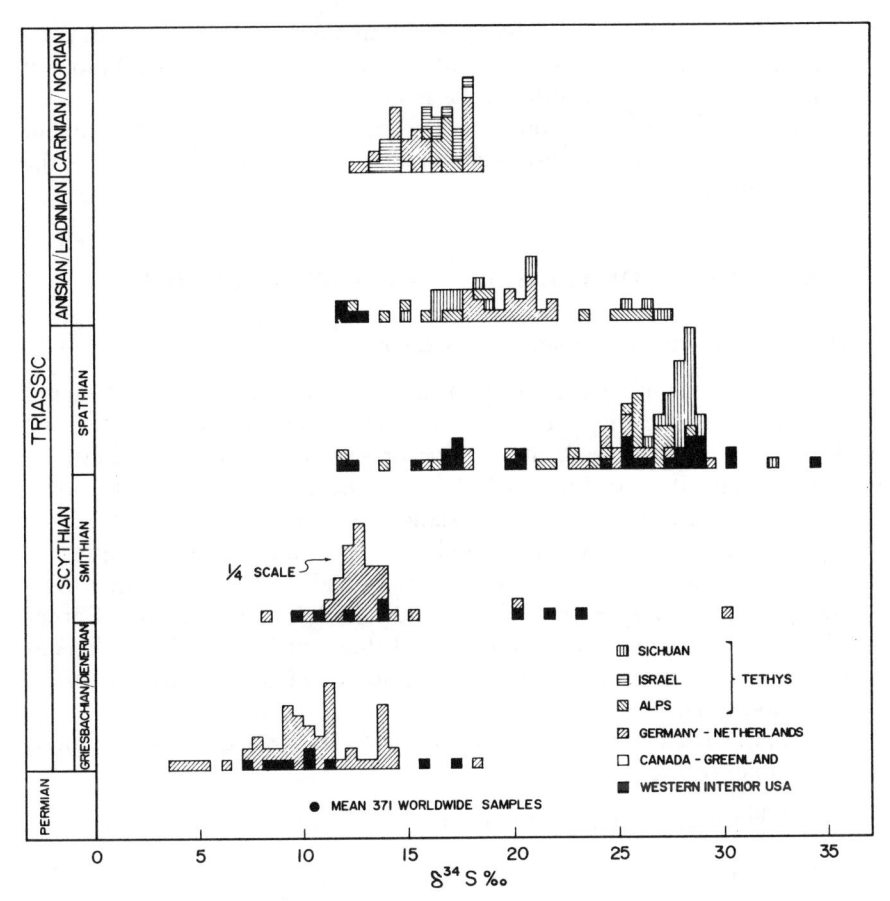

Figure 2.7 The Röt event—a sharp rise of $\delta^{34}S$ (VI) at the end of the early Triassic, after Holser (1984), and Holser et al. (1988), who give sources of data. Each square is one analysis, except for the Smithian, which is increased by a factor of four. Note greatly expanded time scale in the Scythian Stage

Triassic Röt Event has since been documented in detail, as shown in Fig. 2.7, from Holser (1984), who reviewed trends and events in oceanic chemistry. The episodic nature of the marine cycle of sulphur is also demonstrated by the record of 'saline giants' that remove $CaSO_4$ from the ocean at a very high rate (Holser et al., 1980), as discussed in a previous section.

The proximal causes of isotope events like that of $\delta^{34}S$ in the Lower Triassic are speculative. For a positive swing of $\delta^{34}S$ with little or no correlative shift of $\delta^{13}C$, a massive deposition of extra sulphide in a euxinic ocean basin (Leventhal, 1983) would seem to be required. Holser and Magaritz (1987) pointed out that the extraordinary magnitude of the resulting positive swing in $\delta^{34}S$ (Figure 2.6) was facilitated by the low level of sulphate in the oceanic reservoir remaining after the widespread precipitation of sulphate in the saline giants of the Permian.

The short-term events demonstrate a certain independence of the sulphur cycle from the carbon and other cycles, although the specific origin of the events is still obscure.

2.3 MODELS OF THE SULPHUR CYCLE THROUGH TIME

2.3.1 A simple box model and its equations

Modelling of the sulphur cycle starts from a box model (like that of Figure 2.1), an estimated level of reservoirs for the recent past (like those listed in Table 2.1), and an estimate of the present balanced fluxes between reservoirs (Lein, 1983; or Figures 2.3, 2.5). Dynamic operation of the model is designed to account for the past variations of reservoirs such as indicated by Figures 2.2 or 2.6. The dynamics and balances of the system are expressed in a set of equations relating reservoirs, fluxes and their isotopic ratios. The following simple example (based on Garrels and Lerman, 1981; and Berner and Raiswell, 1983) indicates the nature of these analytical relationships as a basis for discussing the inputs, assumptions and outputs calculated by various modellers.

The sulphur (or carbon) system is described by a set of eight equations for the model of Figure 2.1.

Mass balance of the reservoirs of sulphur:

$$S_T = S_{py} + S_{ev} + S_{sw} \tag{4}$$

where the subscripts py = pyrite, ev = evaporite, sw = seawater, and T = total exogenic sulphur.

Steady state seawater condition:

$$F_{py-sw} + F_{ev-sw} = F_{sw-py} + F_{sw-ev} \tag{5}$$

where F_{py-sw} signifies flux from sedimentary pyrite to seawater, etc. This equation requires that S_{sw}, the mass of sulphate dissolved in the oceans, be constant through time, an assumption that is convenient and often made, but this constraint is so unlikely that it should be relaxed if possible.

Reservoir changes:

$$d\,S_{py}/dt = F_{sw-py} - F_{py-sw} \tag{6}$$
$$d\,S_{ev}/dt = F_{sw-ev} - F_{ev-sw} \tag{7}$$

First-order weathering reactions:

$$F_{py-sw} = k_{py} S_{py} \quad (8)$$
$$F_{ev-sw} = k_{ev} S_{ev} \quad (9)$$

where k_{py} = erosion rate for pyrite, etc.

Some modellers have also introduced first-order depositional reactions, but F_{sw-py} and F_{sw-ev} are each likely to be independent of S_{sw}, as discussed in a previous section.

Isotope mass-balance:

$$\bar{\delta}_T S_T = \bar{\delta}_{py} S_{py} + \bar{\delta}_{ev} S_{ev}^+ + \bar{\delta}_{sw} S_{sw} \quad (10)$$

Where the δs are the mean values of sulphur isotope ratio in each reservoir.

Isotope changes in seawater:

$$S_{sw}(d\,\delta_{sw}/dt) = \delta_{py} F_{py-sw} + \delta_{ev} F_{ev-sw} - (\delta_{sw} - \Delta) F_{sw-py}$$
$$-\delta_{sw} F_{sw-ev} \quad (11)$$

where the δs are the sulphur isotope values of deposited sulphur in a given period, and Δ is an average sulphate–sulphide fractionation, $\delta_{sw} - \delta_{py}$. Garrels and Lerman (1984) assume a constant $\Delta = +35‰$ in their modelling.

The constants k_{py} and k_{ev} may be evaluated from modern-day values of these parameters, and in the absence of contrary information, are assumed to have remained unchanged through time. The system is thus described by the eight equations, (4)–(11). It may be used to move backward in time, step by step through short time intervals, assuming that one knows the values for S_{sw}, δ_{sw}, k_{py}, k_{ev}, and Δ, and determine the eight unknowns δ_{py}, δ_{ev}, S_{py}, S_{ev}, F_{py-sw}, F_{sw-py}, F_{ev-sw}, and F_{sw-ev}. Even this simplified model is deceptive. Garrels and Lerman (1984) discovered that the system is not symmetrical in time, so that stepping backward is not exactly the same as stepping forward with the same assumptions and equations.

The more complex models mentioned in a previous secton, involving the carbon and silicate cycles, comprise more equations and more unknowns. Most calculations are done on computer by numerical methods.

2.3.2 Assumptions and results of some previous models

Table 2.4 summarizes some characteristics of several models that have attempted to calculate the course of the sulphur cycle through time. Each of these authors began with a set of equations similar to equations (4)–(11), put in a set of known data, a set of assumptions dependent on their judgement, and calculated the time variance of one or more parameters of the cycle.

Some time ago Nielsen (1965), Holser and Kaplan (1966), Rees (1970) and McKenzie (1972) recognized the potential of the sulphur isotope age

Table 2.4 Models for the sulphur cycle through time

Reference	Input data[a]	Assumptions[a]	Outputs[b]	Comments
Holland (1973)	$\delta^{34}S$	S_{ev}, F_{ri-sw}, fixed	dO_S	Suggests $dO_S = -dO_C$
Garrels and Perry (1974)	$\delta^{34}S$	S_{sw}, F_{ri-sw}, δ_{ri}, $\triangle$, O_{atm} fixed	dS_{py}, dC_{org}	Example calculations
Garrels et al. (1976)	$\delta^{34}S_{ev}$, $\delta^{13}C_{ls}$	6 k's for C, O, S fluxes, S_{ev} fixed	dS, dC, dO	Calculates deviations from present 'steady-state'; varies erosion rate.
Schidlowski et al. (1977)	$\delta^{34}S_{ev}$, $\delta^{13}C_{ls}$	$F_{sw-ev} = kS_{sw}$: S_{sw}/S_{ev} fixed	dO_S	Calculates $dO_S - dO_C$
Claypool et al. (1980)	$\delta^{34}S_{ev}$, $\delta^{18}O_{ev}$	'Symmetrical' net sw fluxes	dO_s	
Garrels and Lerman (1981)	$\delta^{34}S_{ev}$	S_{sw}, C_{sw}, O fixed	dS's, F's, $\delta^{13}C_{es}$	Reproduces $\delta^{13}C$ of Veizer et al. (1980)
Mackenzie and Pigott (1981)	Tectonic events	O fixed	dS, dC	Qualitative analysis of forcing functions
Schidlowski and Junge (1981)	$\delta^{34}S_{ev}$, $\delta^{13}C_{ls}$	As in 1977	dO_S	Recalculation $dO_S - dO_C$
Berner and Raiswell (1983)	$\delta^{34}S_{ev}$	S_{sw}, O fixed	dS's, dC's C_{org}/S_{py}	Garrels and Lerman (1981) model
Garrels and Lerman (1984) (A)	$\delta^{34}S_{ev}$	S_{sw}, O fixed	dS's	Corrected Garrels and Lerman (1981) model, calculated for various S's, F_{py-sw}, F_{ev-sw}.
Garrels and Lerman (1984) (B)	$\delta^{13}C_{ls}$	S_{sw}, C_{sw}, O fixed	dS_{ev}	Similar result for dS_{ev}
Lasaga et al. (1985)	$\delta^{34}S_{ev}$, $\delta^{34}C_{ls}$, tectonic factors	$\triangle$ fixed; various $S_{sw} = f(t)$ assumed	dS_{py}	Last 10^8 years. Elaboration of CO_2 model of Berner et al. (1983)

[a]Subscripts: ev = evaporite; ls = limestone; py = pyrite; ri = river water; sw = seawater; see text, Section 2.3 and Figure 2.1. 'f (t)' means 'function of (geological) time'.
[b]dO_S, dO_C = differential oxygen due to sulphur and carbon cycles, respectively.

curve for calculating time shifts in the sulphur cycle, and made sample calculations. The first comprehensive calculation was by Holland (1973), who assumed a constant river flux and residence time for sulphur in the ocean, and calculated the putative oxygen variation consequent to the variations of $\delta^{34}S$. He expected these variations to be compensated by counterflows in the carbon cycle, and while the data for $\delta^{13}C$ available to him were too diverse to support this assumption, he was able to find tentative corroboration in the averages of S_{py}/C_{org} compiled by Ronov and Migdisov (1971) for geological eras on the Russian Platform.

After Garrels and Perry (1974) elaborated the interconnections of the long-term cycles of sulphur, carbon and silica (Equation 3), Garrels et al. (1976) set up a series of rate equations for the overall system and calculated changes expected (including S_{py} and S_{ev}) in 10 Ma from changes in rates of erosion or photosynthesis. They made two important assumptions that are retained in most later models: (a) that atmospheric oxygen was held exactly balanced through geological time by the compensation of carbon oxidation/reduction for sulphur reduction/oxidation; and (b) that the sulphate reservoir in seawater also remained constant. These two reservoirs probably did not vary during the Phanerozoic by a factor of more than two or three, as attested for (a) by the fossil record and (b) by the mineralogy of evaporites (Holland, 1972), but we think that within these limits they are not very closely controlled. While assumptions of constancy in these reservoirs were necessary to the solution of the models, we should be alert to the applicability of new data to checking their variations.

Schidlowski et al. (1977) worked from a model first applied to the carbon cycle (Junge et al., 1975) that assumed exponential weathering functions and a constant ratio of S_{sw}/S_{ev}. They calculated the oxygen flux from both the sulphur and carbon cycles. Using the isotope age curves then available they were unable to balance that from sulphur with contrary variations from the carbon cycle, and ascribed the differences to oxidation/reduction of iron minerals. Later (Schidlowski and Junge, 1981) recalculated this using the carbon isotope curve of Veizer et al. (1980), and found approximate compensation, but with some possible phase lag between the sulphur and carbon systems that might indicate short-term fluctuations of atmospheric oxygen.

Garrels and Lerman (1981) extended the model of Garrels et al. (1976) described above to get a detailed output of the variations through time of the two reservoirs of sulphur (seawater constant) and their mean isotope ratios, and the fluxes that led to the differences. Assuming equation (3) they used the sulphur results to calculate a corresponding carbon isotope curve, which they found compared favourably with the compilation of Veizer et al. (1980). Berner and Raiswell (1983) made similar calculations to derive S_{py}/C_{org} through time. Garrels and Lerman (1984) have recently improved this model, to remove an inconsistency in the calculation scheme. Their results for the sulphur cycle in terms of S_{ev} are summarized in Figure 2.8, curve A. The same figure compares results from Schidlowski and Junge (1981) (curve C) and two unpublished calculations of Berner using different amounts of S_T (curves B and E). Curve D is an additional calculation by Garrels and Lerman that started with assumed reservoirs at 700 Ma BP (million years before present) and projected forward in time to the present. Interactions of the sulphur and carbon cycles were checked by recalculating the sulphate reservoir, S_{ev}, from the carbon isotope age curve, with similar results.

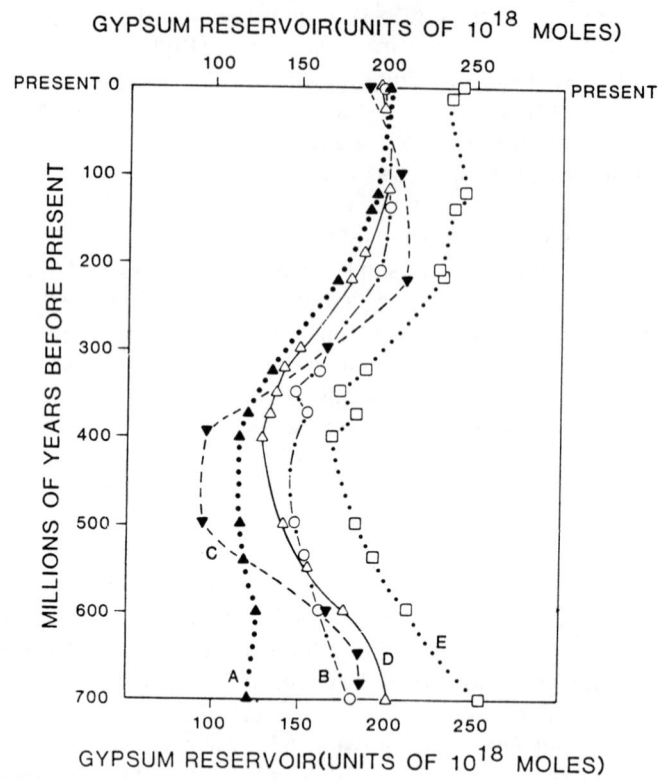

Figure 2.8 The CaSO$_4$ reservoir through time based on models with sulphur isotope input, as calculated by various investigators, after Garrels and Lerman (1984). All calculations assume a constant $\Delta = +35$, constant S_T, constant $S_{sw} = 1280 \times 10^6$ Tg (S), and δ^{34}S from Claypool et al. (1980). Curves A–C calculated backward in time by Garrels and Lerman (1984), R. A. Berner (unpublished), and Schidlowski and Junge (1981); curves D and E calculated forward in time by (D) Garrels and Lerman (1984) and (E) Berner (unpublished). The assumed initial conditions and erosional rate constants differ from model to model, as listed for A, B, D, E in Table 2.5; and for C in Schidlowski et al. (1977)

Garrels and Lerman (1984) state in conclusion, 'The trends deduced by the various modellers are similar. The sulfate reservoir decreased from 700 million years ago to 400 million years ago, then increased strikingly, reaching a maximum between 220 and 100 million. Since then it has declined slightly. The mass is about 80×10^{18} mol in all the models. This range corresponds to a change in the organic carbon reservoir (oxygen production) of 150×10^{18} mol, about 4 times the oxygen in the present atmosphere.' However, the model is insensitive to the size of the initial sulphate reservoir in the Precambrian, and in a later section we will compare these calculated sizes of the evaporite reservoir with those estimated from the geological record.

Another recent calculation, in which Garrels also participated (Lasaga *et al.*, 1985), pursues a more elaborate model first applied to the carbon system (Berner *et al.*, 1983), in which the system is driven by tectonically generated factors such as seafloor spreading. These models are not confined to a closed exogenic cycle, but take into account fluxes to the mantle at mid-ocean ridges, and out of the mantle/crust in volcanism. The effect of the assumption of constant ocean composition was tested by comparing the pyrite flux generated through time in such an ocean with one in which the sulphate content varied linearly with time—the results were not very different. Results for S_{ev} were not given in that paper, so it cannot be compared explicitly with other models.

2.3.3 A new model using reservoir inventory data

All of the previously published models discussed in the last section input a single initial value (at 0 or 700 MaBP), of each of the reservoirs S_{ev} and S_{py}, with S_{sw} constant, and calculate the variation in the evaporite and pyrite reservoirs through time. This procedure throws away such information as we have about the actual distribution of these reservoirs through time. The direct information consists of estimates of the residual masses of evaporite and pyrite sulphur for each geological period remaining after erosion, as listed in Tables 2.2 and 2.3. If we deduce or assume an erosion constant for each of these reservoirs, as was also necessary anyway in the isotope input models, we can calculate the original masses of evaporite and pyrite sulphur deposited during each geological period, independent of the sulphur isotope age curve. If we also estimate the mass of evaporite and pyrite sulphur present at the beginning of the Eocambrian, we can calculate reservoir masses S_{ev} and S_{py} for each period. This is done by letting S_{ev} and S_{py} decay exponentially to account for weathering, and adding each new increment of evaporite or pyrite deposition at the end of its geological period, so that an increment is not exposed to weathering until the end of its deposition. Once the courses of change in S_{ev} and S_{py} have been calculated, changes in the seawater reservoir, S_{sw}, can be estimated by subtraction (equation 4). The entire dynamic model is highly sensitive to the choices of erosion rates and initial (Eocambrian) masses, all of which are only poorly known, so the results of such modelling must be tested by independent criteria.

Table 2.2 implies that only an insignificant mass of evaporites is presently represented in the rocks older than the late Proterozoic, and although the supposition is widely debated we have assumed for this preliminary modelling that the initial evaporite sulphate before the Eocambrian was zero. In contrast, sulphide is abundantly evident in Precambrian rocks. For the initial Precambrian value of S_{py}, the assumption of a small S_{ev} indicates a large

S_{py} (S_T constant). Garrels and Mackenzie (1971, p. 257) calculated a ratio of 0.77 for Precambrian/Phanerozoic sedimentary mass. Applying this ratio to the data of Migdisov et al. (1983), we calculate a preserved mass of Precambrian pyrite sulphur of 3800×10^6 Tg (S).

Other modellers have assumed various erosion rates within a rather restricted range (Table 2.5). In our modelling we initially calculated a smaller erosion rate of $k_{ev} = 0.86 \times 10^{-9}$ a^{-1} from the estimate of Holser et al. (1980) that 18% of Permian $CaSO_4$ has been eroded in the 250 Ma since the close of that period. If this same weathering constant is used for both sulphate and pyrite and for all Phanerozoic time, then very large swings in S_{sw} result, which violate the Holland constraint (Holland, 1972) that limits fluctuations in the concentration of sulphate in seawater to a factor of about two. Accordingly, we ran the model at various erosion rates for sulphide and found that for k_{py} between 1.79 and 1.95×10^{-9} a^{-1} (k_{ev} still 0.86×10^{-9} a^{-1}), the Holland constraint can be met. In all of these cases evaporite sulphate S_{ev} increased monotonically by irregular amounts through the entire Phanerozoic, owing to an erosion rate k_{ev} so low that it never completely compensated for each episode of evaporite deposition. Increasing k_{ev} to 2.0×10^{-9} a^{-1} and varying k_{py} between 1.70 and 1.89×10^{-9} a^{-1} also keeps S_{sw} essentially within those bounds, and these levels of k are more commensurate with those estimated by Garrels and Lerman (1984; see Table 2.5) from the decrement of rock masses with age.

These calculations demonstrate that, even with the large and episodic variations in depositional rate of sulphate and sulphide, and with nominal erosion rate constants, the variations in the sulphur cycle can be modelled within geochemical constraints and independently of the sulphur isotope age curve and equations (10) and (11). We can now refine the inventory model by making *partial* use of the isotope data. We can calculate the mean isotope levels present in the three reservoirs, $\bar{\delta}^{34}S_{ev}$, $\bar{\delta}^{34}S_{py}$, and $\bar{\delta}^{34}S_{sw}$, at the end of each geological period from the results of a run of the inventory model that already satisfies the Holland criterion. The mean isotope value, $\bar{\delta}_T$, is then summed by equation (10). The set of calculated values of $\bar{\delta}_T$ for all periods for that run are then examined for constancy at a value near zero. For example, for $k_{ev} = 2.0 \times 10^{-9}$ a^{-1}, a k_{py} of 1.7×10^{-9} a^{-1} (the lowest value consistent with the Holland constraint) gives the results shown in Tables 2.6–2.8 and illustrated in Figures 2.9–2.11. As listed in Table 2.8, column 10, the resulting $\bar{\delta}_T$ varies only from +0.23 to +2.74, with a mean for all periods of $+1.24 \pm 0.77$. For $k_{py} = 1.89 \times 10^{-9}$ a^{-1} (the highest value consistent with the Holland constraint) $\bar{\delta}_T$ ranges from −1.12 to +1.18. Either of these ranges of values seems close to constancy considering the uncertainties of the inputs, including the assumption of k's constant with time, and no leakage in or out of the exogenic cycle (Section 2.2.6; equation 4). However, the model has not yet been subjected to sensitivity tests, and

Table 2.5 Input conditions for modelling of the sulphur cycle

Modeller	Figure or table	Initial conditions						Erosion rate constants	
		Time (Ma)	S_{ev} (10^6 Tg)	S_{py} (10^6 Tg)	δ_{sw} (‰ moles)	δ_{ev} (‰ moles)	δ_{py} (‰ moles)	k_{ev} (10^{-9} a^{-1})	k_{py} (10^{-9} a^{-1})
Garrels and Lerman	F2.8A	0	6410	6410	+20.0	+18.5	−15.7	5.0	2.5
Berner	F2.8B	0	6410	6410	+19.0	+19.0	−16.0	4.0	2.0
Garrels and Lerman	F2.8D	700	6410	6410	+17.0	+19.0	−16.	5.0	2.5
Berner	F2.8E	700	8180	8180	+17.0	+19.0	−16.0	4.0	2.0
Present work	T2.6–2.8 F2.9	0	2590	5080	—	+22.0	− 9.6	2.0	1.7

Table 2.6 Inventory model: sulphate reservoir (for $k_{ev} = 2.0 \times 10^{-9}$ a^{-1})

Period	Time (Ma[a]) End	Duration	Mass (10^6 Tg) Present	Original	Total[b]	Flux (Tg a^{-1}) To ev	To sw	δ_{ev} (‰)
Neogene	0	23	430	430	2594	18.69	4.43	18.91
Palaeogene	23	42	64	67	2266	1.59	4.59	18.57
Cretaceous	65	75	324	369	2391	4.92	4.36	18.60
Jurassic	140	55	604	799	2350	14.52	3.28	18.98
Triassic	195	35	120	177	1732	5.05	3.22	20.05
Permian	230	50	372	589	1668	11.78	2.27	20.50
Carboniferous	280	65	89	155	1192	2.39	2.21	25.35
Devonian	345	50	184	366	1180	7.33	1.71	26.77
Silurian	395	40	31	68	900	1.70	1.73	28.24
Ordovician	435	65	42	100	902	1.54	1.71	28.54
Cambrian	500	70	220	598	912	8.54	0.68	29.26
Eocambrian	570	130	116	362	362	2.78	0	28.80

[a] Arbitrarily assuming the time scale of Van Eysinga (1975).
[b] Total mass at end of period.

Table 2.7 Inventory model: sulphide reservoir (for $k_{py} = 1.7 \times 10^{-9}$ a^{-1})

Period	Time (Ma^a) End	Duration	Mass (10^6 Tg) Present	Original	Total	Flux (Tg a^{-1}) To py	To sw	δ_{py} (‰)
Neogene	0	23	264	264	8 867	11.47	14.91	−6.84
Palaeogene	23	42	375	390	8 946	9.28	15.08	−6.75
Cretaceous	65	75	802	895	9 190	11.94	15.04	−6.62
Jurassic	140	55	685	869	9 422	15.80	15.24	−5.37
Triassic	195	35	417	581	9 391	16.59	15.43	−5.31
Permian	230	50	462	683	9 351	13.66	15.38	−5.28
Carboniferous	280	65	500	804	9 437	12.38	15.52	−5.33
Devonian	345	50	605	1 087	9 641	21.75	15.18	−5.14
Silurian	395	40	147	287	9 312	7.18	15.88	−5.69
Ordovician	435	65	272	569	9 660	8.76	16.34	−5.48
Cambrian	500	70	427	998	10 150	14.26	16.53	−5.45
Eocambrian	570	130	113	298	10 310	2.29	19.05	−5.76
Precambrian	700	—	3800	12 490	12 490	—	—	−6.00

Table 2.8 Inventory model: sulphur reservoir summary (for $k_{ev} = 2.0 \times 10^{-9}$ a^{-1}; $k_{py} = 1.7 \times 10^{-9}$ a^{-1})

Period (end) (1)	Time (2)	Mass (10⁶ Tg) py (3)	ev (4)	sw (5)	$\bar{\delta}$ (‰) py (6)	ev (7)	sw (calc)[a] (8)	sw(obs)[b] (9)	TC (10)
Neogene	0	8 867	2594	1291	−6.84	+18.91	+20.6	+20.6	+1.18
Palaeogene	23	8 946	2266	1540	−6.75	18.57	21.7	17.7	+0.70
Cretaceous	65	9 190	2391	1171	−6.62	18.60	26.8	16.5	+0.23
Jurassic	140	9 422	2350	980	−5.37	18.98	21.5	16.9	+0.83
Triassic	195	9 391	1732	1629	−5.31	20.05	18.5	16.1	+0.87
Permian	230	9 351	1668	1733	−5.28	20.50	17.4	11.6	+0.39
Carboniferous	280	9 437	1192	2123	−5.33	25.35	16.5	15.9	+1.07
Devonian	345	9 641	1180	1930	−5.14	26.77	17.1	23.5	+2.15
Silurian	395	9 312	900	2540	−5.69	28.24	16.8	24.6	+2.74
Ordovician	435	9 660	902	2190	−5.48	28.54	19.3	22.8	+1.78
Cambrian	500	10 150	912	1687	−5.45	29.26	25.9	29.5	+1.66
Eocambrian	570	10 310	362	2079	−5.76	28.80	30.8	28.8	—

[a] Assuming $\bar{\delta}_T = 0$.
[b] Table 2.2, column 10.
[c] Using δ_{py}, δ_{ev}, δ_{sw} (obs).

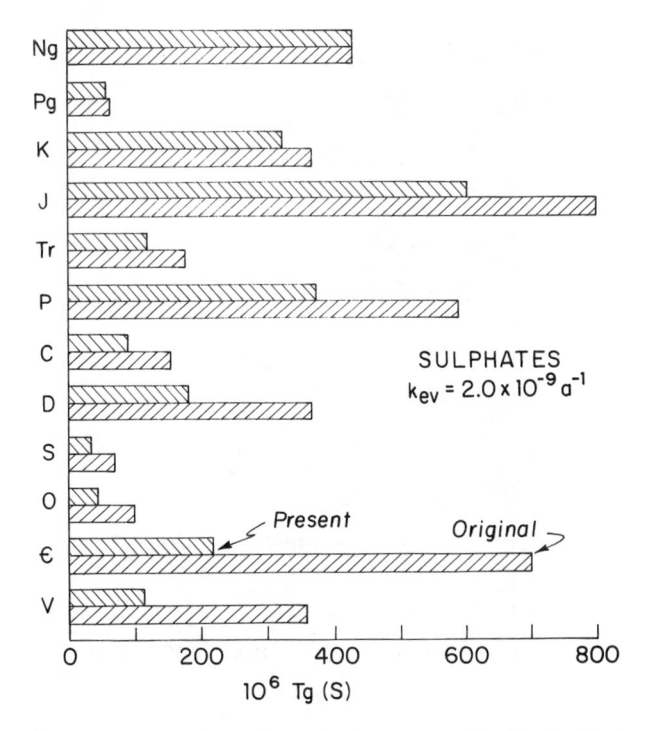

Modelling the Sulphur Cycle Through Phanerozoic Time

Figure 2.9 Mass of sulphate in the sedimentary section (Table 2.6), showing both that presently preserved for each geological period, and an originally deposited mass calculated by an erosional rate constant of $k_{ev} = 2.0 \times 10^{-9}\,a^{-1}$. V = Vendian = Eocambrian = Precambrian

it is evident that the large mass of pyrite relatively constant in $\bar{\delta}_{py}$ (Table 2.7) effectively buffers variations in calculated $\bar{\delta}_T$. Stated otherwise, the $\bar{\delta}_{sw}$ that would be calculated from the inventory model (Table 2.8, column 8) does not track the δ_{sw} observed in evaporite rocks (Table 2.8, column 9).

Figure 2.9 makes clear that, whatever the assumptions of erosion rate, the deposition of massive evaporites (and also of sulphide) is so episodic (Holser et al., 1980) that the widely accepted assumption (Table 2.4) of constant seawater, S_{sw}, is very unlikely to hold. The up-and-down variations of the sulphate reservoir, S_{ev}, given by the inventory-based model (Figure 2.9) contrast with those of the isotope-based model (Figure 2.8, which includes 1280×10^6 Tg (S) in seawater).

2.3.4 Comparison of isotope-based and inventory-based models

The isotope-based model relies on one set of relatively precise input data—the $\delta^{34}S(VI)$ isotope age curve, and on numerous assumptions: constant

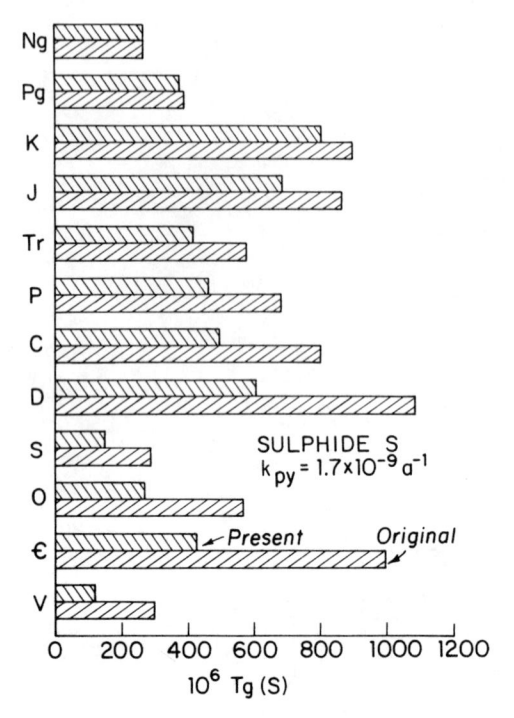

Figure 2.10 Mass of sulphide in the sedimentary section (Table 2.7); erosional rate constant $k_{py} = 1.7 \times 10^{-9}$ a^{-1}. V = Vendian = Eocambrian = Precambrian

mass of exogenic S, mass of seawater S, erosion rates and fractionation factor. The inventory-based model starts from observations of reservoir masses, assumes only constant erosion rates, and uses both isotope age curves, $\delta^{34}S(VI)$ and $\delta^{34}S(-II)$, only in refining the guessed values of erosion rates. We would like to make clear, however, that our estimates of the reservoirs and of the sulphide isotope age curve are for the most part still only provisional and imprecise, so that we cannot clearly say that our new model is already a closer approximation to history. In particular, referring to our basic input displayed in Tables 2.2 and 2.3, the masses of evaporite sulphate (Table 2.2, column 8) and platform sulphide (Table 2.3, column 2) are fairly well based in separate measurements for geological periods. The geosynclinal sulphide (Table 2.3, column 3) and the dispersed sulphate (Table 2.2, columns 4, 5, 6, 7) had to be apportioned among geological periods. We used the masses of rock for this apportionment, while Migdisov et al. (1983; Table 2.15) made a different distribution of sulphide apportioned according to masses of organic carbon measured for each geological period. In addition, we had to make a substantial correction that transferred some

Modelling the Sulphur Cycle Through Phanerozoic Time

measured sulphate to the sulphide reservoir because of oxidation of the sampled material. Finally, we followed Ronov (e.g. 1980) in guessing that the unmeasured composition of the continental shelf and slope and the oceanic sediments were all of a mean composition equivalent to that of the sum of continental platforms and geosynclines, although unlike Ronov we calculated evaporite rocks separately. Is this pile of minor ad-hoc assumptions made in constructing the inventories any better or worse than the gross assumptions made in the isotope models? If one compares the models independently of whether oxidized sulphur is in $CaSO_4$ or seawater, one finds that the mass of oxidized sulphate in the various runs of the inventory-based models is strongly dependent on the erosion rate constants. From the Precambrian to the present, the mass of oxidized sulphate would have either monotonically decreased (k_{ev}, k_{py} = 0.86 × 10^{-9} a^{-1}), monotonically increased (k_{ev} = 0.86 × 10^{-9} a^{-1}, k_{py} = 1.79 × 10^{-9} a^{-1}), or remained roughly constant (k_{ev} = 2.00 × 10^{-9} a^{-1}, k_{py} = 1.70 × 10^{-9} a^{-1}, as in Figure 2.11). These trends may be compared to the isotope-based models of Figure 2.8.

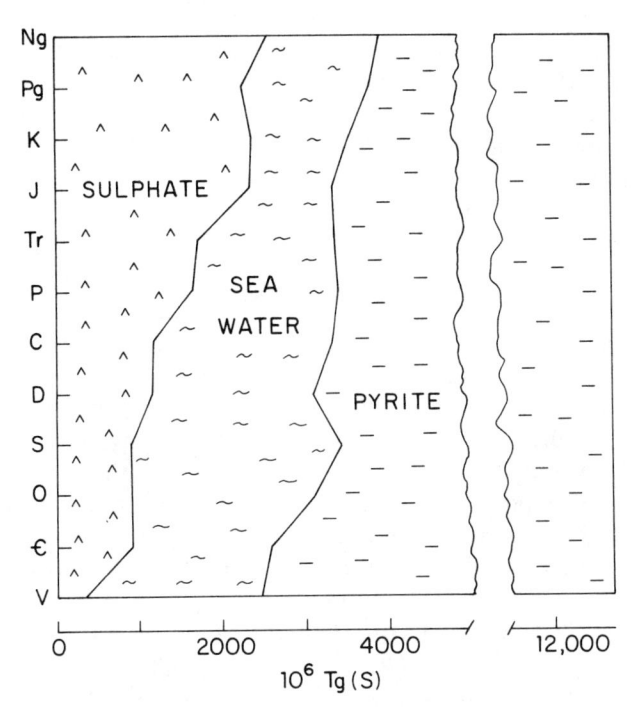

Figure 2.11 Variations of masses of sulphur in three reservoirs through time (Table 2.8), based on an inventory model with erosion constants optimized for Holland criterion (Holland, 1972) and for overall isotope balance through time (k_{ev} = 2.0 × 10^{-9} a^{-1}, k_{py} = 1.7 × 10^{-9} a^{-1}). V = Vendian = Eocambrian = Precambrian

The relative constancy of δ_T that can be modelled from observed inventories of sulphate and sulphide (Table 2.8, column 9) is an indication that the exogenic system is virtually closed (Section 2.2.6), that is, mid-ocean ridge activity, volcanicity and metamorphism have not been an *important* part of the sulphur cycle in the Phanerozoic. This is in agreement with the observation of Migdisov *et al.* (1983, p. 87) that 'A relationship between sulphur and volcanic intensity has not been found, possibly due to the buffering activity of the ocean; nor was it observed when comparing the rate of increase of evaporite sulphur and the rate of accumulation of pyrite sulphur.'

All isotope modellers necessarily assume that the mean effective fractionation in sulphide reduction has remained constant through time, as averaged over all the appropriate facies and rates of sedimentation. Calculation from the inventoried isotope data, however, shows that Δ varies from a low of +16.2‰ (Permian) to a high of +36.9‰ (Silurian), with a mean of +27.6‰ over the geological periods (Table 2.3, column 9). Similar variations in Δ are illustrated in detail by Migdisov *et al.* (1983, pp. 51, 87), who also suggested that at least for the Russian Platform the isotope age curve for $\delta^{34}S$ (–II) is a mirror image of, rather than a parallel to, that for $\delta^{34}S$ (VI). However, the data going into these preliminary estimates of $\delta^{34}S$ (–II are so diverse that it is not yet clear whether these calculated variations of Δ with time are significant. Variations in Δ may be one of the main circumstances accounting for the fact that the isotope-based models do not come close to reproducing inventories based in any regular way (constant k) on the geologically observed time distribution of sulphur in sediments.

2.4 CONCLUSIONS

This comparison and preliminary evaluation of various previous models based on the sulphur isotope age curve, with one based on the present inventories of sulphur, suggest that not only do the former differ among themselves (Figure 2.8), but the two methods also give differing answers. The least known and most critical factors seem to be the rate constants for erosion that are applied to each of the reservoirs. The inventory data suggest that the assumptions of constant seawater composition and constant mean isotopic fractionation, which are primary assumptions of many isotope-based models, may also be far from the mark.

What is needed now is a critical re-evaluation of the system, in which both isotope and inventory data are taken as inputs, each input with an evaluated precision, and the set of equations solved for a best fit to the complete (but generally imprecise) data set. A necessary preliminary step will be a critical evaluation of the precision of the inventory data; such is already estimated (Lindh, 1983; Saltzman *et al.*, 1982) for the sulphate isotope data.

The erosion rate constants (or variables) are apparently critical, and the study of factors that control them should be a primary objective of future work. A first step might be to extend to the early Mesozoic and the Palaeozoic the correction factors, for variations in erosion based on exposed continental area and other tectonic factors, that were applied by Lasaga et al. (1985) to the last 100 Ma.

Most of the isotope-based modelling and all of the inventory-based modelling depends on long-term trends of isotopes and inventories, and does not take much account of the short-term events discussed in Section 2.2.12. A further objective of future work should be an assessment as to whether such events are driven by the same equations and connections as the long-term trends of the cycle.

Finally, an inventory approach to the corresponding variations with time in the carbon cycle should be attempted, integrated with the carbon isotope data, and then the carbon cycle compared with the sulphur cycle to assess the extent of oxidation–reduction compensation between the two cycles (equation 2).

REFERENCES

Andrews, A. J. (1979). On the effect of low-temperature seawater–basalt interaction on the distribution of sulfur in oceanic crust, Layer 2. *Earth Planet. Sci. Lett.* **46**, 68–80.

Arthur, M. A. (1979). Paleoceanographic events—recognition, resolution, and reconsideration. *Rev. Geophy. Space Phys.* **17**, 1484–94.

Berner, R. A. (1972) Sulfate reduction, pyrite formation, and the oceanic sulfur budget. In: Dryssen, D., and Jagner, D. (Eds.) *The Changing Chemistry of the Oceans, Nobel Symposium 20*, Wiley Interscience, New York, pp. 347–61.

Berner, R. A., Lasaga, A. C., and Garrels, R. M. (1983). The carbonate–silicate geochemical cycle and its effect on atmospheric carbon dioxide over the past 100 million years. *Am. J. Sci.*, **283**, 641–83.

Berner, R. A., and Raiswell, R. (1983). Burial of organic carbon and pyrite sulfur in sediments over Phanerozoic time: A new theory. *Geochim. Cosmochim. Acta*, **47**, 855–62.

Burke, W. H., Denison, R. E., Hetherington, E. A., Koepnick, R. B., Nelson, H. F., and Otto, J. B. (1982). Variation of seawater $^{87}Sr/^{86}Sr$ throughout Phanerozoic time. *Geology*, **10**, 516–19.

Claypool, G. E., Holser, W. T., Kaplan, I. R., Sakai, H., and Zak, I. (1980). The age curves of sulfur and oxygen isotopes in marine sulfate and their mutual interpretation. *Chem. Geol.*, **28**, 199–260.

Edmond, J. M., Measures, C., McDuff, R. E., Chan, L. H., Collier, R., Grant, B., Gordon, L. I., and Corliss, J. B. (1979). Ridge crest hydrothermal activity and the balances of the major and minor elements in the ocean: The Galapagos data. *Earth Planet. Sci. Lett.*, **46**, 1–18.

Eriksson, E. (1960). The yearly circulation of chloride and sulfur in nature; meteorological, geochemical, and pedological implications, part II. *Tellus*, **12**, 63–109.

Garrels, R. M., and Lerman, A. (1981). Phanerozoic cycles of sedimentary carbon and sulfur. *Proc. Natl. Acad. Sci. USA*, **78**, 4652–6.

Garrels, R. M., and Lerman, A. (1984). Coupling of the sedimentary sulfur and carbon cycles—An improved model. *Am. J. Sci*, **284**, 989–1007.

Garrels, R. M., Lerman, A., and Mackenzie, F. T. (1976). Controls of atmospheric O_2 and CO_2: past, present and future. *Am. Scient.*, **64**, 306–15.

Garrels, R. M., and Mackenzie, F. T. (1971). *Evolution of Sedimentary Rocks*, Norton, New York, 397pp.

Garrels, R. M., and Perry, E. A. (1974). Cycling of carbon, sulfur, and oxygen through geologic time. In: Goldberg, E. D. (Ed.) *The Sea*, Vol. 5, Wiley, New York, pp. 303–36.

Goldhaber, M. B., and Kaplan, I. R. (1974). The sulfur cycle. In: Goldberg, E. D. (Ed.) *The Sea*, Vol. 5, Wiley, New York, pp. 569–655.

Grinenko, V. A., and Ivanov, M. V. (1983). Principal reactions of the global biogeochemical cycle of sulphur. In: Ivanov, M. V., and Freney, J. R. (Eds.) *The Global Biogeochemical Cycle of Sulphur*, Wiley, New York, pp. 1–23.

Hoefs, J. (1981) Isotopic composition of the ocean-atmosphere system in the geologic past. *Am. Geophys. Union Geodynam. Ser.* **5**, 110–18.

Holeman, J. N. (1968). The sediment yield of major rivers of the world. *Water Resources Res.*, **4**, 737–47.

Holland, H. D. (1972). The geologic history of seawater—an attempt to solve the problem. *Geochim. Cosmochim. Acta*, **36**, 637–57.

Holland, H. D. (1973). Systematics of the isotopic composition of sulfur in the oceans during the Phanerozoic and its implications for atmospheric oxygen. *Geochim. Cosmochim. Acta*, **37**, 2605–16.

Holser, W. T. (1977). Catastrophic chemical events in the history of the ocean. *Nature*, **267**, 403–8.

Holser, W. T. (1979). Mineralogy of evaporites. *Rev. Mineral.*, **6**, 211–94.

Holser, W. T. (1984). Gradual and abrupt shifts in ocean chemistry during Phanerozoic time. In: Holland, H. D., and Trendall, A. F. (Eds.) *Patterns of Change in Earth Evolution*, Dahlem Konferenzen, Springer-Verlag, Berlin, pp. 123–43.

Holser, W. T., and Kaplan, I. R. (1966). Isotope geochemistry of sedimentary sulfates. *Chem. Geol.*, **1**, 93–135.

Holser, W. T. and Magaritz, M. (1987). Chemical events near the Permian-Triassic boundary. *Mod. Geol.*, **11**, 155–80.

Holser, W. T., Kaplan, I. R., Sakai, H., and Zak, I. (1979). Isotope geochemistry of oxygen in the sedimentary sulfate cycle. *Chem. Geol.l*, **25**, 1–17.

Holser, W. T., Hay, W. W., Jory, D. E., and O'Connell, W. J. (1980). A census of evaporites and its implications for oceanic geochemistry. *Geol. Soc. Am. Abstr. Progr.*, **12**, 449.

Holser, W. T., Mackenzie, F. T., Maynard, J. B., and Schidlowski, M. (1988). Geochemical cycles of carbon and sulfur. In: Gregor, C. B., et al. (Eds.) *Chemical Cycles in the Evolution of the Earth*, Wiley, New York, pp. 105–73.

Ivanov, M. V., and Freney, J. R. (Eds.) (1983). *The Global Biogeochemical Sulphur Cycle*, Wiley, New York, 470 pp.

Jorgensen, B. B., Hansen, M. H., and Ingvorsen, K. (1977). Sulfate reduction in coastal sediments and the release of hydrogen sulfide to the atmosphere. In: Krumbein, W. E. (Ed.) *Environmental Biogeochemistry and Geomicrobiology*, Vol. 1, Ann Arbor Science, Ann Arbor, pp. 245–53.

Junge, C. E., Schidlowski, M., Eichmann, R., and Pietrek, H. (1975). Model calculations for the terrestrial carbon cycle: carbon isotope geochemistry and evolution of photosynthetic oxygen. *J. Geophys. Res.*, **80**, 4542–52.

Kovach, J. (1980). Variations in the strontium isotope composition of seawater during Paleozoic time determined by analysis of conodonts. *Geol. Soc. Am. Abstr. Progr.*, **12**, 465.

Krouse, H. R., and Grinenko, V. A. (Eds.) *Stable Isotopes in the Assessment of Natural and Anthropogenic Sulfur in the Environment*, unpublished.

Lasaga, A. C., Berner, R. A., and Garrels, R. M. (1985). An improved geochemical model of atmospheric CO_2 fluctuations over the past 100 million years. *Amer. Geophys. Un. Geophys. Monogr.*, **32**, 397–411.

Lein, A. (1983). The sulphur cycle in the lithosphere. Part II. Cycling. In: Ivanov, M. V., and Freney, J. R. (Eds.) *The Global Biogeochemical Sulphur Cycle*, Wiley-Interscience, New York, pp. 95–127.

Leventhal, J. S. (1983). An interpretation of carbon and sulfur relationships in Black Sea sediments as indicators of environments of deposition. *Geochim. Cosmochim. Acta*, **47**, 133–7.

Li, Y. H. (1972). Geochemical mass balance among lithosphere, hydrosphere and atmosphere. *Am. J. Sci.*, **272**, 119–37.

Lindh, T. B. (1983). Temporal variations in ^{13}C, ^{34}S and global sedimentation during the Phanerozoic. M.S. Thesis, University of Miami.

Mackenzie, F. T. and Pigott, J. D. (1981). Tectonic controls of Phanerozoic sedimentary rock cycling. *J. Geol. Soc. Lond.*, **138**, 183–96.

McDuff, R. E., and Edmond, J. M. (1982). On the fate of sulfate during hydrothermal circulation at mid-ocean ridges. *Earth Planet. Sci. Lett.* **57**, 117–32.

McKenzie, J. A. (1972). A mathematical model for the isotopic balance of sulfur in the ocean. *Geol. Soc. Amer. Abstr. Progr.*, **4**, 197.

Migdisov, A. A., Ronov, A. B., and Grinenko, V. A. (1983). The sulphur cycle in the lithosphere. Part I. Reservoirs. In: Ivanov, M. V., and Freney, J. R. (Eds.) *The Global Biogeochemical Sulphur Cycle*, Wiley-Interscience, New York, pp. 25–95.

Nielsen, H. (1965). Schwefelisotope im marinen Kreislauf und das ^{34}S der früheren Meere. *Geol. Rundsch.*, **55**, 160–72.

Nielsen, H. (1978). Sulfur. Isotopes in nature. In: Wedepohl, K. H. (Ed.) *Handbook of Geochemistry. Part 16B*, Springer-Verlag, Heidelberg.

Rees, C. E. (1970). The sulphur isotope balance of the ocean: An improved model. *Earth Planet. Sci. Lett.*, **7**, 366–70.

Ronov, A. B. (1980). *Osadochnaya Obolochka Zemli*. Nauka, Moscow, 80pp. [Trans. (1982) *Intern. Geol. Rev.* **24**, 1313–88].

Ronov, A. B., Grinenko, V. A., Girin, Yu.P., Savina, L. I., Kasakov, G. A.; and Grinenko, L. N. (1974). Effects of tectonic conditions on the concentration and isotopic composition of sulfur in sediments. *Geokhimiia*, **1974**, 1772–98 [Transl. *Geochem. Intern.*, **11**, 1246–72].

Ronov, A. B., and Migdisov, A. A. (1971). Geochemical history of the crystalline basement and the sedimentary cover of the Russian and North American Platforms. *Sedimentology*, **16**, 137–85.

Saltzman, E. S., Lindh, T. B., and Holser, W. T. (1982). $\delta^{13}C$ and $\delta^{34}S$, global sedimentation, pO_2 and pCO_2 during the Phanerozoic. *Geol. Soc. Am. Abstr. Progr.*, **14**, 607.

Schidlowski, M., and Junge, C. E. (1981). Coupling among the terrestrial sulfur, carbon and oxygen cycles: Numerical modeling based on revised Phanerozoic carbon isotope record. *Geochim. Cosmochim. Acta*, **45**, 589–94.

Schidlowski, M., Junge, C. E., and Pietrek, H. (1977). Sulfur isotope variations in marine sulfate evaporites and the Phanerozoic oxygen budget. *J. Geophys. Res.*,

82, 2557–65.

Schwarcz, H. P., and Burnie, S. W. (1973). Influence of sedimentary environments on sulfur isotope ratios in clastic rocks—Review. *Miner. Deposita*, **8**, 264–77.

Southam, J. R., and Hay, W. W. (1981). Global sedimentary mass balance and sea level changes. In: Emiliani, C. (Ed.) *The Sea*, Vol. 7, Wiley-Interscience, New York, pp. 1617–85.

Vail, P. R., and Mitchum, R. M., Jr. (1979). Global cycles of relative changes of sea level from seismic stratigraphy. *Am. Ass. Petrol. Geol. Mem.*, **29**, 469–72.

Van Eysinga, F. W. B. (1975). *Geological Timetable (3 edn)*. Elsevier, Amsterdam.

Veizer, J., Holser, W. T., and Wilgus, C. K. (1980). Correlation of $^{13}C/^{12}C$ and $^{34}S/^{32}S$ secular variations. *Geochim. Cosmochim. Acta*, **44**, 579–87.

Williams, S. N. (1978). Studies of the base metal sulfide deposits at McArthur River, Northern Territory, Australia: II. The sulfide-S and organic-C relationships of the concordant deposits and their significance. *Econ. Geol.*, **73**, 1036–56.

Wolery, T. J., and Sleep, N. H. (1988). Interactions of geochemical cycles with the mantle. Ch. 3 in: Gregor, C. B. *et al.* (Eds.) *Chemical Cycles in the Evolution of the Earth*, Wiley, New York, pp. 77–103.

Zharkov, M. A. (1981). *History of Paleozoic Salt Accumulation*, Springer-Verlag, Berlin, 308 pp.

Zharkov, M. A. (1984). *Paleozoic Salt-bearing Formations of the World*, Springer-Verlag, Berlin, 427 pp.

Evolution of the Global Biogeochemical Sulphur Cycle
Edited by P. Brimblecombe and A. Yu. Lein
© 1989 SCOPE Published by John Wiley & Sons Ltd

CHAPTER 3
Local and Global Aspects of the Sulphur Isotope Age Curve of Oceanic Sulphate

HEIMO NIELSEN

It appears logical to assume that the total amount of oceanic sulphate and its $\delta^{34}S$ value is controlled by the flux rates of what we call the 'exogenous sulphur cycle'—namely by the inflow of weathering solutions from the continent with preferentially light sulphur on the one hand and by the deposition of S-bearing sediments with a strong S isotope fractionation involved in bacterial sulphate reduction on the other. These two competing main effects keep the $\delta^{34}S$ value of the oceanic sulphate in a steady state situation which is very sensitive to changes in the flux rates. Thus the investigation of the 'age curve' gives us a chance to get information about the variations of these fluxes.

3.1 HOW 'OCEANIC' ARE THE EXISTING EVAPORITE DATA?

Any advanced modelling of the sulphur cycle and its variations during the geological past depends on knowledge of the shape of the sulphur isotope age curve of oceanic sulphate. Several thousand samples of gypsum and anhydrite from evaporation basins of different geological ages all over the world have been measured for this purpose. Thus, from a naive standpoint, one might expect that there exists a sufficiently sound basis for advanced evaluation. In the present chapter we have to discuss how far this optimism is justified.

In this respect we have to remember that no evaporite can be deposited directly within the open ocean but only in a sabkha environment or in a (more or less) closed basin. The accumulation of thick evaporite beds needs a supply of 'fresh' seawater but the mean inflow rate must always remain lower than the maximum evaporation rate. So the isotopic composition within the basin can easily develop in a manner different from the open sea. When the basin becomes completely closed (like the Caspian) the evaporation continues under the regime of the inflowing continental waters.

The question 'open or closed' is still debatable for many fossil evaporite basins. Some authors even argue that all the extreme values expressed in the age curve do not reflect changes in the open ocean but only the local 'isotope history' of the investigated evaporite basins (see, for example, Winogradow and Schanin, 1969).

This dilemma can be ruled out only when each data point employed for the construction of the age curve is verified by evaporite samples from the basins in different continents and in connection with different oceans. Unfortunately this demand is only partly met because of the scarcity of suitable evaporite beds. So the work of the different research groups is restricted to these few well-dated evaporite sequences. No wonder the curves constructed from these data agree so nicely with each other.

Figure 3.1 refers to one of the critical points of the age curve, namely the extreme increase in $\delta^{34}S$ from the Upper Permian (+11‰) to the Röt (above +25‰)* reported from northern central Europe. When both values are considered as properly 'oceanic' the increase in δ would need extremely high flux rates. Therefore we have first to check whether both values are 'oceanic'.

Figure 3.1 Palaeogeographic map of the Upper Permian with the evaporite basins sampled for sulphur isotope measurements

* Upper Lower Bunter, top of the Lower Triassic, Claypool et al. (1980), in their Figure 5, have apparently put these values into the Upper Triassic.

The values reported for the Upper Permian originate from the basins marked by black dots in Figure 3.1. All the values from the Perm basin, USSR, from northern central Europe, from the eastern Alps and Yugoslavia, and from the southern USA range around +11‰ and give the most perfect agreement of the entire age curve. The samples from Brazil have distinctly higher values.

The evaporite basin of northern central Europe was certainly fed from the Nordic Ocean and the Alpine basin from the Tethys. The evaporite basin at Bahia (east coast of Brazil) lies close to the early Atlantic but apparently too far from the 'free' ocean. The Amazon basin extended in an east–west direction through the South American subcontinent and was fed from the Pacific (Szatmari et al., 1979). This basin may be compared in shape and extent with the other large Permian basins and its S isotope range lies only slightly above the European and North American values. Perhaps this $\delta^{34}S$ difference reflects a slight difference in the depositional ages of the two groups.

For the Röt values the situation remains uncertain as well. Until quite recently these high values were known only from Europe, namely from an evaporitic belt ranging from southern Britain through the Netherlands, West Germany, Denmark and East Germany, to Poland, and from two localities in the Italian Alps. These high values were ascribed to extensive bacterial activity in local evaporation basins (Nielsen, 1965).

Quite recently many additional 'typical Röt values' have been published from Sichuan in China, quite on the other edge of the Tethys (Chen and Chu, 1988). The mean $\delta^{34}S$ value of the Sichuan evaporites from the upper Lower Triassic is 1–2‰ higher than that of the European Röt, and this range holds until the lower Middle Triassic (= Lower Muschelkalk in German nomenclature). This argues strongly for a worldwide 'Röt event', as proposed in Holser's (1977) model.

Little is known about the shape of the age curve in the Upper Triassic. Claypool et al. (1980) report values around +14‰ for Keuper sulphate. The complexity of the problem is evident from Figure 3.2: the left side shows a map of northern central Europe with the border lines of the Keuper (Upper Triassic) basin. The shaded area in Denmark denotes the area where the Keuper saline basin reached the stage of potash mineral precipitation. The $\delta^{34}S$ histograms at the right side give the isotopic distribution of gypsum/ anhydrite at the different geographic latitudes. The values decrease systematically from south to north and we have to decide which values are representative of the oceanic value: those in the vicinity of the potash basin or those from southern West Germany.

Figure 3.3 gives an example of how strong changes in the depositional facies may influence the S isotopic record within the same evaporite basin. In this case the lower values of the deeper core section agree with the

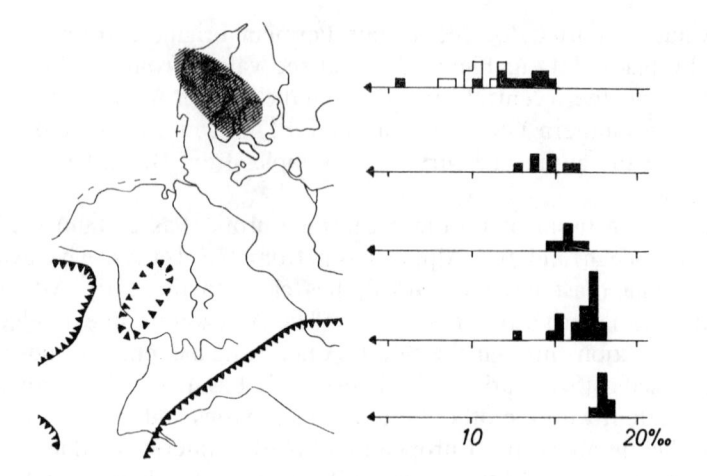

Figure 3.2 Regional trend in $\delta^{34}S$ of Keuper evaporite sulphates from southern West Germany to Denmark (open squares = sulphates of potash facies)

Jurassic values reported from elsewhere, but if drilling had ended above the break in the curve the data would have argued for a marine value above +20‰.

A fundamental difficulty for the recovery of properly dated evaporite sulphates (especially for very old samples) comes from the high mobility of

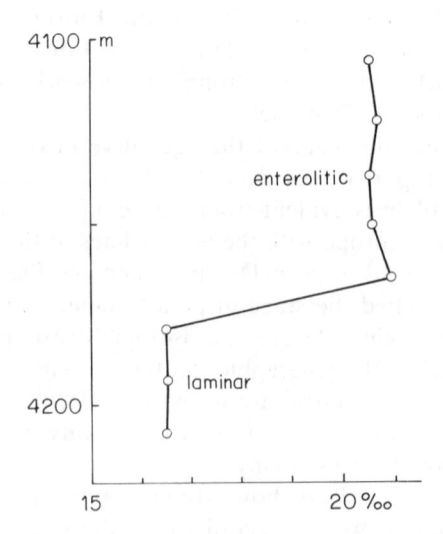

Figure 3.3 $\delta^{34}S$ profile through a Jurassic anhydrite sequence in Israel

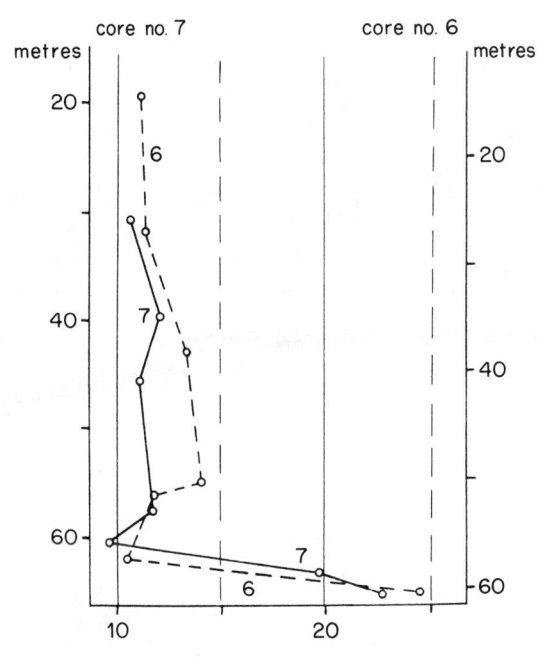

Figure 3.4 δ^{34}S core profiles from Lüthorst with Zechstein evaporite above Röt

salt. In the examples shown in Figures 3.4 to 3.8 the correct age correlation could be established only in connection with a very careful geological field investigation of the sampling sites. It is wrong to date sediments using the δ values of evaporite samples in areas with poor geological background as the isotopic composition can depend on local evolutionary history.

Figures 3.4 to 3.6 refer to a common event of 'salt tectonics'. The two S isotope profiles of Figure 3.4 contain Zechstein (Upper Permian) gypsum/anhydrite overlying Röt sulphate. The explanation of this reversed sequence is given in Figure 3.5: in connection with the tectonic evolution of the whole area the Zechstein salt has overthrusted the younger strata (upper profile) and the gypsum actually observed is the residue from leaching the overthrusted salt mass. Figure 3.6 shows an outcrop of this tectonic contact.

Figure 3.7 shows mobilized Zechstein salt sandwiched as an extended sheet into a sequence of Röt evaporites. The two δ^{34}S profiles are about two kilometres apart but the concordant Zechstein layer has been traced isotopically over several tens of kilometres. Such a layer of the 'wrong' age can confuse the interpretation, and certainly some of the unconformable data in previous publications must be ascribed to this effect. The reverse case—Röt included into Zechstein salt—is shown in Figure 3.8.

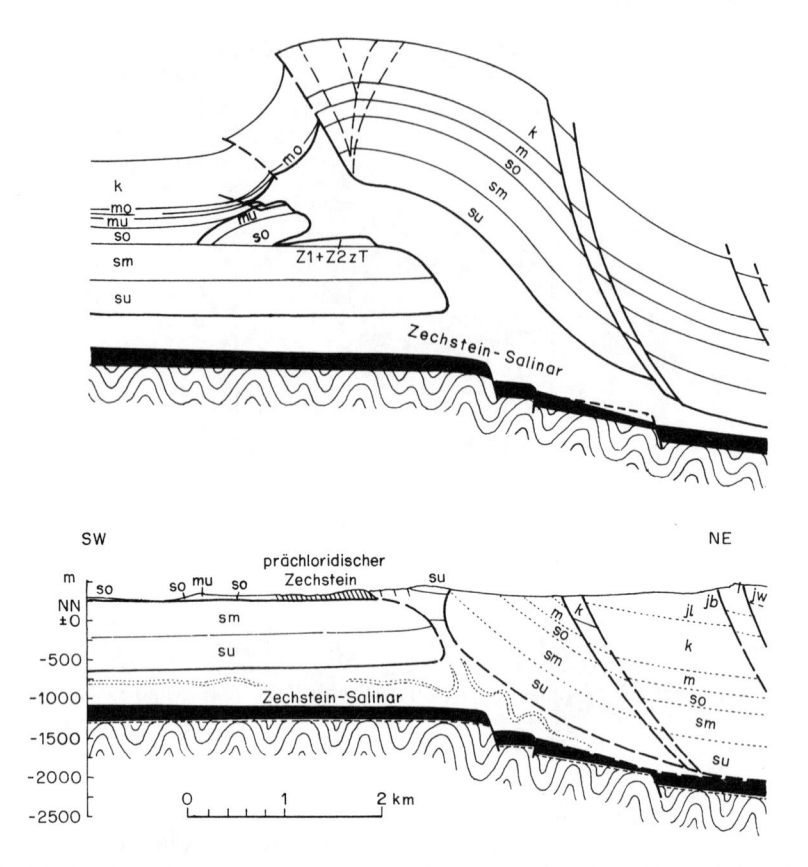

Figure 3.5 Overthrust structure at the western outline of the Hils syncline, about 50 km south of Hannover, West Germany. Upper profile = situation during the tectonic event and lower profile = actual situation. Zechstein Salinar = composite salt bed of the Upper Permian. su, sm = lower and middle section of Lower Triassic ('Bunter'). so = upper section of Lower Triassic ('Röt'). m = Middle Triassic (Muschelkalk). k = Upper Triassic (Keuper). jl, jb, jw = sections of Jurassic (from Hermann et al., 1967)

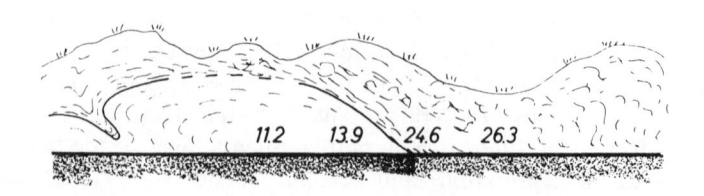

Figure 3.6 Tectonic contact between Röt and Zechstein at the rail cut Giesenberg

The Sulphur Isotope Age Curve of Oceanic Sulphate

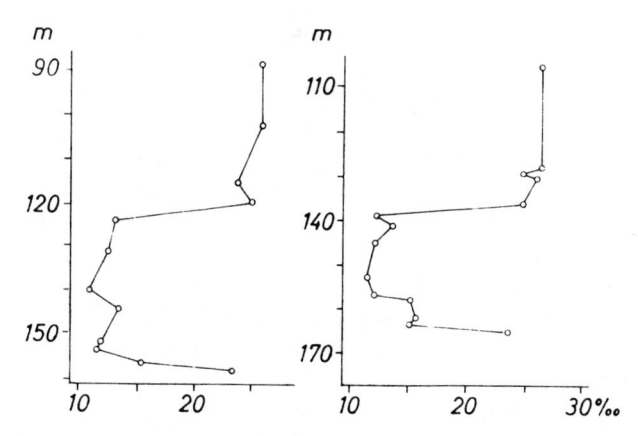

Figure 3.7 $\delta^{34}S$ profiles showing Zechstein sulphate sandwiched into Röt evaporites in the Leine graben, about 30 km north of Göttingen, West Germany. The lateral distance between the two drill holes is 1.8 km

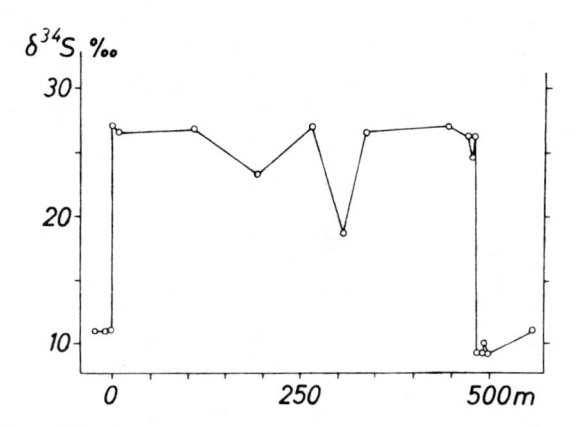

Figure 3.8 $\delta^{34}S$ profile along a mine gallery at the potash mine Adolfsglück. The mushroom-shaped Zechstein salt dome of Hope, 25 km north of Hannover has folded Röt salt into the marginal parts of its 'cap'

3.2 A NEW APPROACH TO THE SULPHUR ISOTOPE AGE CURVE

The examples mentioned above give an idea of the difficulties encountered in the construction of a new age curve. Therefore any new attempt in this direction should be regarded merely as an impetus to promote further discussion of the problem and should not claim to be 'better' than the previously published curves. In any case the large number of published and unpublished S isotope data arising since the appearance of the curve of Claypool *et al.* in 1980 makes it reasonable to risk this. The result is shown in Holser *et al.* (this volume, Figure 2.2). (In order to facilitate the comparison with the previous curves, two of them are inserted in the diagram.)

The new curve is composed of many minor oscillations which had been smoothed out in the previous constructions. The 'oceanic' nature of these oscillations remains debatable, but in my opinion they have the same quality as the strong increase in ^{34}S at the Permian/Triassic boundary. And this renders it necessary to consider their local or global nature in all the future sulphur cycle models.

The new curve omits isotopic evolution during the Precambrian, because this is the most speculative section of the entire age curve. Some Precambrian data support the decision of Claypool *et al.* (1980) to draw their curve for the Precambrian at a range slightly above +15‰ (for example Figure 2.2, Chapter 2, this volume), but there are many samples with much higher values. Examples are the values from the Krol, the Vindhyan and the Grenville (Nielsen, 1978) and a new series of barites with δ^{34}S values in the range from +20 to +35‰ and an age somewhere between 1.3 and 2 Ga in the Aggeneys–Gamsberg area, South Africa (Gehlen *et al.*, 1983). If it turns out that these values have a bearing on the age curve this will certainly be a crucial point in any further attempts to model the early biogenic sulphur cycle, but this question must be held open for further discussion.

REFERENCES

Chen Jin-Shi, and Chu Xue-Lei. (1988). Sulfur isotope composition of Triassic marine sulfates of South China. *Chem. Geol. (Isotope Geosci. Section)*, **72**, 155.

Claypool, G. E., Holser, W. T., Kaplan, I. R., Sakai, H., and Zak, I. (1980). *Chem. Geol.*, **28**, 199.

Gehlen, K. von, Nielsen, H., Chunnett, I., and Rozendaal, A. (1983). *Miner. Mag.*, **47**, 481.

Hermann, A., Hinze, C., Stein, V., and Nielsen, H. (1967). *Geol. Jahrbuch*, **84**, 407.

Nielsen, H. (1965). *Geol. Rundschau*, **55**, 160.

Nielsen, H. (1978). Chapter 16. In: Wedepohl, K.-H. (Ed.), *Handbook of Geochemistry*, Vol. 2, Springer-Verlag, Berlin.

Szatmari, P., Cavazho, R. S., and Simoes, I. A. (1979). *Econ. Geol.*, **74**, 432.

Vinogradow, W. I., and Schanin, L. L. (1969). *Zs. Angew. Geologie*, **15**, 33.

Evolution of the Global Biogeochemical Sulphur Cycle
Edited by P. Brimblecombe and A. Yu. Lein
© 1989 SCOPE Published by John Wiley & Sons Ltd

CHAPTER 4
Contribution of Endogenous Sulphur to the Global Biogeochemical Cycle in the Geological Past

A. YU LEIN AND M. V. IVANOV

4.1 NATURAL FLUX OF ENDOGENOUS SULPHUR IN THE RECENT SULPHUR CYCLE

4.1.1 Sulphur flux during subaerial volcanism

For the sake of simplicity, the evaluation of the natural sulphur cycle described in Chapter 2 examines only three basic reservoirs: the mass of reduced sulphur (S_{py}) in sedimentary rocks, mass of sulphur in evaporites (S_{ev}) and mass of sulphur in seawater (S_{sw}), and four interconnecting fluxes (see Figure 2.1). The model calculations discussed here show how the mass balance and flux parameters must be modified when the sulphur input from endogenous sources (i.e. subaerial volcanism and hydrothermal vents at the sea floor) are taken into account. In most previous publications these contributions have been strongly under estimated. The input of volcanic sulphur to the atmosphere was estimated as only 3 Tg (S) a^{-1} by Granat et al. (1976) instead of the 28 Tg (S) a^{-1} reported by Ivanov (1981). For later estimates see, for example, Lein (1983) and Berresheim and Jaeschke (1983).

A new method based on studies of sulphur concentration in Antarctic and Greenland ice cores (see Chapter 5) was used to estimate sulphur flux rates during catastrophic eruptions. The emission of volcanic sulphur to the atmosphere during the Holocene is estimated as 0.7–3.4 Tg (S) a^{-1}. This value corresponds to our estimates of the sulphur flux rates during catastrophic eruptions (Lein, 1983), since only after powerful eruptions can sulphur reach the middle layers of the troposphere and be transported by convective circulation from tropical latitudes (where most of the active volcanoes are located) to polar regions. Berresheim and Jaeschke (1983) estimated the annual volcanic sulphur flux to the atmosphere as 11.8 Tg (S).

In the present paper, we adopt the values cited above, which range from 11.8 to 28 Tg (S) a^{-1} and accept the average value of the modern sulphur flux from subaerial volcanoes as 20 Tg (S) a^{-1}.

4.1.2 Flux of endogenous sulphur from submarine volcanism

Calculations based on the known concentration data for some key elements in seawater show that the bulk of the oceanic water circulates through the Earth crust in spreading zones once in 7–8 million years. As a consequence of the sharp decrease in solubility of calcium sulphate at elevated temperatures, about 128 Tg (S) are precipitated annually in the form of anhydrite from the underground waters along the spreading axis (Edmond et al., 1979; McDuff and Edmond, 1982, and others). Several researchers believe that this anhydrite will be redissolved as soon as the impregnated rocks move far enough away from the heat source of the spreading centre (Wolery and Sleep, 1976; Mottl et al., 1979; Mottl, 1983). Such episodic trapping of anhydrite would not interfere with the endogenous sulphur cycle and can thus be neglected in further calculations.

On the other hand, an important flux of reduced sulphur to the bottom water occurs in these rift zones. It is partly connected with the outflow of lava and partly with the hydrothermal vents. It is either directly discharged as H$_2$S to the seawater or intermediately deposited as metal sulphide. Following the data given in Chapter 7, this input of hydrothermal reduced sulphur to the exogenous sulphur cycle is taken to be about 30–32 Tg (S) a^{-1}.

The δ^{34} values of hydrothermal hydrogen sulphide emanating along the spreading axis and from sulphide minerals accumulating around the orifices range from 3 to 5‰ (Kerridge et al., 1983; Styrt et al., 1981). Arnold and Sheppard (1981) conclude from these data that up to 90% of reduced sulphur must have been leached out basalt and that only 10% arises from high-temperature inorganic reduction of seawater sulphate. On the other hand, if the S isotope fractionation is estimated to be about 15‰ (Shanks et al., 1981), this process must contribute a substantially higher portion (Holland, 1973). Anyway, a minimum for the endogenic component of around 50% appears realistic, and this means a flux of endogenous sulphur in spreading zones of mid-ocean ridges somewhere between 15 and 30 Tg (S) a^{-1}. Together with the assumed 20 Tg (S) a^{-1} from subaerial volcanism, the total input of endogenous sulphur to the modern global sulphur cycle is 35 to 50 Tg (S) a^{-1}. This is equivalent in magnitude to almost half the present-day natural sulphur flux in river runoff (104 Tg (S) a^{-1}) (Ivanov, 1983).

Evidently, this contribution is not only an important factor in models of the modern global balance of sulphur, but should also be taken into account when we try to understand the variations in the exogenous sulphur cycle

4.2 CONTRIBUTION OF ENDOGENOUS SULPHUR TO THE GLOBAL CYCLE IN THE GEOLOGICAL PAST

4.2.1 Distribution of mass of volcanic and sedimentary rocks during the Phanerozoic

As shown in Figure 4.1 and Table 4.1, the maximum volcanic activity observed was during the entire Ordovician, on the Devonian–Carboniferous boundary, on the Carboniferous–Permian boundary and from the end of the Jurassic through the entire Cretaceous. The mass ratios between sedimentary (Ms) and terrigenic (Mt) rocks demonstrate extensive formation of carbonate rocks and active release of carbon dioxide by volcanoes

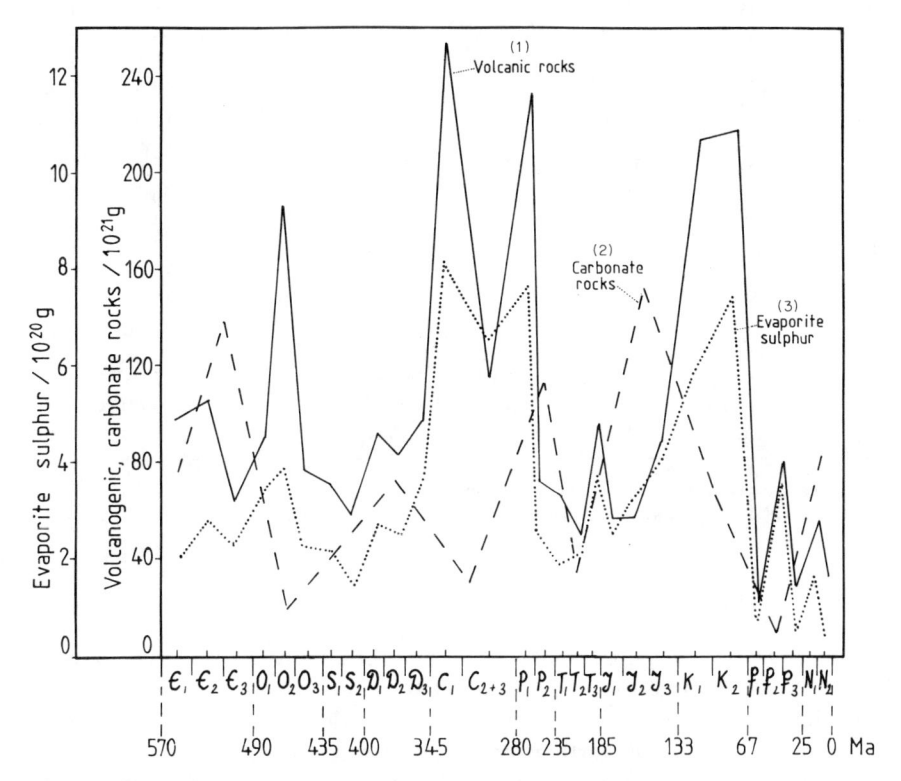

Figure 4.1 Time variations of masses of (1) volcanogenic rocks, (2) ΣCO_2 of carbonate rocks and carbonate admixtures in other rocks (from Budyko et al., 1985), and (3) evaporite sulphur (from Holser et al., Chapter 2, this volume).

Table 4.1. Masses of continental rocks, sedimentary and volcanogenic, by stratigraphic intervals

Stratigraphic interval	Mass* (10^{21}g) sedimentary, Ms	Mass* (10^{21}g) volcanogenic, mv (1)†	Ms/Mv	Mass* (10^{21}g) carbonate, Mc (2)	Mass* (10^{21}g) halogenic, Mh (3)	Mass* (10^{21}g) terrigenic Mt=Ms−(Mc+Mh)	Ms/Mt
N_2	21.16	2.24	9.45	0.45	0.023	20.53	1.03
N_1	40.13	5.57	7.20	4.71	0.92	34.50	1.16
P_3	21.56	2.04	10.56	1.56	1.43	18.57	1.16
P_2	43.62	7.78	5.60	11.57	0.46	31.59	1.38
P_1	13.54	2.46	5.50	3.20	0.07	10.29	1.32
K_2	98.56	22.04	4.47	27.00	0.18	71.38	1.38
K_1	106.81	21.39	4.89	21.40	1.13	83.48	1.27
J_3	55.95	6.55	8.54	16.20	1.66	38.09	1.47
J_2	49.59	5.71	8.68	11.20	0.02	38.38	1.29
J_1	46.53	5.57	8.35	10.90	0.046	35.58	1.31
T_3	59.44	21.17	2.80	11.80	1.012	46.63	1.27
T_2	22.26	11.04	2.01	6.10	0.023	16.14	1.38
T_1	23.37	4.73	4.94	4.00	0.23	19.14	1.22
P_2	37.83	6.97	5.42	4.60	0.805	32.43	1.17
P_1	61.76	23.74	2.60	28.00	3.17	30.59	2.02
C_2+C_3	69.25	10.95	6.32	22.40	0.07	46.78	4.48
C_1	60.48	25.62	2.36	35.20	0.368	24.91	2.42
D_3	54.84	26.46	2.07	19.30	0.322	33.19	1.56
D_2	58.65	22.86	2.57	19.50	0.437	38.71	1.52
D_1	53.22	14.78	3.60	9.30	0.046	43.87	1.21
S	65.06	6.64	9.80	12.70	0.299	52.80	1.25
O	92.55	17.95	5.15	22.90	0.069	69.58	1.33
E_3	44.53	3.47	12.83	16.68	0.046	27.78	1.60
E_2	49.20	5.40	9.11	19.50	0.69	29.00	1.70
E_1	56.25	8.85	6.35	22.40	2.53	31.29	1.80

* Masses are given according to Ronov (1980).
†(1), (2), (3) in Figure 4.1

(Budyko et al., 1985) together with minimum quantities of terrigenic rocks (Table 4.1). No relation was found between the quantity of volcanic rocks and the amount of sulphur in evaporites (Migdisov et al., 1983). This fact is mentioned without further development in Chapter 2. Furthermore, Holser et al. (Chapter 2, this volume) note that Migdisov and co-workers (1983) did not account for the release of sulphur during rock erosion, although this is of particular importance in the evaluation of the actual quantity of halides as the most easily leached deposits. Tables 2.6 and 2.7 in Chapter 2 contain data for the rocks of each geological epoch, including erosion coefficients. The comparison of these new estimates of the masses of evaporite sulphur and of volcanic rocks (Figure 4.1) shows that periods of maximum evaporite formation coincide with periods of strongest volcanic activity along the lines already seen for the interrelation of volcanism with the intensity of carbon dioxide release and accumulation of carbonate rocks (Budyko et al., 1985).

4.2.2 The variation of δ^{34} in seawater, $\delta^{13}C$ of organic C, the isotopic composition of Sr and changes of the ocean level during the Phanerozoic

The 'traditional' explanation for the strong decrease in $\delta^{34}S$ of marine sulphate at the end of the Palaeozoic is assumed to arise from a sharp drop of the ocean level (regression theory) and/or a decrease in seawater salinity through deposition of huge masses of evaporites (Chapter 2). The decrease in $\delta^{34}S$ (Table 4.2, Figure 2.6) would eventually be enhanced by the inflow of light continental sulphur from weathered rocks of the Variscan orogeny.

This very mechanism explains the increase in $\delta^{13}C$ (Figure 2.6) of the organic carbon during the Permian but contradicts the Sr isotope record in seawater during that time. The continental material supplied to the ocean by increased erosion would be expected to have a mean $^{87}Sr/^{86}Sr$ ratio around 0.716. This would raise the $^{87}Sr/^{86}Sr$ minima during the Permian, the Jurassic and Cretaceous, ranging around 0.706, which is closer to the $^{87}Sr/^{86}Sr$ ratio in fresh basalts (0.703) or average values for the oceanic crust (0.704). If increasing erosion has played a role in bringing down the oceanic $\delta^{34}S$ value at the end of the Palaeozoic, its influence on the $^{87}Sr/_{86}Sr$ ratio of ocean water must have been counterbalanced by another component with isotopically light strontium. Evidently, this can only have been 'mantle' material at spreading zones.

Summing up, the following parameters are accessible to the computation of an advanced model for the sulphur cycle during the geological past: (1) the contribution of endogenous sulphur to the global cycle at a higher rate than today (Lein, 1983); (2) correspondence of the distribution curve of evaporite sulphate, masses of volcanic and terrigenic rocks (Table 4.1 and Figure 4.1); (3) correspondence of $\delta^{34}S$ age curves for sulphate and $\delta^{13}C$ for organic carbon and $CaCO_3$, together with the variations in carbon and strontium isotopic composition of seawater (Figure 2.6); (4) variation of ocean level with time, indicating the effect of tectonic factors on biogenic cycles and associated endogenous processes and volcanic gas. The most relevant of these curves are inserted in Figure 2.6 of Chapter 2.

Table 4.2 Mass and isotopic position of sulphur in reservoirs and fluxes of the Phanerozoic

Geol. epoch	t (Ma)	S_{sw_i} (Eg)†	S_{ev_i} (Eg)	$F_{py \to sw_f}$ (Eg)	$F_{py \to sw_f}$ (Eg a^{-1})	$F_{ev \to sw_f}$ $\delta^{34}S$	$F_{ev \to sw_f}$ (Eg a^{-1})	$\delta^{34}S$	$\delta^{34}S_{py}$ (‰)a	$\delta^{34}S_{ev}$ (‰)	S_x (Tg(S) a^{-1})
Ng	23	1300	430	264	14.91	−6.84	4.43	19.91	−9.6	20.6	—
Pg	42	1500	67	390	15.08	−6.75	4.59	18.57	−18.7	17.7	—
K	75	1000	369	895	15.04	−6.62	4.36	18.60	−6.0	16.5	127.3
J	55	1000	799	869	15.24	−5.37	3.88	18.98	−5.8	16.9	25.9
T	35	1700	177	581	15.43	−5.31	3.22	20.05	−4.6	16.1	—
P	50	1800	589	683	15.38	−5.28	2.27	20.50	−7.4	11.6	195.0
C	65	2200	155	804	15.52	−5.33	2.21	25.35	−0.8	15.9	343
D	50	2000	366	1084	15.18	−5.14	1.71	26.77	−12.3	23.5	—
S	40	2600	68	287	15.88	−5.69	1.73	28.24	−6.0	24.6	—
O	65	2200	100	569	16.34	−5.48	1.71	22.8	−2.6	22.8	—
E$_3$	70	1800	698	998	16.53	−5.45	0.68	29.5	+2.4	29.5	—
E$_{1,2}$	130	2100	362	298	19.05	−5.76	0	28.80	—	—	—
*T 2.6		F 2.11	T 2.6	T 2.7	T 2.7	T 2.7	T 2.6	T 2.6	T 2.3	T 2.2	

* Tables and Figures from Holser *et al.* (Chapter 2, this volume).
† Eg = 10^{18} g.

4.3 APPLICATION OF HOLSER'S 'INVENTORY' MODEL IN CALCULATIONS OF THE ENDOGENOUS SULPHUR FLUX DURING GEOLOGICAL EPOCHS

The system of equations for the numerical evaluation of the simple 'inventory' model of the exogenous sulphur cycle with only three reservoirs and four connecting fluxes is described in detail in Chapter 2. The input parameters are the erosion coefficient and the masses of reduced sulphur, evaporite sulphur and seawater sulphur. Our own calculation of the flux of endogenous sulphur to the oceanic sulphate reservoir during each epoch of the Phanerozoic applies a model (Figure 4.2) based on the material isotopic balance equation:

$$S_{sw_t} \cdot \delta^{34}S_{sw_t} = (S_{sw_{t-1}} \cdot \delta^{34}S_{sw_{t-1}}) + [(F_{py \to sw_{t-1}} \cdot \delta^{34}S\ F_{py \to sw_{t-1}};$$
$$+ (F_{ev \to sw_{t-1}} \cdot \delta^{34}S_{ev \to sw_{t-1}}) + S_x \cdot \delta^{34}S_x] - [(S_{ev_t} \cdot \delta^{34}S_{ev_t}) + (S_{py_t} \cdot \delta^{34}S_{py_t})]$$

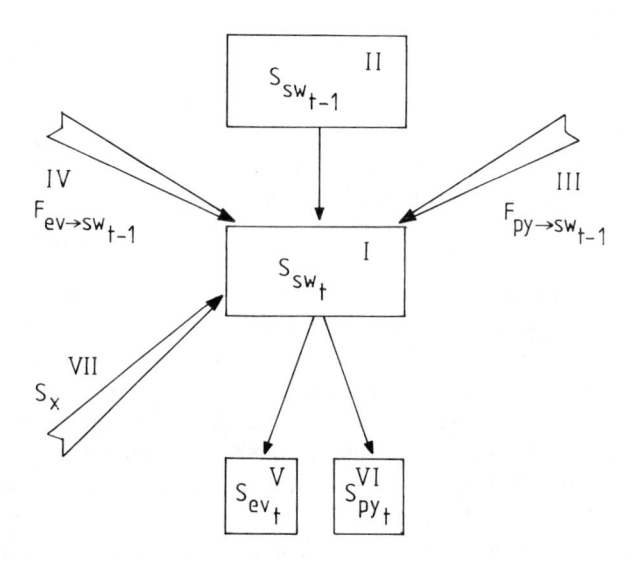

Figure 4.2 Sulphur cycle model.
I, mass of sulphur in oceanic water during time t (epoch).
II, mass of sulphur in oceanic water in the previous epoch (time t-1).
III, flux of pyrite sulphur to ocean during time t from rocks of the previous time (t-1).
IV, flux of evaporite sulphur to ocean of time t from rocks of the previous time (t-1).
V, mass of evaporite sulphur formed during t (deposited).
VI, mass of pyrite sulphur deposited in ocean of time t.
VII, flux of endogenous sulphur to ocean of time t.

where S_{sw_t} is the mass of seawater sulphate sulphur for each epoch (time t); $\delta^{34}S_{sw_t}$ is the isotopic composition of seawater sulphate sulphur for each epoch (time t);

$S_{sw_{t-1}}$ is the mass of seawater sulphate sulphur of any previous epoch (time $t-1$);

$\delta^{34}S_{sw_{t-1}}$ is the isotopic composition of sea sulphate of time $t-1$;

$F_{py \to sw_{t-1}}$ is the flux of sulphur formed on oxidation of sulphide sulphur and getting into the sea sulphate reservoir during the period calculated (time t);

$\delta^{34}S_{F_{py \to sw_{t-1}}}$ is the isotopic composition of sulphur of this flux;

$F_{ev \to sw_{t-1}}$ is the flux of sulphur due to erosion of evaporites into the sea sulphate reservoir during the period calculated (time t);

$\delta^{34}S_{F_{ev \to sw_{t-1}}}$ is the isotopic composition of sulphate sulphur of this flux;

$S_{ev}{}^t$ is the mass of evaporite sulphur deposited from the sea basin during the period calculated (time t);

$\delta^{34}S_{ev_t}$ is the isotopic composition of sulphate sulphur of evaporites deposited from the sea basin during time t;

S_{py_t} is the mass of reduced (pyrite sulphur) deposited in the sea during time t;

$\delta^{34}S_{py_t}$ is the isotopic composition of pyrite sulphur in sea sediments during time t;

S_x is the flux of endogenous sulphur to the ocean during time t;

$\delta^{34}S_x$ is the isotopic composition of endogenous sulphur of this flux, adopted as equal to 1.0%.

All the values used in the equation are listed in Table 4.2.

The calculations show that during the Carboniferous, Permian, Jurassic and Cretaceous periods the flux of endogenous sulphur made up 26–343 Tg (S) a^{-1}. It reached its maximum by the end of the Carboniferous period, namely 343 Tg (S) a^{-1}, which is two times higher than the modern anthropogenic sulphur flux into the biosphere. It is known that the addition of large amounts of anthropogenic sulphur can lead to regional problems of acidification of rainfall. Hence twofold increases in the global flux of sulphur to the biosphere in the geological past may have had catastrophic results.

4.4 CHANGES IN THE GLOBAL SULPHUR CYCLE AND CATACLYSMS IN THE GEOLOGICAL PAST

Since Cuvier's work 150 years ago we know about at least three radical changes in the composition of the fauna and flora of the Earth: late Permian, late Jurassic and on the Cretaceous–Palaeogene boundary. Modern palaeontology rejects the idea of the gradual evolution of the organic world as the only pathway of its development and admits the occurrence of crisis in the biosphere of the geological past (Sokolov, 1976).

When comparing the data on fluxes of endogenous sulphur during different Phanerozoic epochs one can easily see that the increase in the volcanic activity, and specifically a sharp peak in the release of sulphur dioxide gas associated with the global tectonics and activation of rift formation and change of sea level, may be one of the causes of mass extinction of different organisms at the boundaries of Carbonic–Permian, Permian–Triassic and Cretaceous–Palaeogenic periods.

Concluding the discussion of the role of endogenous processes in the global sulphur cycle of the geological past, one should emphasize again that the current idea of the history of our planet is based not only on the consideration of both long-term evolutionary changes but also on such episodical cataclysms.

REFERENCES

Arnold, M., and Sheppard, S. M. F. (1981). East Pacific Rise at latitude 21°N: isotopic composition and origin of the hydrothermal sulphur. *Earth Planet. Sci. Lett.*, **56**, 148–56.

Berresheim, H., and Jaeschke, W. (1983). The contribution of volcanoes to the global atmospheric sulphur budget. *J. Geophys. Res.*, **88** (6), 3732–40.

Budyko, M. I., Ronov, A. B., and Yanshin, A. L. (1985). *History of the Atmosphere*, Hydrometeoizdat, Leningrad, 207 pp. (in Russian).

Edmond, I. M., Measures, C., McDuff, R. E., Chan, L. H., Collier, R., Grant, B., Gordon, L. I., and Corliss, J. B. (1979). Ridge crest hydrothermal activity and the balances of the major and minor elements in the ocean: the Galapagos data. *Earth Planet. Sci. Lett.*, **46**, 1–18.

Granat, L., Rodhe, H., and Hallberg, R. D. (1976). The global sulphur cycle. In: Nitrogen, Phosphorus and Sulphur Global Cycles, SCOPE 7, *Ecol. Bull.* (Stockholm), **22**, 89–134.

Holland, H. D. (1973). Systematics of the isotopic composition of sulphur in the oceans during the Phanerozoic and its implications for atmospheric oxygen. *Geochim. Cosmochim. Acta*, **37**, 2605–16.

Holser, W. T., Hay, W. W., Jory, D. E., and O'Connell, W. I. (1980). A census of evaporites and its implications for oceanic geochemistry. *Geol. Soc. Am. Abstr. Progr.*, **12**, 449.

Ivanov, M. V. (1981). The global biogeochemical sulphur cycle. In: Likens, G. E. (Ed.) *Some Perspectives of the Major Biogeochemical Cycles, SCOPE 17*. John Wiley & Sons, Chichester, pp. 61–81.

Ivanov, M. V. (1983). Major fluxes of the global biogeochemical sulphur cycle. In: Ivanov, M. V., and Freney, J. R. (Eds.) *The Global Biogeochemical Sulphur Cycle, SCOPE 19*, John Wiley & Sons, Chichester, pp. 449–59.

Kerridge, J. F., Haymon, R. M., and Kastner, M. (1983). Sulphur isotope systematics at the 21°N site, East Pacific Rise, *Earth Planet. Sci. Lett.*, **66**, 91–100.

Lein, A. Yu. (1983). The sulphur cycle in the lithosphere, II. Cycling. In: Ivanov, M. V., and Freney, J. R. (Eds.) *The Global Biogeochemical Sulphur Cycle, SCOPE 19*, John Wiley & Sons, Chichester, pp. 95–129.

McDuff, R. E., and Edmond, J. M. (1982). On the fate of sulfate during hydrothermal circulation at mid-ocean ridges. *Earth Planet. Sci. Lett.*, **57**, 117–32.

Migdisov, A. A., Ronov, A. B., and Grinenko, V. A. (1983). The sulphur cycle in the lithosphere. Part I, Reservoirs. In: *The Global Biogeochemical Sulphur Cycle, SCOPE 19*, John Wiley & Sons, Chichester, pp. 25–95.

Mottl, M. J. (1983). Metabasalts, axial hot springs, and the structure of hydrothermal systems at mid-ocean ridges. *Geol. Soc. Am. Bull.*, **94**, 161–80.

Mottl, M. J., Holland, H. D., and Corr, R. F. (1979). Chemical exchange during hydrothermal alteration of basalt by seawater—II. Experimental results for Fe, Mn, and Sulphur species. *Geochim. Cosmochim. Acta*, **43**, 869–84.

Ronov, A. B. (1980). *Sedimentary Envelope of the Earth (Quantitative Regularities of Structures, Composition and Evolution)*, Nauka, Moscow, 79 pp. (in Russian).

Shanks, W. C. III, Bishoff, I. L., and Rosenbauer, R. J. (1981). Seawater sulfate reduction and sulfur isotope fractionation in basaltic systems: interaction of seawater with fayalite and magnetite at 200–350°C. *Geochim. Cosmochim. Acta*, **45**, 1977–95.

Sokolov, B. S. (1976). Organic World of the Earth in Its Way to Phanerozoic Differentiation. *Vestn. AN SSSR*, **1**, 126–43 (in Russian).

Styrt, M. M., Brackman, A. I., Holland, H. D., Clark, B. C., Pistula-Arnold, V., Eldridge, C. E., and Ohmoto, H. (1981). The mineralogy and the isotopic compositions of sulfur in hydrothermal sulfide/sulfate deposits on the East Pacific Rise, 21°N latitude. *Earth Planet. Sci. Lett.*, **53**, 382–90.

Wolery, T. J., and Sleep, N. H. (1976). Hydrothermal circulation and geochemical flux at mid-ocean ridges. *J. Geol.*, **84**, 249–75.

Part II
Evolution of the Sulphur Cycle over Recent Millennia

CHAPTER 5
Human Influence on the Sulphur Cycle

PETER BRIMBLECOMBE, CLAUS HAMMER, HENNING RODHE, ALEXI RYABOSHAPKO AND CLAUDE F. BOUTRON

Before considering recent changes in the sulphur cycle it is necessary to have a clear understanding of the current status. One of the most recent reviews, SCOPE 19, presented a picture of the global biogeochemical sulphur cycle and, in particular, considered the atmospheric component of the cycle (Ivanov and Freney, 1983). This work is based on global sulphur cycles developed by Eriksson (1960). Junge (1963), Robinson and Robbins (1968), Kellogg et al. (1972), Friend (1973) and Granat et al. (1976). This is a rapidly developing field; within a period of just a few years our ideas about the sulphur cycle may change (due to publication of new data) and the cycle itself may be noticeably altered by human activity. This introduction provides a critical review of the SCOPE 19 data on the natural and manmade inputs to the sulphur cycle and introduces the necessary corrections.

The global anthropogenic emission of sulphur to the atmosphere was estimated as 113 Tg (S) a^{-1} in SCOPE 19. This value includes sulphur emission as sulphur dioxide—98 Tg (S) a^{-1}, as sulphate—12 Tg (S) a^{-1} and as reduced sulphur compounds—3 Tg (S) a^{-1}. These estimates were based on extrapolation from data of Cullis and Hirschler (1980) which evaluated oxidized sulphur emission to the atmosphere as 110 Tg (S) a^{-1} in 1980. This value appears to be somewhat overestimated due to high emission factors adopted by the authors. Recently Möller (1984a) has reviewed the emission factors and estimated the anthropogenic flux of oxidized sulphur as 74.9 Tg (S) a^{-1} in 1977 and 90 Tg (S) a^{-1} in 1985. The data given by Möller are in good agreement with results obtained by Varhelyi (1985) who estimated the world anthropogenic emission in 1979 as 80 Tg (S) a^{-1}.

As pointed out in SCOPE 19, part of sulphur dioxide is further oxidized to sulphate, in flues, and subsequently released to the atmosphere. The sulphate fraction in emission depends on the ash content of fuel and combustion technology. In SCOPE 19 the fraction was assumed to be 10%. Adoption of this estimate gave 1985 global sulphur emissions, as sulphur dioxide 80 Tg (S) a^{-1} and that as sulphate 10 Tg (S) a^{-1}. There are no new data for re-evaluating anthropogenic emission of reduced sulphur, 3 Tg (S) a^{-1}, even though the accuracy of the estimate is still extremely low.

On the basis of the above mentioned, the following conclusions can be made: in the mid-1980s total anthropogenic sulphur flux to the atmosphere is 93 ± 14 Tg (S) a^{-1}, the sum including 80 Tg (S) a^{-1} as sulphur dioxide, 10 Tg (S) a^{-1} as sulphate and 3 Tg (S) a^{-1} as reduced sulphur compounds.

A number of estimates of the atmospheric input of reduced sulphur compounds from the ocean, including coastal regions, were published between 1980 and 1984. Barnard *et al.* (1982), Nguyen *et al.* (1983), Cline and Bates (1983), and Andreae and Raemdonck (1983) have convincingly shown that the ocean is an important source of atmospheric reduced sulphur as dimethyl sulphide (DMS). The uncertainty in the estimates is large and the flux value of 20 ± 20 Tg (S) a^{-1} given in SCOPE 19 has to be improved. More recent estimates of the ocean–atmosphere flux of sulphur as DMS range from 20 to 50 Tg (S) a^{-1}, the mean value being 35 Tg (S) a^{-1}.

The flux of gaseous sulphur from coastal waters was assumed to be 5 ± 5 Tg (S) a^{-1} (in SCOPE 19). Varhelyi and Gravenhorst (1981), Nguyen *et al.* (1983), Jaeschke *et al.* (1980), Hitchcock and Black (1984), and Aneja *et al.* (1982) showed the existence of reduced sulphur emisison in coastal waters, but the level of uncertainty remains high. The value of 5 Tg (S) a^{-1} can be taken as an average value, which coincides with the estimate given by Aneja *et al.* (1982). The error in the flux value is probably ± 4 Tg (S) a^{-1}.

The estimates of sulphur flux as H_2S, DMS, CS_2, COS, etc. from the continents remain extremely uncertain. In SCOPE 19 the value of 16 Tg (S) a^{-1} was obtained from a calculation using measured concentrations of reduced sulphur in the continental atmosphere and assuming the mean residence time to be 1 day. Varhelyi and Gravenhorst (1981) believe it to vary from 1 to 8 Tg (S) a^{-1}. In the case of a 12-hour residence time given by Jaeschke *et al.* (1980) the flux should be supposed to be about double this value. Adams *et al.* (1981) estimated the flux as 64 Tg (S) a^{-1}. Delmas and Servant (1983) carried out an interesting study aimed at evaluating soil sulphur emission in humid tropical forests. Extrapolating the values obtained to the total area of tropical forests they estimated a global reduced sulphur flux to the tropical atmosphere as 25 Tg (S) a^{-1}. Thus it seems realistic to assume that the mean atmospheric flux of reduced sulphur over the continents is 20 ± 15 Tg (S) a^{-1}. This gives a total flux of reduced sulphur to the atmosphere of 60 Tg (S) a^{-1}, a value which almost coincides with the estimate of 70 Tg (S) a^{-1} made by Möller (1984b).

Compared with earlier versions of global sulphur budgets, SCOPE 19 provides a significantly increased atmospheric input of volcanic, aeolian and sea salt sulphur. Such increases have been supported by a number of recent publications. Similar sea salt sulphur fluxes to the atmosphere are given by

Petrenchuk (1980), Varhelyi and Gravenhorst (1983), Blanchard and Woodcock (1980). Petrenchuk (1980) and Schütz (1980), and Kushelevsky et al. (1983) present rather increased inputs of aeolian dust of high sulphur content comparied with earlier published data. This enables us to retain an unchanged aeolian emission of 20 ± 10 Tg (S) a^{-1}. The volcanogenic sulphur flux of 28 Tg (S) a^{-1} assumed in SCOPE 19 is probably close to the upper limit for such emissions. Latest publications agree that the flux significantly exceeds the 3 Tg (S) a^{-1} given by Granat et al. (1976) by adding a large fumarolic contribution. In particular, Cadle (1980) believes that volcanic emission varies from 20 to 30 Tg (S) a^{-1}. According to Berresheim and Jaeschke (1983) the mean annual volcanic emission is 11.8 Tg (S) including 7.6 Tg (S) as SO_2, 1 Tg (S) as H_sS and 3.2 Tg (S) as sulphate. It is interesting to note that according to the cited authors SO_2 emission during the intervals between eruptions is 14 times greater than that during eruptions. Berresheim and Jaeschke (1983) estimate from their measurements an average rate of post-eruptive volcanic sulphur emission of about 135 Mg (S) d^{-1} and applied this to 91 volcanoes. For an additional 374 volcanoes they used significantly lower figures for the sulphur emission rate. Lein (1983) used a similar figure for emission but applied it to all 573 active volcanoes. Berresheim and Jaeschke (1983) argue that their estimate is a lower limit. We agree and regard that of Lein as an upper limit, so we are not likely to be much in error if we adopt 20 ± 10 Tg (S) a^{-1} as an average volcanic emission in our calculations. We assume that one-half of the volcanic sulphur is released to the continental atmosphere, while the rest goes into the atmosphere over the ocean. Since COS and CS_2 are relatively uniformly distributed in the Earth's atmosphere, corresponding sulphur fluxes are modelled by distributing them between the continental and oceanic atmosphere in proportion to the area of continents and oceans.

The global sulphur budget has been derived from the values adopted by SCOPE 19. This is desirable because the fluxes were estimated using a large and reliable database based on many years of field observations.

5.1 THE CYCLE TODAY

5.1.1 Introduction

The present day biogeochemical cycle of sulphur, described in the introductory section, is significantly affected by anthropogenic processes (Ivanov and Freney, 1983). Some of these anthropogenic processes are easy to identify and have been reasonably well quantified, e.g. emission of SO_2 to the atmosphere during fossil fuel combustion and the application of sulphur-containing fertilizers to agricultural land. Other processes of potential importance are less well known. The purpose of this section is to discuss, and where possible quantify, the various processes by which man may influence the sulphur cycle.

Table 5.1 summarizes the current global values for the sulphur fluxes to and from the atmosphere. It includes the estimates modified from SCOPE 19. Sulphur is transported from lithosphere to other reservoirs through the natural process of rock erosion and human activity. Erosion processes can only be called 'natural' with reservations. In fact, human activity greatly promotes both wind and water erosion of the surface lithosphere layer. Anthropogenic extraction of sulphur from the lithosphere is carried out in two principal ways: (1) deliberate extraction of sulphur as native sulphur and as sulphides (during metal extraction), (2) as an undesirable contaminant during fossil fuel combustion. Total flux of extracted sulphur is 150 Tg (S) a^{-1}, 93 Tg (S) of which is annually released to the atmosphere, 28 Tg (S) is introduced into soils together with fertilizers and the remaining 29 Tg (S)

Table 5.1 Major fluxes of the global atmospheric sulphur budget, Tg (S) a^{-1} (figures in brackets stand for possible range of flux values)

Flux	SCOPE-19 (1983)	Present work
Anthropogenic emission SO_2, SO_4^{2-}	113 (94–132)	93 (79–107)
Volcanic activity, SO_2[a]	28 (14–42)	20 (10–30)
Aeolian emission, SO_4^{2-}	20 (10–30)	20 (10–30)
COS and CS_2 emission	5 (3.5–6.5)	5 (3.5–6.5)[b]
Natural terrestrial emission, DMS, etc.	16 (4–30)	20 (5–35)
Natural emission in coastal oceanic waters, H_2S, DMS, etc.	5 (0–10)	5 (1–9)
Natural emission from open ocean, H_2S, DMS, etc.	15 (0–30)	35 (20–50)
Sea salt emission, SO_4^{2-}	140 (77–203)	144 (77–203)
Precipitation removal over the continents	51 (37.5–64.5)	51 (38–64)
Dry deposition over the continents, SO_4^{2-}	16 (5–47)	16 (5–47)
Absorption by the continental surface, SO_2	17 (7–27)	17 (7–27)
Precipitation removal over the oceans	230 (160–300)	230 (160–300)
Dry deposition over the oceans, SO_4^{2-}	17 (3–48)	17 (3–48)
Absorption by the ocean surface, SO_2	11 (5–17)	11 (5–17)
Transport from the oceanic to the continental atmosphere	20	20
Transport from the continental to the oceanic atmosphere	100.5	81

[a] Half of volcanic emission is contributed to the oceanic atmosphere.
[b] 1.5 Tg (S) a^{-1} is contributed to the continental atmosphere.

are discharged, as industrial waste waters, directly into rivers. Thus, rivers receive nearly all the sulphur that is deposited from the atmosphere on the continental surface, sulphur from mineral fertilizers, sulphur from waste waters and sulphur contributed through natural erosion by water. However, there may be some delay (decades long) before the sulphur appears in surface waters.

Figure 5.1a schematically presents the current global sulphur cycle (Figure 5.1b shows the prehistoric cycle). It can clearly be seen that the main source of mobile sulphur is the lithosphere and its final sink is the world ocean. Some part of sulphur recycles within the 'continental atmosphere–soil' system, but most of it is transported from the continental atmosphere to the oceanic one. The bulk of both oxidized and reduced sulphur of oceanic origin returns from the marine atmosphere back to the ocean. Judging from a balanced sulphur budget it seems necessary to suppose that sulphur flux from the continental atmosphere to the oceanic one is about 80 Tg (S) a^{-1}, while the opposite flux is 20 Tg (S) a^{-1}. Such an imbalance is certainly the result of anthropogenic emission, which at present accounts for about 55% of total sulphur input to the continental atmosphere.

The anthropogenic influence on the sulphur cycle largely takes place in the following ways:

(a) The emission of SO_2 (and smaller amounts of other sulphur species) into the atmosphere from the use of fossil fuels and industrial processes.
(b) Increased mobilization of sulphate from rocks and soils through mining and agricultural practices which lead to an increased sulphur concentration in river runoff.
(c) Increased aeolian emission of sulphate-containing dust into atmosphere from dry land surfaces caused by farming and animal husbandry practices and by an increased exposure of salty lake sediments.
(d) An increased production of volatile sulphur compounds (e.g. DMS) in coastal seawater due to fertilization by nitrate, phosphate and organic materials.
(e) Changes in the rates of chemical transformation of sulphur compounds in the atmosphere—and thereby also changes in transport and fallout patterns—caused by modifications of the chemical climate, e.g. a changed concentration of hydroxyl radicals over industrialized regions.

We want to emphasize that the increased mobilization of sulphur caused by processes (b), (c) and (d) and the modification according to process (e) should also be regarded as anthropogenic although these processes are not directly caused by industrial emissions. The fourth process is the result of biological processes, but changes in the flux may arise through human activities.

82

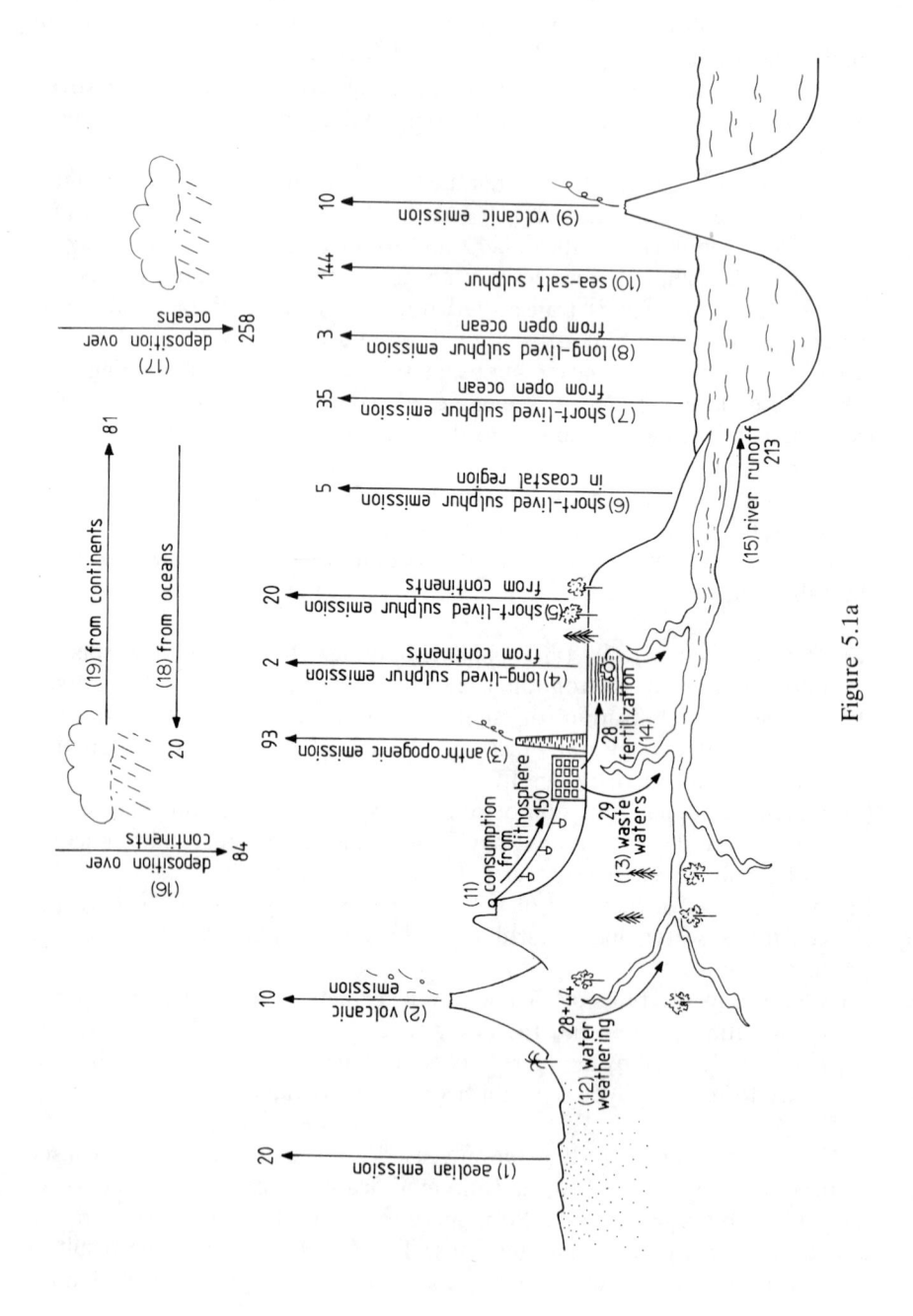

Figure 5.1a

5.1.2 Changes in industrial emission to the atmosphere

Modern man extracts huge amounts of various minerals from the lithosphere to satisfy his economic needs. In most cases minerals contain sulphur either as an impurity or chemically bound with the extracted material. As a rule, industrial activity involves oxidation of the sulphur from sulphides or the elemental form to a four- or six-valent state.

The main way in which sulphur is introduced into the environment is through the combustion of fossil fuels, the production of metal or the extraction of native sulphur. Five main types of human activity can be identified as sources of sulphur in the atmosphere (Afinogenova, 1982):

(a) Fossil fuel combustion for the production of heat or electricity.
(b) Processing sulphide ores of non-ferrous metals.
(c) Ferrous metal production including coke production and use.
(d) Oil refining and treatment of oil products.
(e) Sulphuric acid production from native sulphur.

Figure 5.1a Scheme of the global sulphur cycle for the mid-1980s.
(1) Aeolian emission, SO_4^{2-}, 20 Tg (S) a^{-1};
(2) volcanic emission into the continental atmosphere, SO_2, H_2S, SO_4^{2-}, 10 Tg (S) a^{-1};
(3) anthropogenic emission into the atmosphere, SO_2, SO_4^{2-}, H_2S, 93 Tg (S) a^{-1};
(4) emission of long-lived sulphur compounds into the continental atmosphere, COS, CS_2, 2 Tg (S) a^{-1};
(5) emission of short-lived sulphur compounds into the continental atmosphere, H_2S, DMS and so on, 20 Tg (S) a^{-1};
(6) emission of short-lived sulphur compounds into the atmosphere from coastal regions of the ocean, DMS, H_2S, 5 Tg (S) a^{-1};
(7) emission of short-lived sulphur compounds into the atmosphere from the open ocean, DMS and so on, 35 Tg (S) a^{-1};
(8) emission of long-lived sulphur compounds into the oceanic atmosphere, COS, CS_2, 3 Tg (S) a^{-1};
(9) volcanic emission into the oceanic atmosphere, SO_2, H_2S, SO_4^{2-}, 10 Tg (S) a^{-1};
(10) emission of sea salt aerosol sulphur from the ocean, SO_4^{2-}, 144 Tg (S) a^{-1};
(11) anthropogenic output from the lithosphere SO_4^{2-}, S^{2-}, 150 Tg (S) a^{-1};
(12) weathering and water erosion, SO_4^{2-}, 72 Tg (S) a^{-1};
(13) waste waters, SO_4^{2-}, 29 Tg (S) a^{-1};
(14) mineral fertilizers, SO_4^{2-}, 28 Tg (S) a^{-1};
(15) river runoff into the ocean, SO_4^{2-}, 213 Tg (S) a^{-1};
(16) scavenging from the atmosphere on the continental surface, SO_4^{2-}, SO_2, 84 Tg (S) a^{-1};
(17) scavenging from the atmosphere on the oceanic surface, SO_4^{2-}, SO_2, 258 Tg (S) a^{-1};
(18) transport from the oceanic atmosphere into the continental one, SO_4^{2-}, SO_2, 20 Tg (S) a^{-1};
(19) transport from the continental atmosphere into the oceanic one, SO_4^{2-}, SO_2, S^{2-}, 81 Tg (S) a^{-1}.

Figure 5.1b

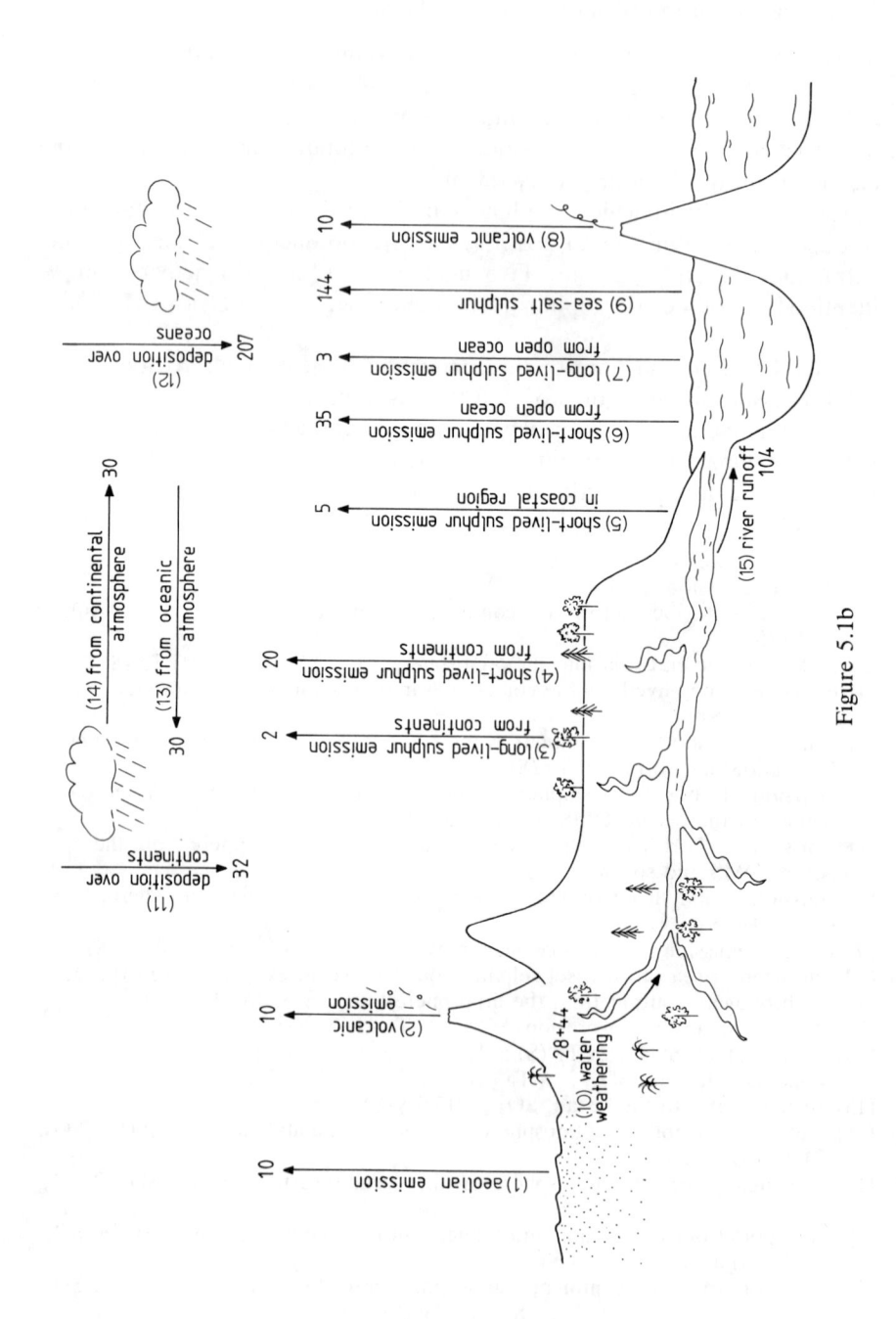

Fossil fuels contain sulphur in the form of organic compounds (in oil, coal, gas, peat, shale and wood), as the metal sulphides (in coal and shale), and as sulphate (in coal, shale and wood). The sulphur content of fuels can vary within wide limits. The cleanest type of fuel is natural gas (sulphur content is about 0.05%), but in contrast, certain kinds of coal contain organic- and pyrite-sulphur up to as much as 7% by weight. During the process of combustion organic- and sulphide-sulphur is oxidized to sulphur dioxide and to a lesser extent sulphur trioxide. On a global scale, the fraction of sulphur that is deliberately removed from flue gases is very small.

The sulphur content of sulphide ores of non-ferrous metals can equal or even exceed the amount of extractable metal. During ore processing and metal smelting sulphide-sulphur is almost completely oxidized to sulphur dioxide. In many cases the sulphur dioxide concentrations in flue gases are so high that it is profitable to collect and utilize it for the production of sulphuric acid. On a global scale, 30–40% of the sulphur from sulphide ore processed in non-ferrous smelters is converted to sulphuric acid.

Emissions in ferrous metallurgy are from the agglomeration processes which involve ore heating and sintering with a simultaneous burn-out of impurities including sulphur. Sulphide-sulphur contained in the ore is almost

---

Figure 5.1b Scheme of natural prehistorical global sulphur cycle.
(1) Aeolian emission, SO_4^{2-}, 10 Tg (S) a^{-1};
(2) volcanic emission into the continental atmosphere, SO_2, H_2S, SO_4^{2-}, 10 Tg (S) a^{-1};
(3) emission of long-lived sulphur compounds into the continental atmosphere, COS, CS_2, 2 Tg (S) a^{-1};
(4) emission of short-lived sulphur compounds into the continental atmosphere, H_2S, DMS and so on, 20 Tg (S) a^{-1};
(5) emission of short-lived sulphur compounds into the atmosphere from coastal regions of the ocean, DMS, H_2S, 5 Tg (S) a^{-1};
(6) emission of short-lived sulphur compounds into the atmosphere from the open ocean, DMS and so on, 35 Tg (S) a^{-1};
(7) emission of long-lived sulphur compounds into the oceanic atmosphere, COS, CS_2, 3 Tg (S) a^{-1};
(8) volcanic emission into the oceanic atmosphere, SO_2, H_2S, SO_4^{2-}, 10 Tg (S) a^{-1};
(9) emission of sea salt sulphur from the ocean, SO_4^{2-}, 144 Tg (S)a^{-1};
(10) weathering and water erosion, SO_4^{2-}, 72 Tg (S) a^{-1};
(11) scavenging from the atmosphere on the continental surface, SO_4^{2-}, SO_2, 32 Tg (S) a^{-1};
(12) scavenging from the atmosphere on the oceanic surface, SO_4^{2-}, SO_2, 207 Tg (S) a^{-1};
(13) transport from the oceanic atmosphere into the continental one, SO_4^{2-}, SO_2, S^{2-}, 20 Tg (S) a^{-1};
(14) transport from the continental atmosphere into the oceanic one, SO_4^{2-}, SO_2, S^{2-}, 30 Tg (S) a^{-1};
(15) river runoff into the ocean, SO_4^{2-}, 104 Tg (S) a^{-1}.

completely transformed to gases in the form of sulphur dioxide. Sulphur dioxide emission from agglomeration processes is mainly dependent on sulphide-sulphur content in ore. In addition to this, there is a contribution from the sulphur released during the production of coke.

Sulphur is mobilized in nearly all stages of oil refining. Malyshev et al. (1963) have shown the sulphur balance in a typical refinery, processing oil with a mean sulphur content of about 2%. The bulk of sulphur (80%) remains in oil products, the sulphur being in much higher concentrations in the high-boiling fractions. About 8% of the sulphur is released from the refinery into the atmosphere as the dioxide. The remaining 12% is partly utilized, partly introduced into solid processed products (coke, asphalt, etc.) and partly discharged into waste waters.

In the production of sulphuric acid, sulphur is oxidized to the dioxide and then to trioxide. Since not all the dioxide is oxidized to trioxide, the former can be partially emitted into the atmosphere. In the past sulphur dioxide emissions of 30 kg per tonne of sulphuric acid produced were typical. However, technical improvements have reduced this emission factor, so that in spite of increased worldwide production of sulphuric acid the atmospheric sulphur oxide emissions from this source are decreasing. Sulphuric acid production currently represents less than 1% of the total anthropogenic sulphur emission into the atmosphere.

Estimates of anthropogenic sulphur dioxide emission into the atmosphere for individual countries or regions and the world as a whole are based on statistical data concerning the use of sulphur-bearing raw materials, production of various kinds of goods, and data on weight emission factors per unit raw material or output. Using such an approach Möller (1984a) estimated that in 1977 the anthropogenic oxidized sulphur flux into the atmosphere was 75 Tg (S) a^{-1}. Varhelyi (1985) studied emission distribution all over the globe. According to her data main emission sources are concentrated in Europe (43.8%) with a significant contribution from North America (24%) and Asia (23%). The estimates placed three-quarters of emission from the USSR in Europe and one-quarter in Asia. As expected the bulk of emission (92%) is associated with the northern hemisphere.

A number of researchers (Robinson and Robbins, 1968; Cullis and Hirschler, 1980; Möller, 1984a; Ryaboshapko, 1983) have tried to assess historical trends of anthropogenic sulphur emission into the atmosphere. Figure 5.2 shows the growth of anthropogenic sulphur emission since the middle of the last century (Möller, 1984a). The data in the figure indicate that global sulphur emission into the atmosphere has increased twentyfold over the past 120 years. Figure 5.2 also shows the trends in atmospheric sulphur emissions in the USA (Husar and Halloway, 1983), Europe without the USSR (Bettelheim and Litter, 1979; Möller, 1984a), and the USSR (present work). It can be seen that emission growth was typical of all

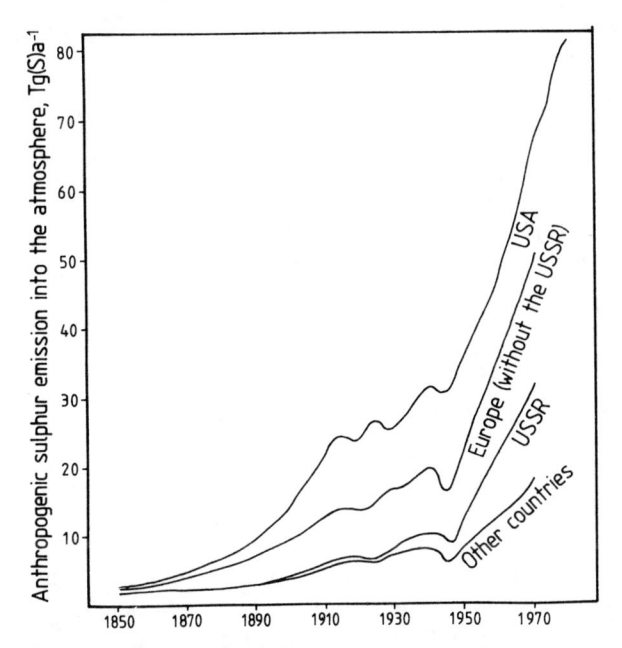

Figure 5.2 Temporal change of the global sulphur emission from anthropogenic sources. Emissions in the USA, in Europe (without the USSR), in the USSR (Russia), and in other countries are shown separately

countries. Until recently it could be assumed that the increase of atmospheric sulphur emission paralleled the economic growth rate of a given country. However, nowadays the situation has changed, and the rate of reduction of sulphur emissions may become a measure of economic development.

5.1.3 Changes in the riverine flux

Sulphur, in both the dissolved state and solid state, is carried by rivers to the ocean. The origin of sulphur in river water is chemical and mechanical weathering of bedrock, atmospheric input due to sea salt sulphur, industrial emissions from power plants, fertilization by superphosphates and ammonium sulphate, mining and processing of ore and municipal and industrial liquid wastes.

The estimates of the input of dissolved (ionic) sulphur to the ocean are based on the product of the mean concentration of SO_4^{2-} in the largest rivers and the discharge of all rivers to the ocean. There are three estimates of particular interest (see Table 5.2). One is by Meybeck (1978) who used data on 40 major rivers and estimated that the annual flux of dissolved sulphur is 105 Tg (S) a^{-1}. His estimate is based on a total water discharge

Table 5.2 A comparison of recent estimates of riverine sulphur flux

	Meybeck (1978)	Ivanov et al. (1983)	Husar and Husar (1985)
Number of rivers considered	40	****	54
Total water discharge 10^{16} l a^{-1}	3.74	4.24	4.1
Mean ionic sulphur concentration mg (S) l^{-1}	2.8	4.9	3.2
Natural sulphur flux via rivers Tg (S) a^{-1}	****	104	46–85
Anthropogenic sulphur flux via rivers Tg (S) a^{-1}	****	93	46–85
Total flux via rivers Tg (S) a^{-1}	****	208	131–170

**** = not determined.

of 3.74×10^{16} l a^{-1} and a mean concentration of sulphur equal to 2.8 mg (S) l^{-1}. The concentration is lower than that calculated by Livingstone (1963) (3.7 mg (S) l^{-1}) and the more recent data of Ivanov et al. (1983) and Husar and Husar (1985) (3.2 mg (S) $^{-1}$).

The estimate by Ivanov et al. (1983) was based on the data of Livingstone (1963), Alekin and Brazhikova (1964), Gibbs (1972) and Depetris (1976). They estimated that the flux of ionic sulphur excluding manmade pollution is 104 Tg (S) a^{-1} by examining the SO_4^{2-}/Cl^- ratio. The recent estimates of the sulphur concentration are a little different, but they consider the annual flow of water to be about 4.2×10^{16} l a^{-1} after the work of Korzon et al. (1974).

Ivanov et al. (1983) then estimated the manmade contribution to the transport of dissolved sulphur to the ocean. They considered annual emissions of anthropogenic sulphur into the atmosphere and the consumption of sulphur by industry, including production of fertilizers to derive an annual flux of anthropogenic sulphur to rivers. They conclude that 104 Tg (S) a^{-1} is supplied by man to rivers and transported to the ocean. Therefore, 208 Tg (S) a^{-1} dissolved sulphur is transported from continents to the ocean. If the discharge by world rivers used by Ivanov et al. (1983) is taken, then the mean concentration of sulphur dissolved in rivers today is calculated to be 4.9 mg (S) l^{-1}. The concentration of sulphur in smaller rivers draining industrial parts of the world has increased significantly during the last century; for example, the increase is five fold in the Elbe river which drains central Europe. Such data are not available for the largest of the world's rivers. The increase in the concentration of sulphur in those rivers during this century should be carefully evaluated so that we can understand the importance of both the direct anthropogenic inputs and the indirect inputs from such processes as erosion resulting from human activity. This is especially necessary because not all of the sulphur consumed by various industries is directly transferred to rivers. Ivanov et al. (1983) estimate that on a global scale 104 Tg (S) a^{-1} arises from human activities. This may be nearer the present situation than the slightly lower values of Husar and Husar (1985) which are based on older river water analyses. The correction may be particularly important for Europe and North America.

Besides the dissolved sulphur, rivers transport particulate sulphur to the ocean. Meybeck (1977) estimated that 28 Tg (S) a^{-1} is transported in suspended state. This estimate is based on the data for the total solid load in rivers (1.55×10^4 Tg a^{-1}), the mean contents of sulphur in shales (2×10^{-3} g/g), and the mineral fraction of total suspended solids, 90%. The new data on the transport of sediment by rivers are 1.35×10^4 Tg a^{-1} for suspended solids and 1 to 2×10^3 Tg a^{-1} for bedload and flood discharges (Milliman and Meade, 1983). This suggests that the original estimate by Meybeck is reasonable. Even if we assume that soil rather than shale is a typical representative of the particulate matter in rivers, the estimate would not change much because the mean concentration of sulphur in soils is about 3×10^{-3} g/g.

The above data indicate that the contribution of particulate sulphur to its global flux from continents to the ocean is small compared to the transport of dissolved sulphur. This is probably the reason why we have so little knowledge about the forms of particulate sulphur and their origin.

In conclusion, there are very few reliable data on the natural and anthropogenic fluxes of sulphur from continents to the ocean via rivers. One possible way to determine the anthropogenic influences is to compare the historical and present chemical analyses of river water and study the fluxes of sulphur in small representative catchments which would represent various climatic, morphologic, geological and anthropogenic conditions. Such studies may give quantitative indications of the balance between geochemical and biochemical natural processes and anthropogenic processes that control various stages of the global cycle of sulphur.

5.1.4 Changes in aeolian contribution

The atmospheric transport of sulphur-bearing soil dust particles due to wind is usually a characteristic of arid regions of the globe. Wind erosion accounts for not less than 5000 Tg of dust annually released into the atmosphere (Schütz, 1980), the dust containing about 20 Tg of suphur (Ryaboshapko, 1983). The aeolian input of dust (including sulphur) to the atmosphere is commonly considered a natural flux. Nevertheless, the impact of human activity on the flux magnitude is beyond doubt.

The earliest anthropogenic impact on aeolian emission and biogeochemical sulphur cycle as a whole was probably connected with the development of grazing, particularly with sheep and goat breeding. Intensive pasturing in dry regions (possibly coupled with climatic changes) results in the degradation of the plant cover and upper soil layer promoting erosion by the wind (Prospero and Nees, 1977). This process, which began two to three thousand years ago, led to continued desertification of vast areas of central Asia, the Middle East and northern Africa. The conditions prevailing in the above regions are characterized by frequent dust storms which lift large amounts of sulphur-bearing soil dust into the air and transport the particles great distances.

Another important anthropogenic factor that increases aeolian emissions is the development of irrigative farming in arid zones. Ploughing in itself promotes wind erosion. Besides this, inadequate irrigation techniques result in salinization of the upper soil horizons and an increase in the sulphur content of the soil dust layer. According to Takhtamyshev (1971) the sulphur concentration in the upper soil layer and soil dust can increase fivefold, as the land is brought under cultivation and reach up to 10 000 ppm. The process of soil salinization was even observed in the ancient civilizations of the Middle East and central Asia. Modern irrigation techniques make it possible to avoid salinization or at least slow down the process. However, huge amounts of sulphur in the form of soil dust are still lifted to the atmosphere today, over the vast areas of saline soils that have appeared as a result of human activity.

The Aral Sea region (Kazakhstan, USSR) is a vivid example of the anthropogenic impact on aeolian emission (Grigoryev, 1985). The water of rivers draining into the Aral Sea is so intensively used for irrigation that the flow into the sea has been drastically reduced. Exposed bottom areas are covered with a porous layer of sediment with a high salt content. The atmosphere now receives from 0.1 to 1 Tg (S) a^{-1} along with the dust from such areas. The Aral Sea will probably disappear completely within 20–30 years. As a result, 10^4 Tg of sea salt containing about 300 Tg (S) will be available for aeolian transport.

It is difficult to give a quantitative assessment of anthropogenic impact on aeolian sulphur emission. Since over the human history more than half the area of the world's arid zone has been, to various degrees, subject to human activity, so it is probable that about a half of aeolian sulphur flux to the atmosphere is related to man-induced changes. The review SCOPE 19 estimates the present aeolian sulphur flux as 20 Tg (S) a^{-1} (Ryaboshapko, 1983). In prehistoric times the value may have been 10 Tg (S) a^{-1}. However, this estimate, based on the above, is purely speculative at the moment.

5.1.5 Changes in emission of reduced sulphur compounds

It is known that discharges of nutrients, mainly phosphate and nitrate, from municipal and industrial sources to lakes and coastal waters lead to an increased rate of primary production in these water bodies. Since there are some indications of a positive relation between the production of DMS in surface water and the primary productivity (Andreae and Barnard, 1984), man could influence the emission of DMS from these fertilized water bodies. It is not certain whether there is a linear relationship between primary productivity and DMS release, because the distribution of phytoplankton species is an important consideration. According to our estimates it is unlikely that man's influence on the production of DMS from seawater through the input of nutrients is significant for the global sulphur budget. More subtle interactions are conceivable as a significant fraction of the DMS produced by microorganisms in the ocean is oxidized in photosensitized reactions (Brimblecombe and Shooter, 1986). If anthropogenic activities increase the concentration of photosensitizers in coastal waters through agricultural runoff of organic compounds, then the effective flux of DMS to the atmosphere could be reduced.

5.1.6 Changes in chemical processes in the atmosphere

Emissions of air pollutants have substantially affected the chemical climate within the industrialized regions of the world. The most obvious changes are a lowering of the pH in cloud and rainwater caused by sulphur and nitrogen compounds and an increased concentration of oxidant (mainly ozone and probably hydroxyl radical) caused by emission of NO_x, hydrocarbons and carbon monoxide. On the other hand, if the concentrations of NO_x are high (above a few ppb) they will actually decrease the concentration of hydroxyl radical (Rodhe et al., 1981). These changes in chemical climate are likely to influence the rate and oxidation of various trace species of both anthropogenic and natural origin. Thereby the rate of removal and the transport scales could be affected. It should also be pointed out that in remote parts of the troposphere where the concentration of NO_x is very low the atmospheric concentrations of methane and other hydrocarbons may lower the concentration of the hydroxyl radical.

It has been suggested that the increased concentration of hydroxyl radical in the atmosphere over Europe, and other industrialized regions, would cause a rapid oxidation of DMS emitted from the oceans and transported in over the industrial region. Part of the observed peak in sulphur deposition could then be due to an anthropogenic influence on the natural sulphur flux: an anthropogenic *sink* rather than an anthropogenic source.

A rough estimate of the magnitude of this additional sulphur deposition over Europe can be made as follows. Let us accept that an upper limit for the average concentration of DMS in marine air at the European west coast is taken to be 200 ng (S) m^{-3}. Such concentrations are higher than those reported earlier for the Atlantic (Barnard et al., 1982) and the Gulf of Mexico (Andreae and Barnard, 1984) but in agreement with recent measurements from the remote Pacific (Andreae and Raemdonck, 1983). With a scale height of 1000 m, a length of the coastline of 2×10^6 m and a mean westerly airflow in the 0–1000 m layer of 5 m s^{-1}, the annual flux of DMS in over the European continent would be less than about 0.06 Tg (S). This flux is insignificant in comparison with the European anthropogenic emission of 28 Tg (S) a^{-1} (Dovland and Saltbones, 1979) and the total deposition of about 20 Tg (S) a^{-1}.

Thus, it does not seem likely that DMS from the ocean contributes significantly to the peak in sulphur deposition over Europe. However, the flux estimated above (0.06 Tg (S) a^{-1}) can not be taken to represent the total contribution of DMS to the European sulphur budget. This is because a substantial fraction of the SO$_2$ and sulphate carried by the winds from the Atlantic may well be secondary products of oceanic DMS emissions. Despite this it is possible for DMS emissions to be very high in some coastal regions in late spring and early summer. This could have some impact on the local sulphur concentration (Turner and Liss, 1985).

5.2 RECORDS OF PAST CHANGES—ANTHROPOGENIC AND NATURAL

5.2.1 Introduction

An alternative way of assessing the anthropogenic influence on the present sulphur cycle is to study available records of past fluxes. Such fluxes are mainly associated with the accumulation of sulphate in ice and lake sediments (MARC, 1985). Some direct measurements of past sulphate concentrations in rainwater and river water also exist. Direct measurements go back to the mid-19th century, but reliable data covering a wide area are confined to the second half of the 20th century (see Section 5.2.2). The information on sulphur fluxes contained in ice cores from the central regions of polar ice sheets and some ice caps is direct, in the sense that they consist of well-preserved and well-layered frozen past precipitation. These layers can be dated with great certainty (Hammer et al., 1978).

The natural sulphur fluxes can be studied from such records over a ten thousand year time span, but in order to use the records as indicators of the variability and range of natural sulphur fluxes, it is necessary to select a time period, which shows a reasonable resemblance to present climatic conditions. It is desirable, however, that the time period include the climatic variations, which have occurred since man changed from his hunting habits to more permanent settlement.

Most of the last inter-glacial period (Holocene) covering some 11 000 years may serve well as such a reference period. It represents our own climatic age, apart from the first few millennia after the last glacial, and offers some of the most detailed records of Quaternary phenomena: ice sheets, ice caps, lake sediments, soils, peat bogs, tree rings, etc. Only some of these records are useful for studies of past sulphur fluxes. Among the latter type of records ice sheets and ice caps are of particular interest. These are restricted, almost entirely, to the polar regions of the Earth and are situated remote from major population centres. The latter factor makes them excellent data sources for studying the extent to which sulphur pollution reaches the low and mid-troposphere at remote places (see Section 5.2.3).

The composition of lake sediments is less directly related to the atmospheric sulphur concentration, but they are much more widespread and many are situated in areas where the impact of local or regional pollution is a matter of great concern.

To what extent the different source regions for atmospheric sulphur contribute to the sulphur concentrations found in the various records depends of course on the geographical site as well as on the type of record.

5.2.2 Historical analyses

The earliest systematic measurements of solute in precipitation were started at Rothamsted in England in 1853. There are indications that the sulphate levels may have doubled in the century since 1880, but there is no way of checking their reliability (Brimblecombe and Pitman, 1980). Early analysts had great difficulty in determining sulphate and this is illustrated by the variability in Figure 5.3 which shows deposition estimated from early precipitation analyses from the USA (Data sources are given in the similar figure for nitrate in Brimblecombe and Stedman (1982)). Measurements of sulphate in precipitation in central Russia at the beginning of this century indicate concentration levels that are about a factor of three lower than those measured in the same area during the 1970s (Bulatkin and Maksimova, 1978). Some early data are summarized in the review of Eriksson (1952).

Since 1955 many stations in the European Air Chemistry Network (EACN) have monitored the ionic composition of precipitation (Rodhe et al., 1984). The data from EACN indicate that the sulphate levels in precipitation in Scandinavia increased by about 50% between the late 1950s and the late 1960s. Since the early 1970s the levels probably have decreased by some 20% (Rodhe and Granat, 1984). A comparison between measurements of sulphate in precipitation in the USA carried out in 1957 with similar measurements in the late 1970s shows no clear trend.

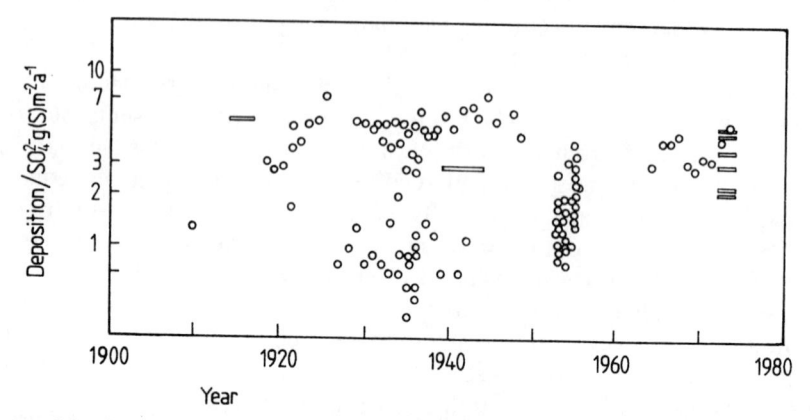

Figure 5.3 Determinations of sulphate deposition in eastern North America showing wide variability. Circles, annual measurements; rectangles, longer averages

There are numerous early analyses of river composition, but as yet these have not been widely used in studies of historical change, e.g. early US Geological Survey data of the type that can be found in Clarke (1924). Beeton (1965) used data for the Great Lakes of North America to show a doubling in sulphur concentrations in some of these inland waters over the last century.

5.2.3 Ice cores

Storage of information on the past composition of the atmosphere in ice cores requires a number of favourable conditions. Survival depends, most importantly, on preservation of the ice. A prerequisite for deducing ancient levels of atmospheric sulphur concentrations (or impurities in general) from ice cores is that local 'in air' and 'in ice' concentrations can be related. This is only possible under certain glaciological and meteorological conditions (Junge, 1975; Hammer, 1985a). These conditions are only satisfied in polar regions so it is not surprising that most data on sulphur and other trace substances in ice cores have been obtained from the more central parts of Antarctica, the Greenland Ice Sheet and various ice caps in the Arctic.

There are now numerous available data on the sulphate content of the ice sheets of Antarctica and Greenland, though most of the early data must be used with care. For Antarctica, see Lorius *et al.* (1969), Doronin (1975), Cragin *et al.* (1977), Delmas and Boutron (1978, 1980), Korotkevitch *et al.* (1978), Delmas *et al.* (1982), Aristarain *et al.* (1982), Herron (1982), Legrand and Delmas (1984) and Palais and Legrand (1985). For Greenland, see Koide and Goldberg (1971), Weiss *et al.* (1975), Cragin *et al.* (1974, 1975), Herron *et al.* (1977), Busenberg and Langway (1979), Davidson *et al.* (1981),

Herron (1980, 1982), Herron and Langway (1985) and Neftel et al. (1985). There are presently less available data on glaciers outside the polar regions; Davies et al. (1982), Wagenbach et al. (1983), Punning et al. (1984) and Astratov et al. (1986).

It should be emphasized that the sulphate concentrations to be measured are very low, especially in the remote regions of Antarctica and the Greenland Ice Sheet, where background values are in the 6–35 ng (S)/g (H_2O) range. Contamination problems during 'in field' sampling, transportation, storage and laboratory analysis must then be carefully taken into account by various methods; e.g. cross-sectional analysis of the cores may be necessary in order to check the penetration of outside contamination, especially in the upper 60 m of ice sheets, which is very porous. It is also important to check the reliability of the sulphate data by other means. One way to do this is measuring all the other major ions simultaneously (H^+, Na^+, HN_4^+, $K+$, Mg^{2+}, Ca^{2+}, Cl^-, NO_3^- and organic ions such as methane sulphonate) and verifying that the ion balance is satisfied (see Figure 5.4a, Legrand and Delmas, 1984). However, this procedure does not guarantee against contaminated samples, as ion balance must also exist in these ices. Establishing the ion balance is also an important step in order to suggest in which chemical form sulphur was present in the atmosphere (see Figure 5.4b). Most sulphate appears to be present as H_2SO_4.

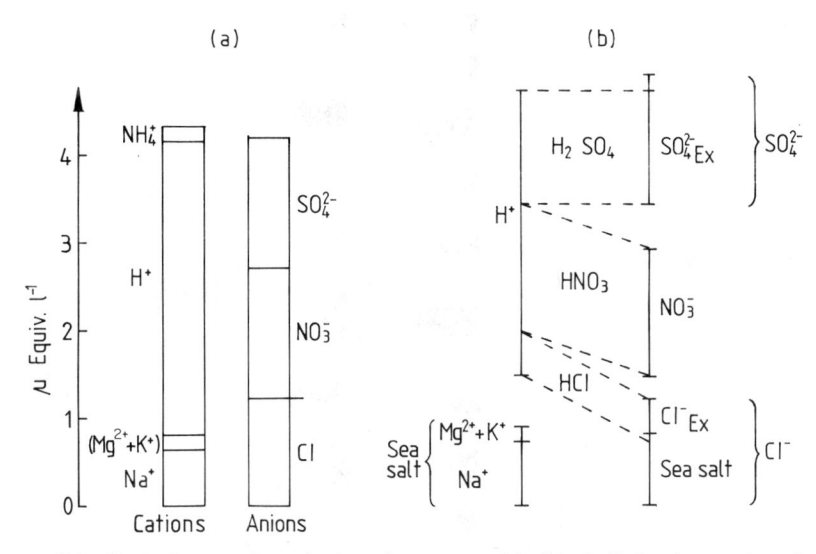

Figure 5.4 Snow layers, deposited at the geographic South Pole, Antarctica, from 1959 to 1969
(a) Ion balance,
(b) Reconstruction of the different contributions (acidity and sea salt). From Legrand and Delmas (1984)

The simultaneous determination of all major ions has been greatly improved by advances in ion chromatography, though the important H^+ concentration must be determined by other techniques. The H^+ concentration can be determined by pH related measurements, (e.g. via acid titration (Legrand et al., 1982), correcting for ambient air CO_2 (Hammer, 1980), avoiding ambient air) or by solid electrical conductivity measurements on the ice (Hammer, 1980). The solid conductivity technique has the advantage of being performed direclty on the cleaned ice core surface, thus reducing contamination problems. Furthermore it is fast and has a spatial resolution of a few millimetres along the core. The method measures only the strong acids in the core and ambient air problems are absent.

As can be seen in Figure 5.4, H^+ is an important cation and its concentration is generally (though not always) quite close to the sum of the nitrate and sulphate concentrations (see Figure 5.5, Hammer, 1983). The latter holds only for the Holocene and not for the whole glacial period. In

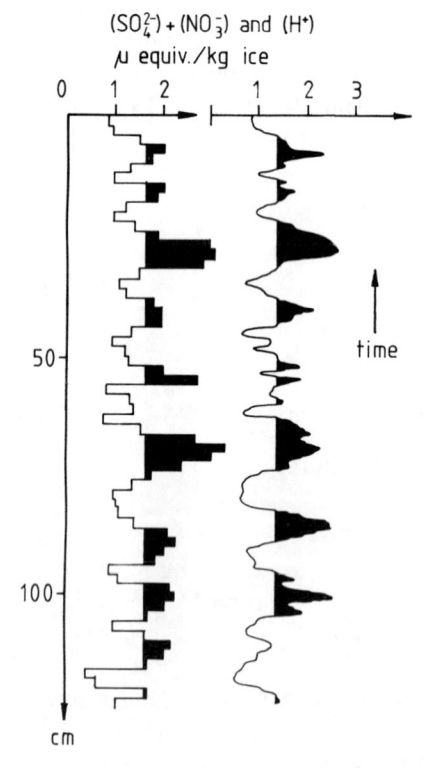

Figure 5.5 Seasonal variations in $(SO_4^{2-}) + (NO_3^-)$ and (H^+) as measured by ECM (electrical conductivity method—Hammer, 1980, 1983) on deep ice from the antarctic Byrd core. Age approximately 1000 years. (from Hammer, 1983)

principle it is recommended that H⁺ be determined by more than one method, but for practical reasons, that can only be accomplished on selected samples. Some ice cores are several kilometres long and the present techniques only allow a few trace substances to be measured both with a high resolution and in a continuous way.

The early data on sulphur concentrations in ice cores were hampered by unrecognized contamination problems, but even more recent data are not free from such problems. Most of the data published in the 1980s are no doubt correct, but that does not *per se* secure a true picture of the sulphur flux during past times: sampling frequency, glacio-meteorological noise and 'in ice' to 'in air' concentrations are problems which have to be dealt with, for each drill site, before a 'true' local air concentration can be inferred from ice cores.

The final step, which is to relate local air concentrations of sulphur to changes in the source regions and source strength, is more complicated and modelling of atmospheric transport processes must in principle be undertaken. For some purposes such a strict procedure is not necessary and from already published studies we may conclude:

(a) The mid-tropospheric natural sulphur load only varied within a factor of approximately 2–3 during the last 10 000 years (in 5–20 year averages). The published data are too few to allow a precise estimate of the variability; also, most authors do not state how many years their samples represent. Cragin *et al.* (1975), de Angelis *et al.* (1984), Palais and Legrand (1985) and Herron and Langway (1985).

(b) Volcanism played an important role in determining the variations of the mid-tropospheric sulphur load during the past 10 000 years (see Figure 5.6, Hammer *et al.* (1980) and Hammer (1985a)).

(c) Anthropogenic sulphur has increased the northern hemisphere mid-tropospheric sulphur load 2–4 times the natural load since last century. The increase is most evident in the last two decades (Neftel *et al.*, 1985); see Figure 5.7. No similar increase has been observed in Antarctic ice cores (Delmas and Boutron, 1980); see Figure 5.8.

(d) During the last glacial period the mid-tropospheric sulphur load was strongly increased in both hemispheres (as compared to the Holocene), but a quantitative assessment of the various causes and sources is complex.

In order to elucidate the above findings, the more recent data from Antarctica and Greenland ice cores will be discussed in the following sections. Sampling sites are shown in Figures 5.9a and 5.9b.

Central East Antarctica is a low accumulation area (less than approximately 5–10 cm (H_2O equiv.) a^{-1}, while the higher accumulation area in West Antarctica corresponds to a low accumulation area in Greenland.

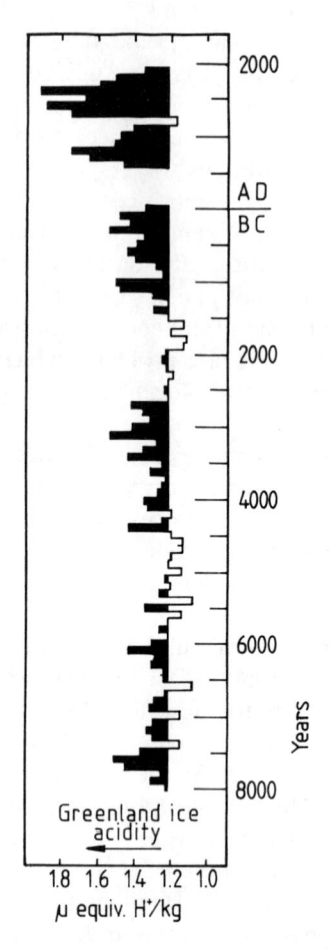

Figure 5.6 The influence of volcanism on mid-tropospheric acidity (set out in black) over the northern hemisphere, as derived from the Greenland Camp Century ice core (see Figure 5.9b). The measurements are based on *ca* 10 000 individual acidity measurements but are presented in 100 year samples. From Hammer (1985a)

Antarctica
Low accumulation of snow in Central East Antarctica means that the data covering the last century have been obtained from the upper very porous firn, but as extreme care was taken during sampling and measuring (Delmas and Boutron, 1980) we can rely on the findings: sulphate is found to be deposited both as a primary and secondary aerosol. Primary aerosol is sea salt, but it accounts for less than 10% of the total sulphate. The secondary aerosol source is H_2SO_4 ('excess sulphate'), and may be of biogenic and/or

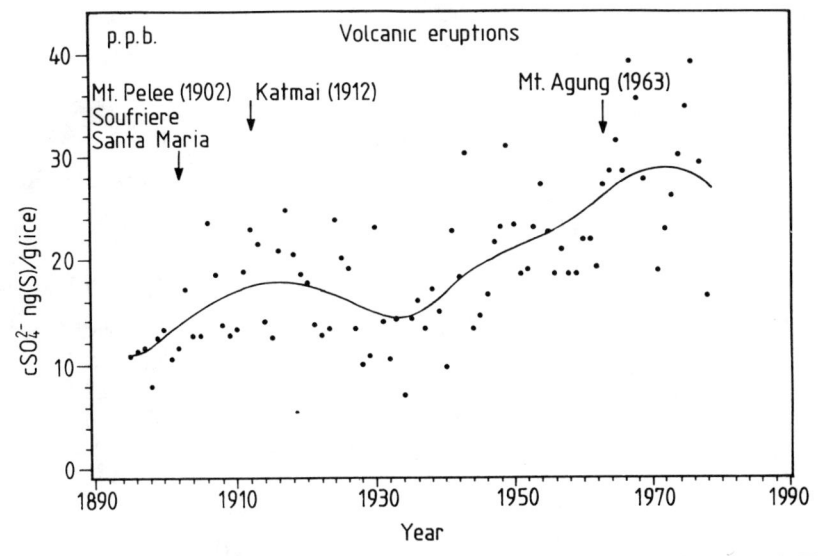

Figure 5.7 Annual (SO_4^{2-}) in a shallow core from Dye 3, South Greenland. The general trend is shown with a spline-smoothed curve. From Neftel *et al.* (1985). Neftel *et al.* claim that the Katmai eruption in Alaska is clearly seen in the acidity record of Dye 3; this is not true. This is indirectly demonstrated by the authors' own sulphate data. The Katmai eruption is clearly seen only in Central and North Greenland ice cores (Hammer, 1984)

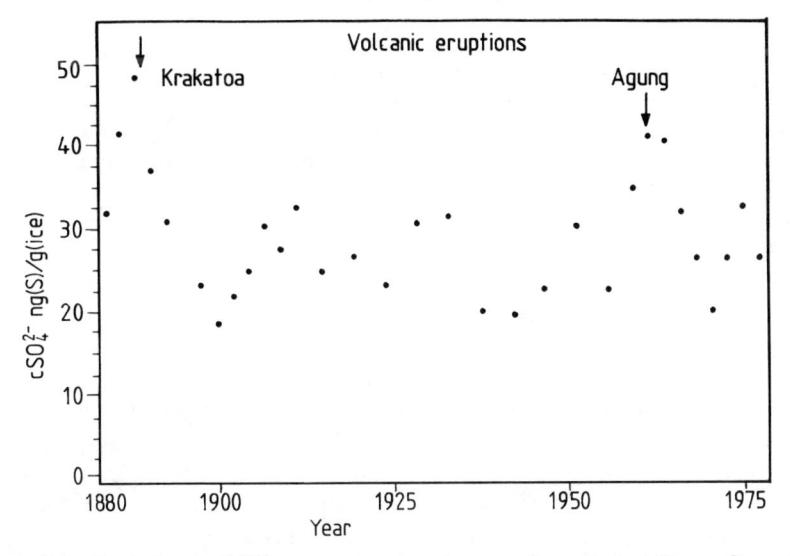

Figure 5.8 Variations of SO_4^{2-} concentrations in snow deposited at Dome C, central East Antarctica, during the last century. Agung and Krakatoa eruptions are clearly detected (modified from Delmas and Boutron, 1980)

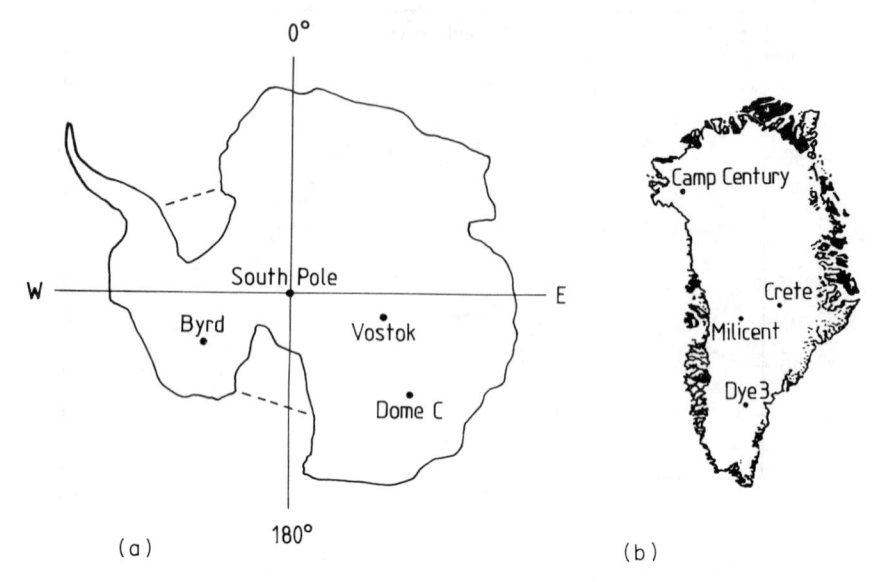

Figure 5.9 Location of important drill sites.
(a) four in Antarctica
(b) four in Greenland

Table 5.3. Geographic South Pole, Antarctica: estimation of the relative contributions to sulphate deposition on snow for a volcanic and a non-volcanic time period. Concentrations are given in μ equiv l^{-1}. Percentages are relative to total sulphate. From Legrand and Delmas (1984)

Years	1959–1963 non-volcanic period	1964 volcanic period
Total SO$_4^{2-}$ (mean)	0.95	3.03
Sea salt $_4^{2-}$	0.08 (8%)	0.08 (2.5%)
Excess SO$_4^{2-}$		
volcanic	< 0.09 (< 9%)	2.15 (71%)
non-volcanic	0.8 (> 83%)	0.8 (> 26.5%)

volcanic origin (Delmas and Boutron, 1980; Legrand and Delmas, 1984). The relative contributions of the sources are illustrated in Table 5.3.

Antarctic data (Delmas and Boutron, 1980) clearly indicate that the influence of anthropogenic sources of sulphur is negligible in the most remote regions of the southern hemisphere. This is to be expected because the largest part of anthropogenic emissions occur in the northern hemisphere. Various major volcanic eruptions such as Agung Bali 1963, Krakatoa 1883 and Tambora 1815 left the Antarctic ice cores with enhanced sulphate concentrations (Delmas and Boutron, 1978, 1980; Legrand and Delmas, 1984, see Figure 5.8). Similar increases in sulphuric acid concentration in Greenland ice cores have been observed after major volcanic eruptions, though HCl and HF sometimes contributed to the volcanic acid fallout (Hammer, 1977; Hammer et al., 1980; Herron, 1982). This raises the question as to what extent volcanic activity determines the natural sulphur load of the mid-troposphere. Both in Central Antarctic and Central Greenland ice cores sea-salt sulphate contributes little to the total sulphate concentration during the Holocene. Volcanic and biogenic sulphate should therefore account for a large part of the natural sulphate in ice cores. As major volcanic eruptions are so clearly detected by sulphate peaks in ice cores it is reasonable to assume that more moderate and minor eruptions must add substantially to the sulphur load of the mid- and upper atmosphere. Presently the precise nature of the sulphate background in ice cores can only be roughly assessed. The changing climate can also have an influence on the sulphur concentrations in ice cores, but for the Holocene we need much more data before an accurate assessment can be undertaken.

A particularly interesting case of climatic influence on sulphur fluxes is registered in the ice cores, which reach back into the last glacial period. The Antarctic deep ice cores of Byrd, Vostok and Dome C demonstrate an increase of sulphur in precipitation, as compared to the Holocene (Korotkevitch et al., 1978; De Angelis et al. 1984; Palais and Legrand, 1985). The sulphate concentrations are found to be especially high during the last stages of the Wisconsin ice age. The corresponding enhancement factors (as compared to the Holocene) are estimated to be about 1–5 (Byrd) and 2–3 (Vostok and Dome C). The causes of these high late glacial sulphate values are still not well understood. It has been suggested that they could be due to an increase of volcanic or biogenic productions, but it is more likely that they are due to enhanced glacial aridity, a stronger atmospheric circulation (Petit et al., 1981) and/or a 2–3 times lower precipitation in the polar regions (Hammer, 1985a, 1985b). However, it must be remembered, that the data obtained in East Antarctica come from a region of very low precipitation and therefore dry fallout may also have contributed to the sulphur concentrations (Legrand and Delmas, 1984).

Greenland
Until recently the sulphur data covering the 20th century, from the Greenland Ice Sheet, suffered from either inadequate sampling frequency or contamination problems. One of the reasons for the lack of data over the 20th century was, in fact, a fear of contamination in the porous ice cores covering this time period; hence most data were obtained from *in situ* pit studies. Such pits only cover a few very recent years, but the risk of contamination in pits is lower. The data of Herron (1982) suggested a 2–3

fold increase of sulphate over Greenland, compared to the natural sulphate level of the mid-troposphere, but until recently (Neftel et al., 1985) this finding was controversial. The continuous data now available (1 year averages) covering the time period 1895–1978 (see Figure 5.7) clearly indicate an anthropogenic influence. The obvious increase during the late sixties and in the seventies can hardly be explained solely by volcanic eruptions. There was no obvious climatic trend in the period, hence a climatic cause for the increasing deposit of sulphate can be ruled out. The data are from the Dye 3 region in South Greenland where the annual precipitation varies around 40–70 cm (H_2O equiv.) a^{-1} as shown in Figure 5.9b. One would expect a similar increase in the more isolated parts of Central Greenland, but no similar data on sulphur concentrations are available at present.

The continuous acidity profile from the Central Greenland Crete ice core, covering the years 553–1972 AD (Hammer et al., 1980, see Figure 5.10), shows no increase in acidity due to anthropogenic activity, but the influence of volcanism is evident. Figure 5.11 shows examples of enhanced sulphate concentrations due to volcanic eruptions for the Dye 3 core (Herron, 1982). It is unlikely that the mid-tropospheric sulphate is present in the Crete core as neutralized sulphuric acid, because neither NH_3 nor other neutralizing trace substances are sufficiently abundant to accomplish this. The lack of a continuous sulphur profile over the 20th century part of the Crete core forces us to look for an explanation for the 'missing' anthropogenic acidity trend. (See Table 5.4 for typical Greenland ice core impurity composition.) The fact that Crete is more isolated and at a higher elevation (3170 m above sea level) than Dye 3 (2490 m above sea level) can hardly explain the 'missing' trend. The most probable explanation seems to be a combination of three facts: at Dye 3 the increase of sulphur is most clearly seen when the years 1965–1981 are included—the Crete core ends in 1972; earlier in the century the anthropogenic trend may be masked by short-term climatic changes and varying volcanism; finally the acidity (from pH measurements) data over the 20th century were difficult to correct for ambient CO_2 influence (Hammer et al., 1980). New ice cores from Central Greenland, obtained in 1984, suggest that an increase in acidity can, in fact, be seen over the years 1970–1983. However, without the latter period included, this is less evident.

Of the anthropogenic sulphur releases in the lower troposphere only a small fraction reaches the fairly well-mixed troposphere, but even small volcanic eruptions add to the mid-tropospheric sulphur load by direct injection, whereas seawater sulphate contributes little. Thus the main part of the natural sulphur concentration in polar ice cores (Holocene) can be ascribed to two sources—volcanism and biogenic activity, as in the case of Antarctic ice cores. Volcanic activity may be responsible for a large fraction of the natural sulphur in ice cores during the Holocene. This is suggested by the fairly stable climate over the past 8–9000 years, the low average

Human Influence on the Sulphur Cycle

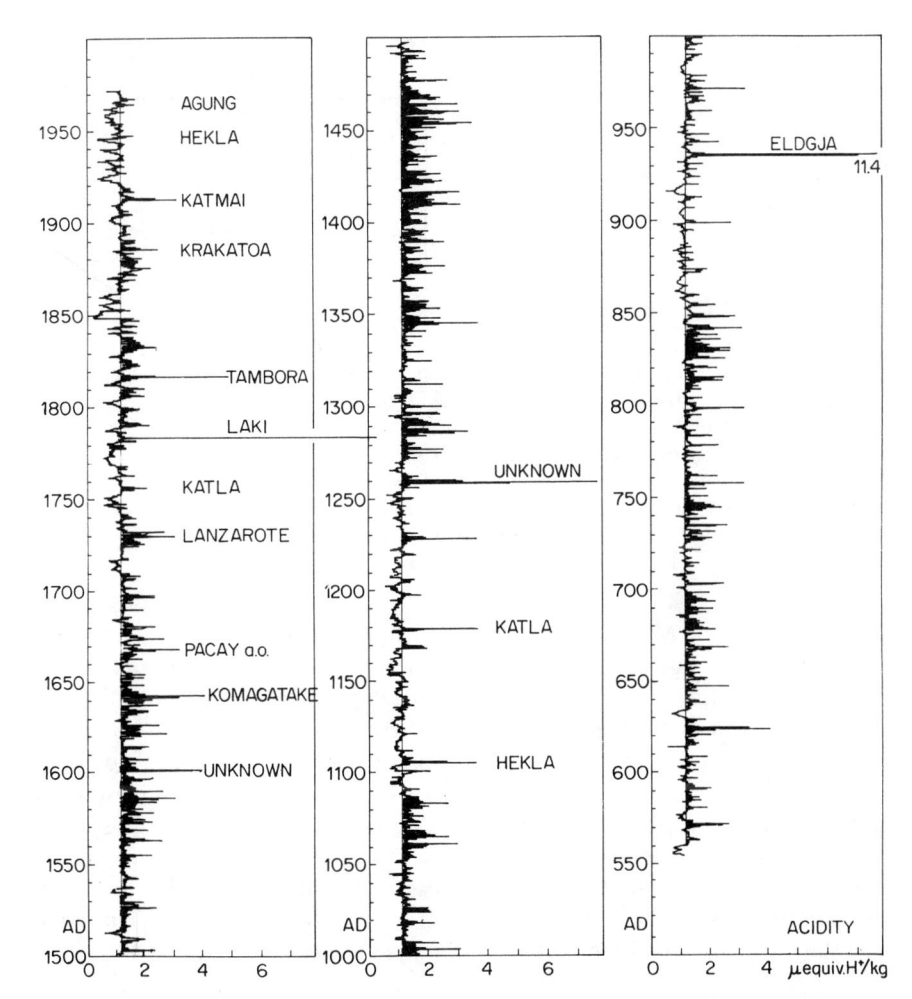

Figure 5.10 Mean acidity of annual layers from AD 1972 to 553 in the ice core from Crete, Central Greenland. Acidities above the background, 1.2 ± 0.1 μ equiv H^+ kg^{-1} (ice), are due to fallout of volcanic acids, mainly H_2SO_4, from eruptions north of 20°S. The ice core is dated with an uncertainty of ± 3 years at AD 553, which makes possible the identification of several large eruptions known from historical sources, and the accurate dating of the Icelandic Eldgja eruption shortly after the settlement that was completed AD 930. From Hammer et al. (1980)

concentrations of dust in Greenland ice cores over the same period and the data in Figure 5.10.

In Figure 5.6, 10 000 measurements of acidity in the Camp Century deep ice core (northwest Greenland, see Figure 5.9b), shown as 100 year averages, may be used to estimate the average contribution of volcanic sulphur to the

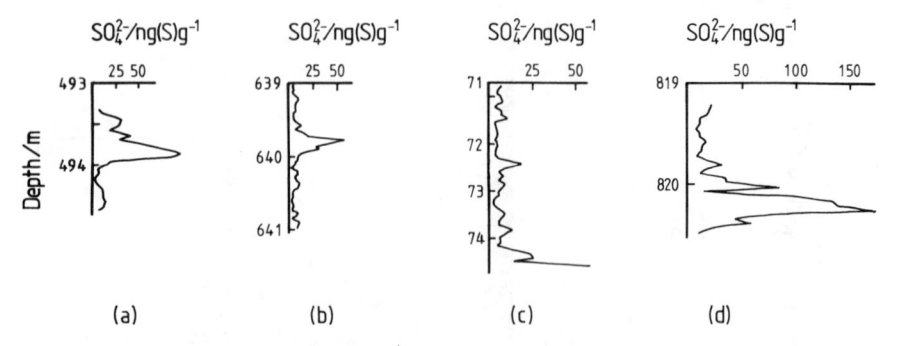

$SO_4^{2-}/ng(S)g^{-1}$ (a), $SO_4^{2-}/ng(S)g^{-1}$ (b), $SO_4^{2-}/ng(S)g^{-1}$ (c), $SO_4^{2-}/ng(S)g^{-1}$ (d)

Figure 5.11 Dye 3, Greenland: detailed variations of the concentrations of sulphate in four sections of the 901 m deep ice core. The major concentration peaks are interpreted as follows:
(a) eruption of Eldgja, Iceland, estimated at 934 AD.
(b) eruption of unknown origin estimated at 535 AD.
(c) Krakatoa eruption 1883 AD in lowermost interval.
(d) eruption of unknown origin estimated at approximately 40 BC. From Herron (1982)

atmospheric load. More data are needed in order to use the ice core for such a purpose, but if the Camp Century data are used as 'a rule of the thumb', an estimate can be given. Volcanism contributed an estimated 0.7–3.0 Tg (S) a^{-1} to the whole atmosphere. This estimate is based on Greenland ice core data and it holds whether one uses the procedure of Hammer (1977; Hammer et al., 1981), or uses ice concentrations of sulphur and northern hemisphere average precipitation to estimate mid-tropospheric sulphur fluxes. In Figure 5.6 a large part of the acidity (80%) is due to HNO_3, but NO_3^- concentrations show less variability than SO_4^{2-}, probably due to a more uniform stratospheric source (Risbo et al., 1981). The large uncertainty in the estimate is due to the uncertainty of where the eruptions

Table 5.4 Typical bulk impurity composition of the Greenland Ice Sheet[a]. From Hammer (1983)

dust[b]	50		
anions[c]		cations[c]	
NO_3^-	1.0	H^+	1.2
SO_4^{2-}	0.5	NH_4^+	0.3
Cl^-	0.5	Na^+	0.4
		$Mg^{2+}+Ca^{2+}+K^+$	0.1
sum of anions	2.0	sum of cations	2.0

[a] Post Wisconsin ice excluding organic material.
[b] μg kg^{-1} of ice.
[c] μequiv kg^{-1} of ice.

took place, climatic changes, etc. It is, however, important to notice that if volcanism contributed more than 0.7–3.0 Tg (S) a^{-1} over the last 10 000 years, a large fraction of the sulphur must have remained in the low troposphere. Ice core data give an uper limit to the change in sulphur budget in the troposphere.

As in the case of Antarctic ice cores, Greenland samples suggest high sulphate concentrations (Cragin et al., 1974; Herron and Langway, 1985), the main differences between the two hemispheres being due to the larger land masses and continental shelves in the northern hemisphere; hence the enhancement factors are higher for Greenland. Although the results from Central Greenland are less clear than South Greenland a threefold increase appears to arise through anthropogenic activities. The picture in the Antarctic remains uncertain.

5.2.4 Lake sediments and soils

A major problem with these sources of information on changes in the sulphur budget is the difficulty in relating atmospheric sulphur concentrations to those measured in cores or pitwall samples. It is a more complex problem than is the case for ice and precipitation samples. In the latter case we are dealing with direct atmospheric precipitates, thus avoiding the influence from complex sedimentation processes, exchanges within the sediments and the complicated changes induced by geological, ecological and environmental conditions (see, for example, Wright, 1982). Even under favourable conditions it is not clear if observed sulphur concentrations in the sediments can be quantitatively related to atmospheric deposition rates. In addition the changes in sediments may sometimes reflect not only changes in the atmospheric flux, but also changes in surface water chemistry, brought about by industrial and agricultural developments.

The influence of anthropogenic pollution has been observed in numerous records from lake sediments, soils and peatbogs, but most data are from indirect or direct acidification studies or trace metal analyses of, for example, Pb, Hg, Cu and Zn (MARC, 1985). Indirect data on acidity can in some cases indicate anthropogenic influence or natural variations thousands of years back in time (Renberg and Hedberg, 1982; Battarbee et al., 1985), but it requires additional models and assumptions to derive sulphur records from such data.

Another problem is the dating of these records: the dating cannot compete with the precise dating of ice cores, but for many purposes the Pb^{210} method (covering the last two centuries) or the ^{14}C method are satisfactory. The dating of soils is usually less precise, due to the low sedimentation rates and the mixing of various components within it. For all three types of sedimentary records it is important to note that the dating of a certain stratigraphical level only means that certain components of that level have been dated; exchanges between layers via mobile phases of sulphur components have to be taken into consideration.

Sulphur accumulation in lake sediments due to anthropogenic activity can be demonstrated in a number of ways:

Comparison of sedimentary sulphur content in lakes from watersheds in regions with high and low sulphur deposition rates.
Comparison of sulphur content of younger sedimentary layers exposed to high sulphur deposition rates with that of older sedimentary layers from periods with low sulphur deposition rates. The sedimentary layers have been accurately dated by fission products (bomb test period) and ^{210}Pb.
Comparison of isotopic composition of sulphur from sediments in zones of high and low sulphur deposition rates or from sediments from different layers in a zone of current high deposition rates.

Data from a number of lake sediments in Canada and the eastern USA show a marked enrichment of sulphur in the most recent sediments (Cook, 1980; Nriagu and Coker, 1983; Nriagu, 1984; Thode and Dickman, 1983; Mitchell et al., 1984; Holdren et al., 1984; Fry, 1986). Enrichment factors (present-day concentrations divided by pre-industrial values) of up to 9 have been reported for lakes near the smelters at Sudbury, Ontario, and 2–5 in areas with no local sources of pollutant sulphur. The few data available also show that the flux of excess sulphur into the lake sediments is accompanied by a significant shift in the isotopic composition of the sulphur (Nriagu and Coker, 1983; Thode and Dickman, 1983). The onset of accelerated deposition of sulphur in the sediments has generally been dated to between 1850 and 1890, corresponding to the beginning of industrialization of the source regions.

The available information thus points to the fact that the sulphur enrichment of the recent lake sediments in North America is related to the general environmental pollution with sulphur. However, it is not clear whether the enrichment factors observed can be related quantitatively to changes in atmospheric deposition rates. This arises because it is difficult to relate the sediment data to a particular time period for the following reasons:

Different sedimentation rates in different watersheds.
Sediment focusing: the finer particles of sediment with the highest sulphur content are deposited in the deepest part of the lake.
Variability within layers may be greater than differences between layers; in principle this does not exclude a valid record.
Productivity of lakes may vary due to inputs of other nutrients at different times and so influence the accumulation of sulphur in the sediments.

Human Influence on the Sulphur Cycle

Oxidized surface layer can influence the transport of sulphur across the sediment–water interface.

Similar problems, as for lake sediments, arise in the determination of atmospheric pollution by analysing a time series of soils for total sulphur content:

Unless the position of soil sampling is accurately defined, it is difficult to take corresponding soil samples years apart.

Even if the original site can be accurately determined, natural spatial variability may be so great that increases due to atmospheric sulphur fluxes are not apparent.

Comparison between samples analysed by workers many years apart may be invalidated by the methods of analysis. Early methods were unreliable and usually resulted in low sulphur concentrations being obtained.

Increased deposition rates can be shown along a transect from a source of high pollution. These observations can be invalidated if changes in soil type are not taken into consideration. This problem can be overcome by studying the isotopic composition of the soil and pollutant sulphur.

Some of these problems may be overcome by studying stable sulphur isotopes in the soil system because isotopic information may allow corrections to be made for the variable residence time of sulphur in different soils and soil horizons (Krauss and Grinenko, 1988). As vegetation under stress from high sulphur concentrations tends to emit isotopically light sulphur, the foliage and ultimately the soil become isotopically heavy. Such observations may aid the reconstruction of past sulphur fluxes to soils. However, we conclude that presently it is only from ice cores that reasonably lengthy, reliable records are obtainable. Currently most of these are for very remote areas, but it is possible that cores from high altitude temperate glaciers may provide useful data in the future.

5.3 SULPHUR CYCLE PRIOR TO HUMAN ACTIVITY

In the preceding sections we considered the observed environmental changes caused by anthropogenic disturbance to the sulphur cycle. A question arises: are these data sufficiently reliable to reconstruct an unperturbed natural global sulphur cycle? At present the answer must be no: the available data are sparse and attempts to answer the question will necessarily be speculative.

This section makes an attempt to assess the anthropogenic contribution to each of the fluxes in the cycle considered in the introduction to this chapter. There are several fluxes that have probably not changed much over the historical period: volcanic emission, emission of sulphate with sea salt and emission of short-lived (H_2S, DMS, etc.) and long-lived reduced sulphur compounds (COS, CS_2) in the open regions of the world ocean has probably remained constant averaged over a number of years. The change (if any) in some other fluxes is probably insignificant. Among such fluxes are the emission of short-lived and long-lived reduced sulphur compounds on land and in coastal ocean regions (see Section 5.1.5). This category also includes sulphur flux from the oceanic atmosphere into the continental one though it may have increased slightly. Let us assume that the flux related to natural water erosion is unchanged as well. As was supposed in Section 5.1.4, the aeolian flux due to anthropogenic activity had doubled, i.e. it may have been 10 Tg (S) a^{-1} in prehistoric times.

Thus, during the last few centuries atmospheric sulphur emissions have grown by 103 Tg (S) a^{-1} and river discharge by 57 Tg (S) a^{-1}. Besides, the part of anthropogenic sulphur which is removed from the atmosphere over the continents is ultimately found in river runoff.

If we assume that the modern continental atmosphere is a well-mixed block, we can easily find that out of 103 Tg (S) a^{-1} anthropogenic sulphur about 51 Tg (S) a^{-1} is transported to the marine atmosphere and 52 Tg (S) a^{-1} returns to the continental surface (see Figure 5.1a). It can be assumed that the relative proportions of sulphur in the continental atmosphere removed over the land surface and that transported to the marine atmosphere remained constant. Thus in the prehistoric period the annual sulphur deposition on the continental surface was 84–52=32 Tg (S) a^{-1} and the annual transport to the marine atmosphere 81–51=30 Tg (S) a^{-1}. Consequently the deposition over the ocean surface was 258-51=207 Tg (S) a^{-1} in total.

The above considerations are illustrated in Figure 5.1b, which schematically presents the sulphur cycle prior to the beginning of human activity. Comparing the data given in Figures 5.1a and 5.1b we can see that during the historical period river runoff has increased from 104 Tg (S) a^{-1} to 213 Tg (S) a^{-1}, i.e. mean sulphur concentration in river water has doubled, and shows reasonable agreement with data summarized in Section 5.1.3. Sulphur input from the atmosphere to the continents has increased over the period by 2.6 times. It should be noted that, since anthropogenic emission is mainly as sulphur dioxide, it is the absorption of sulphur of sulphur dioxide by the continental surface that is largely responsible for the increase rather than an increase in the scavenging of sulphur by precipitation. On the whole, the continental and oceanic atmospheres in the prehistoric period were nearly balanced with regard to sulphur fluxes.

5.4 TRANSPORT SCALES AND MODELLING

The average residence times for the various sulphur compounds in the atmosphere differ substantially—from less than a day for DMS in the tropical

Table 5.5 Rough estimates of average residence times (t) and transport distances (L) in the atmosphere for different sulphur compounds (Rodhe, 1978; Rodhe and Isaksen, 1980; Möller, 1984b)

	Planetary boundary layer		Free troposphere		
	t/day	transport distance $L/10^3$ km	t/day	transport distance longitudinal 10^3 km	distance latitudinal 10^3 km
H_2S	1	0.5	2	4	1
DMS	<0.5	<0.2	<1	<2	<0.5
COS	500	global	500	global	global
CS_2	10	5	20	global	5
SO_2	1	0.5	10	synoptic	3
SO_4^{2-}	4	2	15	global	4

marine boundary layer to probably over a year for COS. As a result, the average transport distances vary from a few hundred kilometres to global. Table 5.5 contains rough estimates of these time and space scales for the most important sulphur compounds. The average transport distances for the free troposphere have been separated into longitudinal and latitudinal components.

It is clear from the table that anthropogenic emissions of SO_2 near the surface will be mainly confined to the nearest few thousand kilometres. Only the fraction that escapes the boundary layer and is mixed into the free troposphere will be spread over distances of a global scale. On the other hand, SO_2 emitted from eruptive volcanoes 'starts off' in the free troposphere, or even in the lower stratosphere, and thus will be more globally dispersed.

In general, the deposition from an anthropogenic source of sulphur will dominate over the natural sulphur deposition in a sufficiently small area surrounding the source. According to Brimblecombe (1985), the anthropogenic sulphur emission in Europe in the year 1840 amounted to 0.43 Tg (S) a^{-1}, or about 1.5% of the 1980 value. Application of a simple but well-verified model for atmospheric sulphur transport (Eliassen and Saltbones, 1983) to the 1840 emission data shows that the deposition from these sources exceeds 0.2 g (S) $m^{-2}a^{-1}$ in a small area covering England, part of Belgium and the Netherlands, thus probably exceeding the deposition from natural sources in this area. It seems that under rather special meteorological conditions significant and noticeable quantities of air pollutants could be deposited in regions remote from pollution sources at an early date (Brimblecombe et al., 1986).

By 1980, the European anthropogenic emissions had increased to about 28 Tg (S) a^{-1} (Dovland and Saltbones, 1979). The deposition from these sources exceeds 0.2 g (S) m^{-2}a^{-1} over most of Europe and 12 g (S) m^{-2}a^{-1} in central Europe, far exceeding the deposition from natural sources. The changes in North America have been similar (Welpdale, 1978; Möller, 1984a).

As was discussed in a previous section, data from Greenland ice cores indicate an increase by about a factor of two or three in the deposition of sulphate during the 1970s compared to the average pre-industrial value. However, the increase is not much greater than the range of variability of the pre-industrial values so the statistical significance of the increase is hard to establish outside southern Greenland. A natural question to ask is whether the indicated increase in sulphur deposition on Greenland is consistent with our notion about anthropogenic and natural sulphur sources. This question is not easy to answer since most models used to study the transport and deposition of sulphur are either limited to the industrialized regions and their immediate surroundings (e.g. Eliassen and Saltbones, 1983) or global but restricted by only representing zonal average values (e.g. Rodhe and Isaksen, 1980). None of these models take proper account of topography.

It is likely that a substantial fraction of the sulphur that is deposited on the Greenland ice has been transported to the site above the planetary boundary layer (most Greenland ice cores have been obtained at an elevation of about 2000–3000 m). The relative importance of the different sources will then depend on the ability of their emissions to penetrate into the mid-troposphere. Eruptive volcanoes will evidently be relatively more important than industrial SO$_2$ emissions. If one assumes that natural and anthropogenic surface sources (other than eruptive volcanoes) are distributed at similar distances from Greenland, then one could assume that the natural/anthropological ratio for sulphur deposition in Greenland is typical of the ratio for the total northern hemisphere. With the northern hemisphere anthropogenic emissions amounting to 60–80 Tg (S) a^{-1} (Möller, 1984a) and the natural emissions to 30–50 Tg (S) a^{-1} (Andreae and Barnard, 1984), the ratio of the present to the pre-industrial flux would be in the range of 2.2–3.7.

It could be argued that the bulk of the natural emissions is likely to occur in tropical regions and thus would be further away from Greenland than the anthropogenic emissions. On the other hand, tropical emissions are more likely to end up in the free troposphere than are high latitude emissions because of the prevalent convective mixing in cumulonimbus clouds that takes place in the tropics (Chatfield and Crutzen, 1984).

Comparing the ratio 2.2–3.7 with the factor of two to three increases, one may conclude that there is reasonable agreement. However, in view of the uncertainty of the ice core data, as well as our limited knowledge about transport and deposition processes, no definite conclusions can be drawn.

The modest increase observed in Greenland ice cores does not seem to agree very well with the postulate that the present Arctic haze is caused mainly by anthropogenic emissions. However, this haze has a large maximum in March and April. During this time of the year there is little precipitation in the Arctic, and generally a stable boundary layer, implying very inefficient deposition processes. At the same time, the atmospheric circulation pattern is generally favourable for northward transport of European anthropogenic sulphur. The impact of this sulphur on the deposition pattern in the Arctic will be relatively smaller, since one reason for the occurrence of this very long-range transport is inefficient deposition processes. Greenland receives deposition not only from the boundary layer, but the free troposphere as well.

An attempt to model the global distribution of sulphur compounds from both natural and anthropogenic sources has indicated that the anthropogenic contribution to the sulphate concentration in the free troposphere varies from about 80% in the mid-latitudes of the northern hemisphere to about 10% in the tropics (Rodhe and Isaksen, 1980). This result must be regarded as tentative since the concentrations of SO_2 calculated by their model in the upper troposphere were much lower than those that have been observed. The origin of these relatively high concentrations of SO_2 and sulphate in the troposphere are not well understood, although it has been suggested that transport of 'polluted' boundary layer air to the upper troposphere by large cumulonimbus clouds could be an important factor, at least in tropical latitudes (Chatfield and Crutzen, 1984).

5.5 ENVIRONMENTAL IMPACT OF CHANGES IN THE SULPHUR CYCLE

In recent times the atmospheric sulphur cycle has undergone the greatest changes as a result of anthropogenic activity. Hence it is here that the earliest observations of environmental impact are found. These are related to increased atmospheric concentrations of sulphur compounds. Obviously local effects (in the vicinity of emission sources), were the first to be described (e.g. Evelyn, 1661). The impact of increased sulphur dioxide and sulphate concentrations on health can sometimes be found in regional medical statistics. Certain meteorological situations produce conditions which aid the formation of sulphuric acid smog. Long-term exposure to such smogs has important implications for urban health. The effect of sulphur dioxide on vegetation in the vicinity of emission sources is also marked. Lower plants (mosses and lichens), as well as coniferous tree species are the most sensitive. High concentrations of sulphur compounds damage metal and stonework, and artistic and cultural objects through corrosion and structural weakening.

On a regional scale, the most striking manifestation of sulphur cycle disturbance is the phenomenon of environmental acidification. The phenomenon, described in detail in scientific and popular literature, has spread by now over vast areas of northern Europe and eastern North America. It has also been observed in Japan and may appear in a number of other regions around the globe. Regional impact may include health effects of high particulate sulphate concentrations—tens of micrograms of sulphate sulphur per cubic metre. In addition, the presence of particles in the atmosphere reduces the level of solar radiation. Observations from satellites indicate that sulphate haze can cover large areas in the North American continent and stretch far into the Atlantic.

Global effects include possible growth of sulphate particle concentration in the stratosphere. This can have an impact on the global climate of the Earth. The increase in stratospheric sulphate could be caused by the growth of tropospheric concentrations of long-lived anthropogenic sulphur compounds (COS and CS_2) and, consequently, by their increased penetration into the stratosphere. It should be noted that the issue still remains almost uninvestigated. At present there is no firm evidence for increasing tropospheric concentrations of COS or CS_2. However Hofmann and Rosen (1980) noticed the growth of stratospheric concentrations of sulphate particles.

The impact of anthropogenic perturbations of the sulphur cycle on the hydrosphere can be of a non-global nature only. Even if man extracts from the lithosphere and uses all sulphur-bearing minerals (coal, oil, sulphide ores, native sulphur), the sulphur content in the world ocean will increase by no more than 0.001%. On the other hand, increases in the sulphate concentration of rivers and lakes in industrialized regions were noticed by the beginning of the present century. We do not know whether the increased sulphate concentration has affected the freshwater biota.

Since the sulphur cycle in the hydrosphere and sediments is closely connected with the cycles of organic carbon and nutrients, anthropogenic disturbances of these factors can lead to sulphur cycle perturbation. The discharge of large amounts of organic carbon and nutrients into closed freshwater bodies stimulates the development of anaerobic sulphate reducing microorganisms. This results in the formation of free H_2S in lower water layers and disappearance of higher life forms in a given water body.

It is alarming that the sulphur cycles of the Black and Baltic Seas are also disturbed due to the input of organic substances with river runoff. As a result, anaerobic conditions have formed in the Baltic Sea depressions and the level of the 'dead' H_2S zone in the northwestern part of the Black Sea has risen. The incidents of H_2S destroying mussel fields and rising to surface water layers have become more frequent here (Kosarev, 1984).

In the lithosphere the sulphur cycle is very closely connected with the cycles of non-ferrous metals. Intensive and ever increasing extraction of non-ferrous metals could lead to almost complete depletion of sulphide ores in the next 30–50 years. It is worth noting yet another aspect of the anthropogenic impact on the lithospheric sulphur cycle. The commonly applied technique for water pumping into oil-bearing strata results in an intensive growth in sulphate-reducing bacteria. The organic substances in oil products act as a substrate and sulphates in pumped water serve as a source of oxidant. As the process of sulphate reduction develops, H_2S is released and the extracted oil becomes contaminated with sulphur. In conclusion one should note that there are other interesting effects such as the microbial acidification of mine waters and the spontaneous combustion of sulphide-containing waste dumps.

5.6 FUTURE SCENARIOS AND DIRECTIONS

5.6.1 Future perspectives

In the coming fifty years the main problems concerning the anthropogenic influence on the sulphur cycle are likely to be the emissions of SO_2 and associated NO_x to the atmosphere in highly industrialized continental areas. Here the effects of corrosion of building materials and cultural artefacts, damage to human health and injury to vegetation will continue to be a problem in the areas near cities and industrial complexes. In sensitive regions the acidification of soils and surface waters provides considerable challenge in terms of control measures which must necessarily be international rather than local. Beyond the problem of acidification comes the question of the role of sulphur compounds in the complex issue of 'novel forest decline'. The near future requires us to achieve and act on a clear understanding of these issues.

Scenarios for future anthropogenic sulphur emissions from Europe have been worked out at the International Institute for Applied Systems Analysis. The scenarios suggest that these emissions will not change substantially over the next 50 years (see Figure 5.12). This is probably also the case for North American anthropogenic emission. However, it is likely that a peak in the effect of deposited sulphur, such as the leaching of base cations from soils, may lie in the future. In some countries, often in tropical and subtropical regions, large increases in emission are to be expected in the future (e.g. Venezuela, Nigeria, China, etc.) (SCOPE 36) so the global picture will be a little different (see Figure 5.13). In tropical and subtropical countries with a high growth in the utilization of fossil fuel anthropogenic sources of sulphur emission to the atmosphere may become more important

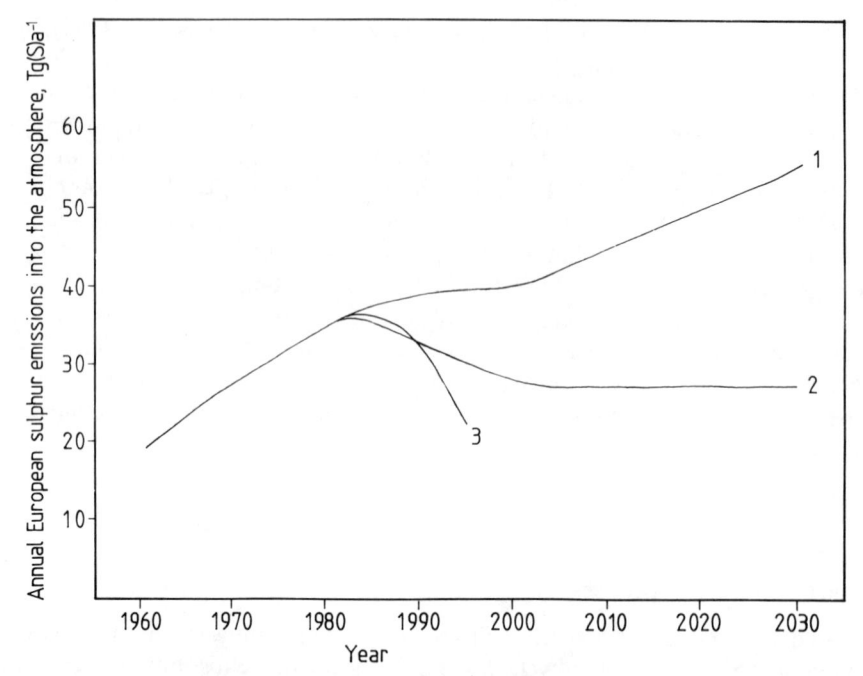

Figure 5.12 The long-term trend in European sulphur emissions prepared by the International Institute of Applied Systems Analysis and modified by Eliassen (1988).
1 Without emission control measures (highest energy use scenario).
2 Without emission control measures (lowest energy use scenario).
3 With 30% reduction in emissions by 1993 (lowest energy use scenario).

than natural sources. One should add that the control of sulphur emissions may not be free of associated environmental problems. Large amounts of contaminated gypsum are a likely product of an extensive programme to reduce sulphur emission to the atmosphere.

Over a thousand year time scale predictions are much more uncertain. It is important to note that only 7% of the available sulphur in fossil fuels has been released. Long-term acidification could affect geological processes such as weathering. Paces (1985) has argued that there is a sharp transition point in catchments as the alkaline materials are removed by acidification. Widespread loss of buffering capacity in soils could be very serious.

5.6.2 Future work

In order to understand the changes underway it is clear that a number of projects need to be undertaken. In establishing the long-term changes we need a better knowledge of the past levels of sulphur compounds in the

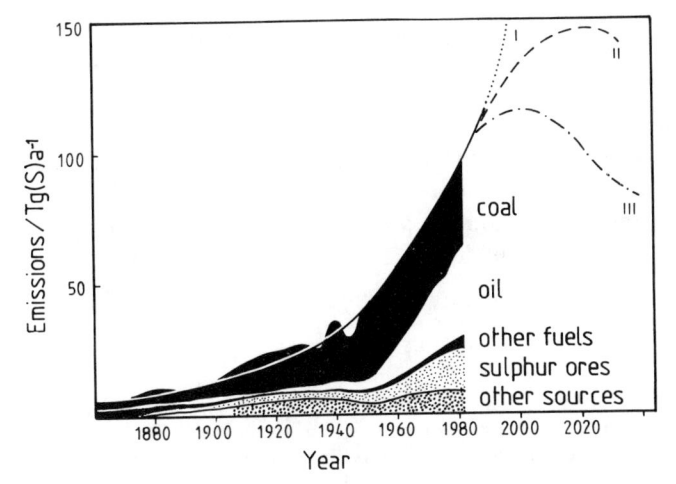

Figure 5.13 Long-term global anthropogenic sulphur emission into the atmosphere (Ryaboshapko, 1985).
I   Exponential increase at the present rate.
II Scenario with no emission control measures.
III Scenario with emission control measures.

atmosphere. In current ice core work there is little knowledge of the changes in isotopic composition which would add a new dimension to our understanding of the origin of the sulphur compounds deposited in polar regions. New solid source techniques for mass spectroscopy may allow isotopic analysis of nanogram quantities of sulphur compounds and aid such work. However, isotopic studies of polar ice need to be coupled to a thorough understanding of atmospheric transport to polar regions.

Isotopic studies of soils may prove valuable in trying to establish past sulphur fluxes. In particular it should be possible to gain useful information by examining the distribution of sulphur isotopes between the inorganic sulphate pool and the larger more slowly replaced organic sulphur pool. Sampling for such studies should proceed on the basis of soil horizons and not to arbitrarily defined depths.

We need a better understanding of the anthropogenic and natural emissions of small organo-sulphur compounds. The anthropogenic emissions of COS and CS_2 are very poorly understood. The fate of DMS and the relative importance of various oxidative pathways to methane sulphonic acid, dimethyl sulphoxide and sulphate is still unclear. Further we need more extensive information on the distribution, both temporal and spatial, of these compounds and their sources. The stable and less soluble of the sulphur gases (e.g. COS) may be important in influencing the concentration of the sulphate aerosol in the stratosphere in epochs of low volcanic activity.

There is a need to examine the older measurements and analyses of the composition of waters and soils, or to find archived samples or to used layered materials such as sediments or tree rings to build up a better picture of century long changes in the cycling of sulphur. In particular it may be that much can be learnt from analyses of river water and a detailed comparison between present and past sulphur fluxes. In addition we need a well-organized comparative programme of representative drainage basins to learn about the individual contributions of riverine sulphur to the global cycle.

REFERENCES

Adams, D. F., Farwell, S. O., Robinson, E., Pack, M. R., and Bamesberger, W. L. (1981). Biogenic sulfur source strength. *Environ. Sci. Technol.*, **15** (12), 1493–8.

Afinogenova, O. G. (1982). Integrated approach to the estimation of the total anthropogenic emission of sulfur dioxide on the territory of an individual country. In: Galperin, M. V. (Ed.) *Methodological and Systemotechnical Questions of Environment Pollution Monitoring*, Annls. Inst. Appl. Geophys., **48**, Moscow, Gidrometizdat, pp. 70–80 (in Russian).

Alekin, O. A., and Brazhnikova, L. V. (1964). *Runoff of Dissolved Compounds from the Territory of the USSR*, Nauka, Moscow (in Russian).

Andreae, M. O. and Barnard, W. R. (1984). The marine chemistry of dimethylsulfide. *Mar. Chem.*, **14**, 267–79.

Andreae, M. O., and Raemdonck, H. (1983). Dimethyl sulfide in the surface ocean and the marine atmosphere: a global view. *Science*, **221** (4612), 755–47.

Aneja, V. P., Aneja, A. P., and Adams, D. F. (1982). Biogenic sulfur compounds and the global sulfur cycle. *J. Air Pollut. Contr. Assoc.*, **32** (8), 803–7.

Aristarain, A. J., Delmas, R. J., and Briat, M. (1982). Snow chemistry on James Ross Island, Antarctic Peninsula. *J. Geophys. Res.*, **87**, 11004–12.

Astratov, M., Zaveryaeva, I., and Ryaboshapko, A. (1986). Variations of sulfate contents in Arctic atmospheric precipitation during the last 90 years. *Meteorology and Hydrology*, no. 1, 43–6.

Barnard, W. R., Andreae, M. O., Watkins, W. E., Bingemer, H., and Georgii, H. W. (1982). The flux of dimethylsulfide from the oceans to the atmosphere. *J. Geophys. Res.*, **87** (C11), 8787–93.

Battarbee, R. W., Flower, R. J., Stevenson, A. C., and Rippey, B. (1985). Lake acidification in Galloway: a palaeoecological test of competing hypotheses. *Nature*, **314** (6009), 350–2.

Beeton, A. M. (1965). Eutrophication of the St. Lawrence Great Lakes. *Limnol. Oceanogr.*, **10**, 240–54.

Berresheim, H., and Jaeschke, W. (1983). The contribution of volcanoes to the global atmospheric sulfur budget. *J. Geophys. Res.*, **88** (C6), 3732–40.

Bettelheim, J., and Litter, A. (1979). Historical trends of sulfur oxide emissions in Europe since 1865. Central Electricity Generating Board, UK, Report PL-GS/E/1/79.

Blanchard, D. C., and Woodcock, A. H. (1980). The production, concentration, and vertical distribution of the sea-salt aerosols. In: Kneip, Th.J., and Lioy, P. J. (Eds.) *Aerosols, Anthropogenic and Natural Sources and Transport Ann. N.Y. Acad. Sci.*, **338**, 330–47.

Brimblecombe, P. (1985). Is acid rain a new phenomenon? *Proceedings of the Acid Rain Inquiry*, Scottish Wildlife Trust.

Brimblecombe, P., Davies, T. D., and Tranter, M. (1986). Nineteenth century black Scottish showers. *Atmos. Environ.*, **20**, 1053–7.

Brimblecombe, P., and Pitman, J. I. (1980). Long term deposit at Rothamsted, England. *Tellus*, **32**, 261–7.

Brimblecombe, P., and Shooter, D. (1986). Photo-oxidation of dimethylsulphide in aqueous solution. *Marine Chemistry*, **19**, 343–53.

Brimblecombe, P., and Stedman, D. H. (1982). Historical evidence for a dramatic increase in the nitrate component of acid rain. *Nature*, **298**, 460–2.

Bulatkin, G. A., and Maksimova, Yu. A. (1978). Contents of some trace elements and sulphate in precipitation over the central part of the Russian Plain. In: *Lichens as Indicators of Environmental Conditions*. Tallinn Botanical Garden of the Academy of Sciences of the Estonian SSR, pp. 61–3.

Busenberg, E., and Langway, C.C. Jr. (1979). Levels of ammonium, sulphate, chloride, calcium and sodium in snow and ice from southern Greenland. *J. Geophys. Res.*, **87**, 1705–9.

Cadle, R. D. (1980). A comparison of volcanic with other fluxes of atmospheric trace gases. *Rev. Geophys. Space Phys.*, **18**, 746–52.

Chatfield, R. B., and Crutzen, P. J. (1984). Sulphur dioxide in remote oceanic air, *J. Geophys. Res.*, **89**, 7111–32.

Clarke, F. W. (1924). *The Data of Geochemistry*, USGS Bulletin 770, Washington.

Cline, J. D., and Bates, T. S. (1983). Dimethyl sulfide in the equatorial Pacific ocean: a natural source of sulfur to the atmosphere. *Geophys. Res. Lett.*, **10**, 949–52.

Cook, R. B. (1980). Thesis. Columbia University, New York.

Cragin, J. H., Herron, M. M., and Langway, C. C. Jr. (1975). The chemistry of 700 years of precipitation at Dye 3, Greenland. Res. Rep. 341, U.S. Army Cold Region Res. and Eng. Lab., Hanover, N.H.

Cragin, J. H., Herron, M. M., Langway, C. C. Jr., and Klouda, G. (1974). Interhemispheric contribution to the changes in the composition of atmospheric precipitation during the late cenozoic era. Conference on Polar Oceans, Montreal, May 1974.

Cragin, J. H., Herron, M. M., Langway, C. C. Jr., and Klouda, G. (1977). Interhemispheric comparison of changes in the composition of atmospheric precipitations during the late Cenozoic era. In: Dunbar Maxwell, J. (Ed.) *Polar Oceans (Proceedings of the Polar Oceans Conference, Montreal, 1974)*, Arctic Institute of North America, pp. 617–31.

Cullis, C. F., and Hirschler, M. M. (1980). Atmospheric sulphur: natural and man-made sources. *Atmos. Environ.*, **14**, 1263–78.

Davidson, C. I., Chu, L., Grimm, T. C., Nasta, M. A., and Quamoos, M. P. (1981). Wet and dry deposition of trace elements onto the Greenland Ice Sheet. *Atmos. Environ.*, **15**, 1429–37.

Davies, T. D., Vincent, C. E., and Brimblecombe, P. (1982). Preferential elution of strong acids from a Norwegian ice cap. *Nature*, **300**, 161–3.

De Angelis, M., Legrand, M., Petit, J. R., Barkov, N. I., Korotkevitch, Y. S., and Kotlyakov, V. M. (1984). Soluble and insoluble impurities along the 950 m deep Vostok ice core, Antarctica. Climatic Implications. *J. Atmos. Chem.*, **1**, 215–39.

Delmas, R. J., and Boutron, C. F. (1978). Sulfate in Antarctic snow: spatio-temporal distribution. *Atmos. Environ.*, **12**, 723–8.

Delmas, R. J., and Boutron, C. F. (1980). Are the past variations of the stratospheric sulfate burden recorded in central Antarctic snow and ice layers? *J. Geophys. Res.*, **85**, 5645–9.

Delmas, R. J., Briat, M., and Legrand, M. (1982). Chemical composition of south polar snow. *J. Geophys. Res.*, **87**, 4314.

Delmas, R., and Servant, J. (1983). Atmospheric balance of sulfur above an equatorial forest. *Tellus*, **35B** (2), 110–20.

Depetris, P. J. (1976). Hydrochemistry of the Paraná River. *Limnol. Oceanogr.*, **21**, 736–9.

Doronin, A. N. (1975). Chemical composition of snow near Vostok Station and along the axis Mirny–Vostok. *Bull. Sov. Antarct. Exp.*, **91**, 62–8.

Dovland, H., and Saltbones, J. (1979). Emission of sulphur dioxide in Europe in 1978. Norwegian Institute for Air Research, Rpt EMEP/OCP/2/79.

Eliassen, A. (1988). Pattern of anthropogenic sulphur deposition in Europe, past present and future. In: *Long Range Transport of Atmospheric Pollutants. Annls. Inst. Appl. Geophys.*, *Vol.* **71**, Moscow, Gidrometizdat, pp. 59–64 (in Russian).

Eliassen, A., and Saltbones, J. (1983). Modelling of long range trajectories of sulphur over Europe. *Atmos. Environ.*, **17**, 1457–73.

Eriksson, E. (1952). Composition of atmospheric precipitation II. Sulphur, chloride, iodide compounds. *Tellus*, **4**, 280–303.

Eriksson, E. (1960). The yearly circulation of chloride and sulphur in nature: meteorological, geochemical and pedological implications. Part 2. *Tellus*, **12**, 63–109.

Evelyn, J. (1661). *Fumifugium*, Gabriel Bedel and Tomas Collins, London.

Friend, J. P. (1973). The global sulphur cycle. In: Rasool, S. I. (Ed.) *Chemistry of the Lower Atmosphere*. Plenum Press, New York, London.

Fry, B. (1986). Stable isotope distributions and sulfate reduction in lake sediments of the Adirodack Mountains, New York. *Biogeochemistry*, **2**, 329–43.

Gibbs, R. J. (1972). *Geochim. Cosmochim. Acta*, **36**, 1061–6.

Granat, L., Rodhe, H., and Hallberg, R. O. (1976). The global sulfur cycle. In: Nitrogen, phosphorus and sulfur—global cycles. SCOPE Report 7, *Ecol. Bull.*, Stockholm, **22**, 89–134.

Grigoryev, A. A. (1985). *Anthropogenic Impact on the Environment from Space Observations*. Leningrad, Nauka.

Hammer, C. U. (1977). Past volcanism revealed by Greenland ice sheet impurities. *Nature*, **270**, 482–6.

Hammer, C. U. (1980). Acidity of polar ice cores in relation to absolute dating, past volcanism and radio-echoes. *J. Glaciol.*, **25**, 359–72.

Hammer, C. U. (1983). Initial direct current in the buildup of space charges and the acidity of ice cores. *J. Phys. Chem.*, **87**, 4099–103.

Hammer, C. U. (1984). Traces of Icelandic eruptions in the Greenland Ice Sheet. *Jökull*, **34**, 51–6.

Hammer, C. U. (1985a). The influence on atmospheric composition of volcanic eruptions as derived from ice core analysis. *Ann. Glaciol.*, **7**, 125–9.

Hammer, C. U. (1985b). The Byrd ice core: continuous acidity measurements–solid conductivity ECM. Progress report presented at Symposium on snow and ice and the atmosphere: 19–24 Aug. 1984. *Ann. Glaciol.*, **7**, 214.

Hammer, C. U., Clausen, H. B., and Dansgaard, W. (1980). Greenland Ice Sheet evidence of post-glacial volcanism and its climatic impact. *Nature*, **288**, 230–35.

Hammer, C. U., Clausen, H. B., and Dansgaard, N. (1981). Past volcanism and climate revealed by Greenland ice cores. *J. Volcanol. Geotherm. Res.*, **11**, 3–10.

Hammer, C. U., Clausen, H. B., Dansgaard, W., Gundestrup, N., Johnsen, S. J., and Reeh, N. (1978). Dating of Greenland ice cores by flow models, isotopes, volcanic debris and continental dust. *J. Glaciol.*, **20** (82), 3–26.

Hammer, C. U., Clausen, H. B., Dansgaard, W., Neftel, A., Kristinsdottir, P., and Johnson, E. (1985). Continuous impurity analysis along the Dye 3 deep core. In: The Greenland Ice Sheet Programme. *AGU Monograph*, **33**, 90–4.

Herron, M. M. (1980). The impact of volcanism on the chemical composition of Greenland ice sheet precipitation. Ph.D. thesis, State University of New York at Buffalo, USA.

Herron, M. M. (1982). Impurity sources of F^-, Cl^-, NO_3^- and SO_4^{2-} in Greenland and Antarctic Precipitation. *J. Geophys. Res.*, **83**, 3052–60.

Herron, M. M., and Langway, C. C. Jr. (1985). Chloride, nitrate and sulfate in the Dye 3 and Camp Century, Greenland ice cores. In: The Greenland Ice Sheet Programme. *AGU Monograph*, **33**, 77–84.,

Herron, M. M., Langway, C. C. Jr., Weiss, H. V., and Cragin, J. H. (1977). Atmospheric trace metals and sulfate in the Greenland Ice Sheet. *Geochim. Cosmochim. Acta*, **41**, 915–20.

Hitchcock, D. R., and Black, M. S. (1984). $^{34}S/^{32}S$ evidence of biogenic sulfur oxides in a salt marsh atmosphere. *Atmos. Environ.*, **18** (1), 1–12.

Hofmann, D. J., and Rosen, J. M. (1980). Stratospheric sulfuric acid layer: evidence for an anthropogenic component. *Science*, **208**, 1368–70.

Holdren, G. R. Jr., Brunelle, T. M., Matisoff, G., and Wahlen, M. (1984). Timing the increase in atmospheric sulphur deposition in the Adirodack Mountains. *Nature*, **311** (5983), 245–8.

Husar, R. B., and Halloway, J. M. (1983). Sulfur and nitrogen over North America. In: *Ecological Effects of Acid Deposition*, Nat. Swedish Environ. Prot. Board. Rep. PM1636.

Husar, R. B., and Husar, J. D. (1985). Regional river sulfur runoff. *J. Geophys. Res.* **90** (C1), 1115–25.

Ivanov, M. V., and Freney, J. R. (Eds.) (1983). *The Global Biogeochemical Sulfur Cycle, SCOPE19*. John Wiley & Sons, Chichester, 470 pp.

Ivanov, M. V., Grinenko, V. A. and Rabinovich, A. P. (1983). In: Ivanov, M. V., and Freney, J. R. (Eds.) *The Global Biogeochemical Sulphur Cycle, SCOPE19*. John Wiley & Sons, Chichester, pp. 331–56.

Jaeschke, W., Claude, M., and Herrman, J. (1980). Sources and sinks of atmospheric H_2S. *J. Geophys. Res.*, **85** (C10), 5639–44.

Junge, C. E. (1960). Sulfur in the atmosphere. *J. Geophys. Res.*, **65**, 227–37.

Junge, C. E. (1963). *Air Chemistry and Radioactivity.* Academic Press, New York, London, 381 pp.

Kellogg, W. W., Cadle, R. D., Allen, E. R., Lazrus, A. L., and Martell, E. A. 1972. The sulfur cycle. *Science*, **175**, 587–96.

Koide, M., and Goldberg, E. D. (1971). Atmospheric sulfur and fossil fuel combustion. *J. Geophys. Res.*, **76**, 7689–96.

Korotkevitch, Y., Petrov, V. N., Barkov, N. I., Sukhonosova, L. I., Dmitriyev, D. N., and Portnov, V. G. (1978). Results of the study of the vertical structure of Antarctic ice sheet in the vicinity of Vostok. *Antarctica, Inf. Bull. Sov. Ant. Exp.*, **97**, 135–48.

Korzon, V. I. et al. (1974). *World Water Balance and Water Resources of the Earth*, Hydrometeoizdat, Leningrad (in Russian).

Kosarev, A.N. (1984). *The Southern Seas*, Znanije, Moscow, (in Russian).

Kushelevsky, A., Shani, G., and Haccoun, A. (1983). Effect of meteorological conditions on total suspended particulate (TSP) levels and elemental concentration

of aerosols in a semi-arid zone (Beer-Sheva, Israel). *Tellus,* **35B**, 55–65.
Legrand, M. R., Aristarain, A. S., and Delmas, R. J. (1982). Acid titration of polar snow. *Anal. Chem.,* **54**, 1336–9.
Legrand, M. R., and Delmas, R. J. (1984). The ionic balance of Antarctic snow: a 10 year detailed record. *Atmos. Environ.,* **18** (9), 1867–74.
Lien, A. (1983). The sulphur cycle in the lithosphere, II Cycling. In: Ivanov, M. A., and Freney, J. R. (Eds.) *The Global Biogeochemical Sulphur Cycle, SCOPE 19.* John Wiley & Sons, Chichester, pp. 95–129.
Livingstone, D. A. (1963). *Geol. Surv. Prof. Paper 440–G,* Washington.
Lorius, C., Baudin, G., Cittanova, J., and Platzer, R. (1969). Impuretés solubles contenues dans la glace de l'Antarctique. *Tellus,* **21**, 136–48.
Malyshev, Yu. A., Shikhov, V. V., and Shmataov, V. F. (Eds.) (1963). *Economics of Processing and Consumption of Sulfury Oil and Oil Products.* Bashkir Branch of the USSR Academy of Sciences, Ufa, 152 pp. (in Russian).
MARC (1985). *Historical Monitoring.* Monitoring and Assessment Research Centre, University of London.
Meybeck, M. (1977). Manuscript for UNESCO meeting in Amsterdam, 1976.
Meybeck, M. (1978). *Proc. UNESCO/SCOPE Workshop, Melreux,* UNESCO Press, 1978.
Milliman, J. D., and Meade, R. H. (1983). *J. Geol.* (Chicago), **119**, 1–21.
Mitchell, M. J., David, M. B., and Uutala, A. J. (1984). Sulphur distribution in lake sediment profiles as an index of historical depositional patterns. *Hydrobiology,* **121** (2), 121–7.
Möller, D. (1984a). Estimation of the global man-made sulfur emission. *Atmos. Environ.* **18**, 19–27.
Möller, D. (1984b). On the global natural sulfur emission. *Atmos. Environ.,* **18**, 29–39.
Neftel, A., Beer, J., Oeschger, H., Zürcher, F., and Finkel, R. C. (1985). Sulphate and nitrate concentrations in snow from South Greenland 1895–1978. *Nature,* **314** (6012), 611–13.
Nguyen, B. C., Bonsang, B., and Gandry, A. (1983). The role of the ocean in the global atmospheric sulfur cycle. *J. Geophys. Res.,* **88** (C15), 10903–14.
Nriagu, J. O. (1984). Role of inland water sediments as sinks of anthropogenic sulphur. *Sci. Total Environ.,* **38** (Sept.), 4–13.
Nriagu, J. O., and Coker, R. D. (1983). Sulphur in sediments chronicles past changes in lake acidification. *Nature,* **303** (5919), 692–4.
Paces, T. (1982). *Ambio,* **11**, 206–8.
Paces, T. (1985). Sources of acidification in Central Europe estimated from elemental budgets in small basins. *Nature,* **315**, 31–6.
Palais, J. M., and Legrand, M. (1985). Soluble impurities in the Byrd Station ice core. Antarctica: their origin and sources. *J. Geophys. Res.,* **90** (C1), 1143–54.
Petit, J. R., Briat, M., and Royer, A. (1981). Ice Age aerosol content from East Antarctic ice core samples and past wind strength. *Nature,* **293** (5831), 391–4.
Petrenchuk, O. P. (1980). On the budget of sea salts and sulfur in the atmosphere. *J. Geophys. Res.,* **85** (C12), 7439–44.
Prospero, J. M., and Nees, R. T. (1977). Dust concentration in the atmosphere of the equatorial North Atlantic. *Science,* **196**, 1196–8.
Punning, Y. A., Punning, K., and Tyugu, K. (1984). Sulphur concentration in atmospheric precipitation. Paper presented at the SCOPE workshop on sulphur cycle, Tallinn, USSR, 27 July–3 August, 1984.
Renberg, I., and Hedberg, T. (1982). The pH history of lakes in SW Sweden as calculated from the subfossil diatom flora of the sediments. *Ambio.,* **1** 30–3.

Risbo, T., Clausen, H. B., and Rasmussen, K. L. (1981). Supernovae and nitrate in the Greenland Ice Sheet. *Nature*, **294**, 5842, 637–9.
Robinson, E., and Robbins, R. C. (1968). *Sources, Abundance and Fate of Gaseous Atmospheric Pollutants. Final Report*, Project PR-6755. Standford Res. Inst., Menlo Park, Calif., 110 pp.
Rodhe, H., Crutzen, P., and Vanderpol, A. (1981). Formation of sulfuric and nitric acid in the atmosphere during long-range transport. *Tellus*, **33**, 132.
Rodhe, H., and Granat, L. (1984). An evaluation of sulphate in European precipitation 1955–1982. *Atmos. Environ.*, **18**, 2627–39.
Rodhe, H., Granat, L., and Soderlund, R. (1984). Sulphate in Precipitation, Rept. CM64 University of Stockholm, Department of Meteorology.
Rodhe, H., and Isaksen, I. (1980). Global distribution of sulphur compounds in the troposphere estimated in a height-latitude transport model. *J. Geophys. Res.*, **85**, 7401–9.
Ryaboshapko, A. G. (1983). The atmospheric sulfur cycle. In: Ivanov, M. A., and Freney, J. P. (Eds.) *The Global Biogeochemical Sulphur Cycle, SCOPE 19*. John Wiley & Sons, Chichester, pp. 203–96.
Ryaboshapko, A. G. (1985). Sulphur in the biosphere. *Priroda*, **N7**, 42–50 (in Russian).
Schütz, L. (1980). Long-range transport of desert dust with special emphasis on the Sahara. In: Kneip, Th. J., and Lioy, P. J. (Eds.) *Annl. N.Y. Acad. Sci.*, **338**, 525–32.
Takhtamyshev, Ch. N. (1971). On sulfate salinization of soils in AkDarja region of Samarkand district. *Pochvovedenie*, **11**, 160–4 (in Russian).
Thode, H. G., and Dickman, M. (1983). Unpublished report.
Turner, S. M., and Liss, P. S. (1985). Measurements of various sulphur gases in a coastal marine environment. *J. Atmos. Chem.*, **2**, 223–32.
Varhelyi, G. (1985). Continental and global sulfur budgets—I Anthropogenic SO_2 emissions. *Atmos. Environ.*, **19**, 1029–40.
Varhelyi, G., and Gravenhorst, G. (1981). An attempt to estimate biogenic sulfur emission into the atmosphere. *Időjaras*, **85** (3), 126–33.
Varhelyi, G., and Gravenhorst, G. (1983). Production rate of air-borne sea-salt sulfur deduced from chemical analysis of marine aerosols and precipitation. *J. Geophys. Res.*, **88** (C11), 6737–51.
Wagenbach, D., Gorlach, U., Hoffa, K., Junghans, H. C., Munnich, K. O., and Schotter, U. (1983). A long term aerosol deposition record in a high altitude alpine glacier. Paper presented at WMO Technical Conference on Observation and Measurement of Atmospheric Contaminants, Vienna, Austria, 17–21 October.
Welpdale, D. M. (1978). Large scale atmospheric sulphur studies in Canada. *Atmos. Environ.*, **12**, 661–70.
Weiss, H. V., Bertine, K. K., Koide, M., and Goldberg, E. D. (1975). The chemical composition of a Greenland glacier. *Geochim. Cosmochim. Acta*, **39**, 1–10.
Wright, H. E. Jr. (1982). Lake and wetland sediments as records of past atmospheric composition. In: Goldberg, E. D. (Ed.) *Atmospheric Chemistry*, Report of the Dahlem Workshop on Atmospheric Chemistry, Springer-Verlag, Berlin, pp. 135–57.

Part III

Interaction of Sulphur and Carbon Cycles in some Modern Ecosystems

Evolution of the Global Biogeochemical Sulphur Cycle
Edited by P. Brimblecombe and A. Yu. Lein
© 1989 SCOPE Published by John Wiley & Sons Ltd

CHAPTER 6
Interaction of Sulphur and Carbon Cycles in Marine Sediments

M.V. Ivanov, A. Yu. Lein, M.S. Reeburgh and G.W. Skyring

6.1 QUANTITATIVE RELATIONSHIPS BETWEEN SULPHATE REDUCTION AND CARBON METABOLISM IN MARINE SEDIMENTS*

6.1.1 Introduction

Measurements of the quantitative relationships between the oxidation of organic carbon and sulphate reduction in marine sediments are fundamental to the understanding of the biochemical cycling of C and S. These relationships may also reflect the form and amount of these elements at various stages of cycling or the rates at which their diagenesis occur.

A general conclusion of several investigators is that a major portion of the organic carbon deposited in sediments is rapidly oxidized to CO_2 by the sulphate-reducing bacteria (SRB) and that 2 moles of organic carbon are oxidized for every mole of sulphate reduced. However, the degree to which the data that have prompted these conclusions represent the true situation, is constrained. These constraints involve the methodology and the reliability of measured amounts and rates, the complexity of ecosystems, and the variations imposed by seasons, environmental perturbation and patchiness. These topics will be reviewed with the aim of drawing together the data which are concerned with the biogeochemical dynamics of C and S cycling.

6.1.2 Organic substrates for the sulphate-reducing bacteria in marine sediments

Recent studies of the anaerobic oxidation of organic matter in sediments indicate that fatty acids are quantitatively the most significant penultimate end-products (Balba and Nedwell, 1982; Chambers, 1985; Christensen, 1984; Mountfort and Asher, 1981; Sansone and Martens, 1982; Skyring, 1987,

*G.W. Skyring.

1988; Sørensen et al., 1981). Sulphate reduction is the most important process in the complete oxidation of these fatty acids in anoxic marine sediments where sulphate is not limiting. Widdel (1980) gives the general equation for the complete oxidation of fatty acids by the SRB as follows:

$$4H(CH_2)_nCOO^- + (3n+1)SO_4^{2-} + H_2O \rightarrow (4n+4)HCO_3^- \\ + (3n+1)HS^- + OH^- + nH^+$$

For example, when acetate is oxidized completely, the atomic ratio of organic carbon oxidized to sulphate-S reduced is 2 : 1. This is the ratio which is usually used to calculate the quantity of organic-C required to support measured sulphate reduction rates. However, as 'n' increases, the C : S ratio decreases and this may affect such calculations in some unusual circumstances. The presently available data suggest that, in most marine environments, acetate is the dominant natural substrate (Balba and Nedwell, 1982; Banat et al., 1981; Chambers, 1985; Christensen, 1984; Sansone and Martens, 1982; Skyring, 1988; Plumb and Reichstein, 1984; Plumb et al., 1983; Sorensen et al., 1981). A consequence of this observation is that the acetate-oxidizing SRB must be ecologically the most important group of SRB in marine environments. However, other low molecular weight fractions of the dissolved organic carbon (DOC) reservoirs have been shown to be quantitatively important substrates for the SRB. Oremland and Silverman (1979) found that sulphate reduction in San Francisco Bay (USA) sediments was stimulated by lactate but not acetate, and lactate may also be an important natural substrate for the SRB in the sediments of the Ems–Dollard estuary, Holland (Laanbroek and Veldkamp, 1982). From molybdate inhibition experiments Sorensen et al., (1981) and Christensen (1984) considered that propionate and butyrate were quantitatively significant substrates for the SRB in European coastal sediments. However, Skyring (1988), on the basis of the observed stoichiometry between acetate oxidation and sulphate reduction in a coastal lake (Lake Eliza; South Australia) sediment, concluded that propionate was probably not a major substrate for the SRB even though it accumulated in molybdate-treated samples. Succinate (Balba and Nedwell, 1982) and formate (Barcelona, 1980) may also be important natural substrates for the SRB in some marine environments.

Skyring (1984) found that sulphate reduction in the cyanobacterial mats of Spencer Gulf, was stimulated twofold by a complex mixture of growth substrates, but not by acetate or lactate. In respect of alternative energy substrates for the SRB, Skyring et al. (1977) and Stams et al. (1985) described the utilization of amino acids by several strains of marine SRB, and Smith and Klug (1981) suggested that amino acids may be important substrates for the SRB in freshwater sediments. Amino acids excreted into the rhizosphere by plant roots may be important natural substrates for the SRB in sea grass beds, coastal marshlands and mangroves (Jorgensen et al., 1981).

Aizenshtat *et al.* (1984) found that the DOC in the porewaters of Solar Lake (Sinai, Egypt) is mostly amino acids and the concentration increases with depth in the sediment reaching 50 mM. Why these amino acids do not appear to be suitable substrates for the SRB in Solar Lake (since sulphate reduction rates decreased rapidly with the depth of sediment) was enigmatic. The fact that free amino acids accumulate in this environment indicates that there must be very severe restrictions other than substrate availability on microbial activity, including that of the SRB.

Substantial methane oxidation rates have been attributed to the SRB in the subsurface sediments of Saanich Inlet (Devol and Ahmed, 1981; Devol *et al.*, 1984) and other investigators have also suggested that the SRB are involved in methane oxidation in various subsurface marine sediments (Iversen and Blackburn, 1981; Kosiur and Warford, 1979; Martens and Berner, 1977; Reeburgh, 1976, 1980; Zehnder and Brock, 1980). While the geochemical evidence for the coupled methane oxidation and sulphate reduction appears to be convincing, pure cultures of methane-oxidizing SRB, or even active consortia, have not been isolated and recently Reeburgh (see Section 6.3) concluded that the presently known SRB could not be responsible for methane oxidation unless alternative oxidants are involved.

6.1.3 Methods and variability: primary productivity

Hall and Moll (1975) and Bunt (1975) reviewed many of the methods used for assessing aquatic primary productivity and it is evident from recent publications on this topic that the principal methodologies have not changed significantly over the last 10 years*. However, satellite imaging methods for determining pelagic chlorophyll concentrations have greatly improved the database for global oceanic photosynthesis. The synthesis of organic matter by primary producers is the first step in most concepts of biogeochemical cycles since it is this organic matter which is the energy source for the various organisms involved (Trudinger *et al.*, 1979). A review of primary production in aquatic ecosystems is beyond the scope of this contribution. Therefore attention has been restricted to the following ecosystem in which quantitative relationships between the cycling of C and S have been investigated: cyanobacterial and other microbenthic mats, coastal salt marshes, sea grass beds (meadows), and sediments receiving planktonic rain.

Cyanobacterial and microbenthic mats
Cyanobacterial mats were chosen by this SCOPE workshop as model ecosystems for studying the quantitative relationships between the carbon

---

*Since preparation of this paper, Roberts *et al.* (1985) have extensively reviewed measurement of plant biomass and net primary production.

and sulphur cycles. Cyanobacterial mats are benthic ecosystems, of which many types are present in modern aquatic environments (see Cohen et al., 1984). Some are very simple structures such as the 1 to 2 mm smooth mats in Spencer Gulf (South Australia) in which *Microcoleus* colonizes carbonate sediments (Bauld et al., 1979, 1980; Bauld, 1984; Skyring and Johns, 1980; Skyring et al., 1983). On the other hand, the microbial components of the mats of Solar Lake are multi-layered and quite complex biologically, but are relatively free of clastic material (Krumbein et al., 1977).

One of the most convenient methods for determining primary productivity is to use ^{14}C-bicarbonate as a tracer to monitor the rate of incorporation of ^{14}C into photosynthate. In order to calculate a primary productivity of a mat ecosystem, it is essential to know the concentration of the bicarbonate being assimilated. The simplest way would be to determine the ^{14}C specific activity of the bicarbonate in the water surrounding the mat. However, Revsbech et al. (1981) found that, even after two hours, a ^{14}C-bicarbonate tracer had not equilibrated with the bicarbonate in a diatomaceous mat covering the sediments of Randers Fjord (Denmark). They pointed out that specific activities calculated from the concentration of the bicarbonate in the overlying water would have resulted in underestimation of the primary productivity of the mat. Joint (1978) also used a ^{14}C method to measure primary productivity of the benthic phototrophs of estuarine mud flats of the Lyhner River (UK) in which *Enteromorpha* contributed a large amount of organic matter to the sediments. Bauld (1984) and Bauld et al. (1979) used a ^{14}C-bicarbonate tracer and determined the bicarbonate concentration in Spencer Gulf seawater by titration with standard acid to determine the primary productivity of cyanobacterial mats. The results were consistent with those determined for other cyanobacterial mat ecosystems (Bauld, 1984). However, in hypersaline environments, an accurate determination of bicarbonate by titration may be complicated by the effect of salt on pH measurements. There are, however, other methods which do not rely on accurate pH measurements. Horner and Smith (1982) described a method for estimating the CO_2 bicarbonate concentration in water by ^{14}C dilution. This method is independent of salinity for the calculation of bicarbonate concentration. Carbon dioxide-specific membrane electrodes are suitable for determining the total CO_2-bicarbonate in natural waters and CO_2 may also be measured by compact, dedicated infrared spectrophotometers or by gas chromatography.

Javor and Castenholz (1984) used slurry suspensions of cyanobacterial mat from Laguna Negro (Mexico) to estimate primary productivity and pointed out that, with this method, light penetration causes some problems in relating experimentally determined productivity with that of the *in situ* mat. In a method developed by Bauld et al. (1979, 1984), discs of intact cyanobacterial mat were incubated so that the original orientation of the mat to light was always preserved. Thus, if it is possible to use discs, this would be the preferred method for estimating cyanobacterial mat primary productivity under conditions that would approximate the *in situ* conditions. In some of the complex mat structures there are layers of photosynthetic sulphur bacteria below the cyanobacteria (Castenholz, 1984; Stolz, 1984). Estimates of the non-oxygenic photosynthesis by these bacteria may be estimated by preferentially inhibiting the cyanobacterial photosystem II with 3-(3,4-dichloropheny)-1,1-dimethyl-urea (DCMU). However, some of the cyanobacteria such as *Oscillatoria limnetica* also oxidize sulphide phototrophically (Cohen, this volume, Section 8.1), and Revsbech and Ward (1984) pointed out that CO_2 fixation in the presence of DCMU does not provide a measure of the importance of anoxygenic phototrophy in the intact oxygen-producing mat. Illumination of the samples with far red (800–900 nm) light specific for bacteriochlorophyll would be preferable for determining bacterial phototrophy (Cohen, this volume, Section 8.1).

The methods for estimating the primary productivity of cyanobacterial mat communities discussed above were generally based on the fixation of $^{14}CO_2$ by small representative samples. Kinsey (1983, 1985) reviewed the monitoring of CO_2 flux in a body of water (marked with dye) for estimating the net carbon metabolism over large areas of various ecosystems of a coral reef. The method relied on the measurement of fluxes of bicarbonate (determined by alkalinity measurements) and oxygen in a marked water body as tidal currents moved the water over metabolizing benthic communities which included extensive cyanobacterial mats occuring in various parts of coral reefs.

Usually, the quantity of ^{14}C fixed as particulate organic carbon (POC) in the photosynthesizing community is used to estimate productivity. However, Plumb *et al.* (1982) estimated that the net quantity of ^{14}C appearing in the dissolved organic fraction (DOC) associated with a cyanobacterial mat was 1–3% of the total productivity. Photo-assimilation and heterotrophic assimilation may complicate the calculation for gross DOC production by cyanobacterial mats (Plumb *et al.*, 1983). Clearly DOC should be estimated when calculating total productivity.

Oxygen production rates, based on bulk-volume analyses, have been used to estimate primary productivity of several cyanobacterial ecosystems. However, Javor and Castenholz (1984) considered that there were many problems associated with this method, mainly because respiration rates in the light and dark are not equal. Recently developed micro-electrode techniques have provided data on the *in situ* rate of O_2 production in intact mats (Jørgensen *et al.*, 1983; Revsbech *et al.*, 1981) and Revsbech *et al.* (1983) comprehensively reviewed the use of both oxygen production and bicarbonate fixation methods for measuring primary productivity. They concluded from their experiments on bacterial and diatomaceous mats, that the ^{14}C-bicarbonate method was the most reliable because the oxygen method was complicated by the formation of bubbles. Oxygen production by cyanobacterial mats in Solar Lake was further discussed by Jørgensen et al. (1983). A summary of the primary production measurements for various microbenthic mats is given in Table 6.1.

Coastal Salt Marsh Ecosystems
Schubauer and Hopkinson (1984) comprehensively reviewed methods for measuring above and below ground productivity by salt marsh macrophytes of which *Spartina*, *Juncus* and *Salicornia* are the main genera. They harvested complete plants at various times of the year to calculate productivity. They concluded from their investigations with *Spartina* (Georgia, USA) that the statistical variations (standard error or deviation) for the living and dead biomass were ± 10–20% and ± 15–27% for above and below ground, respectively. They concluded that various methods for measuring above ground productivity were satisfactory. However, measuring below ground production by separating live and dead tissue was difficult, tedious and expensive. They also concluded that, since root mass was much less than rhizome or dead material, the process could be made more efficient without sacrificing too much accuracy by including the root mass with the dead material.

Schubauer and Hopkinson (1984) found that the below ground production was around 1.6 times the above ground production and that the total plant production was very high at 7620 g (635 mol (C)$m^{-2}a^{-1}$) for *Spartina alterniflora* and *S. cynsuroides*. These values are the highest productivities reported for salt marsh ecosystems. However, they considered that, if anything, the *in situ* production was underestimated. Livingstone and Patriquin (1981) found that the ratio between below and above ground production in a Nova Scotia (Canada) *Spartina* salt marsh was around 1.75, which was very close to that calculated by Schubauer and Hopkinson (1984). The latter also measured turnover rates (productivity/mean biomass) at 5.09 to 5.35 a^{-1} for *Spartina*. Estimates by several methods of above and below ground production for a variety of the salt marsh ecosystems of the eastern USA are available and for detailed information, reference may be made to Pomeroy et al. (1981), Valiela et al. (1982), Wiegert et al. (1981), and Woodwell et al. (1973).

High productivity characterizes coastal salt marsh estuarine ecosystems and, consequently, very high rates of microbial activity are characteristic of the sediments (Table 6.1). In addition to the high productivity of the marshland grasses, phytoplankton and benthic algae make a significant contribution to the total productivity of salt marshes; these have been calculated at around 6% and 10%, respectively for Georgia salt marshes (Pomeroy et al., 1981).

Sea grasses

Sea grasses are frequent colonizers of the lower slope of estuarine sediments and shallow seas. Zieman and Wetzel (1980) reviewed the methodology for estimating the productivity of sea grasses and they recommended the method developed by Jacobs (1979) in which production is estimated from various leaf components. However, estimates of net productivities, like those for salt marsh grasses, are complicated by the inadequacies for measuring below ground productivities. Estimates of net annual productivities range from 4 to 8 mol (C) m^{-2} with an exceptionally high value of 83 mol (C) m^{-2} being estimated for the *Zostera* meadows of the Bering Sea (Hood, 1983). Moriarty *et al.* (1985) measured leaf primary productivity for *Zostera capricornia* in subtropical Moreton Bay (Queensland, Australia) and calculated an annual rate of around 40 mol (C) m^{-2}. Jørgensen (1977) estimated (methods not specified) that the organic detritus in Limfjorden (Denmark) was derived mainly from phytoplankton and the eelgrass, *Zostera marina*, and that the average daily input to the sediments was 0.1 mol (C) m^{-2}. The productivity of *Zostera capricornia* was estimated by the rate of leaf elongation by Moriarty *et al.* (1985) at a similar average daily rate of 0.11 mol (C) m^{-2}. Below ground productivity was not measured. Moriarty *et al.* (1985) amended the daily productivity to 1.6 g (0.13 mol) (C) m^{-2} to account for DOC excretion and epiphyte production. Burkholder and Doheny (1968) found that the ratio of leaf to rhizome plus roots of *Zostera marina* was 2 : 3 in sand and 10 : 3 in mud, and Sand-Jensen (1975) calculated that above ground productivity was 2–6 times below ground productivity. In this respect, the above ground production of sea grasses was relatively greater than above ground production in salt marsh grasses.

Planktonic ecosystems

Much of the organic matter in the bottom sediments of estuaries, enclosed seas, unconfined continental shelves, continental slopes, and abysses of the deep ocean is derived from phytoplankton, zooplankton and faecal debris. However, it is probably the phytoplankton (Meadows and Campbell, 1978) which is the largest carbon source for resident heterotrophic microbial populations. Meadows and Campbell (1978) briefly described the various methods used for measuring phytoplankton productivity. The measurements are relatively simple compared to those required for marsh and sea grasses and the standard errors are considerably lower. Forsberg (1985) briefly reviewed the ^{14}C methods for estimating primary productivity of phytoplankton. Problems he discussed were concerned with ^{14}C equilibration, cellular carbon pools, light saturation and culture growth rate. In some instances, primary productivity may be overestimated by as much as 70% although considerable underestimates are also known. Phytoplankton are patchy (see

Table 6.1 Quantitative relationships between primary productivity and sulphate reduction

Ecosystem	Primary productivity mmol day^{-1} m^{-2}	Primary productivity mol a^{-1}	Sulphate reduction mmol day^{-1} m^{-2}	Sulphate reduction mol a^{-1}	Percent carbon oxidized by SRB	References
Cyanobacterial mats						
Solar Lake (Sinai)						
Microcoleus	700–1000		70		14–20	Cohen et al., 1980
Oscillatoria						Jørgensen and Cohen, 1977
						Krumbein et al., 1977
Spencer Gulf (Australia)						
Microcoleus	17–258	13.5	2–104	6.5	80–100	Skyring et al., 1983
Lyngbya	170–241[1]		2–21[2]		12	Bauld, 1984[1]
						Skyring, 1984[2]
Salt marshes						
Great Sipiwissett (USA)						
Spartina alterniflora		17[3]				Valiela et al., 1982[3]
Above ground		58–75[4]	60[5]	18[5]	54[5]	Howes et al., 1985[4]
Below ground			75[6]			Howes et al., 1984[5]
						Howarth and Teal, 1979[6]

Sea grass meadows				
Limfjorden (Denmark) *Zostera marina*	14.2	9.5	2.6	53	Jørgensen, 1978
Moreton Bay (Australia) *Zostera capricornia*	107 (aboveground)	10.5		80	Moriarty et al., 1985
Planktonic					
Long Island Sound (USA)	17.4		29–39	11	Aller and Yingst, 1980
The Baltic Sea	5.3 (to sediment)			36	
(Gdansk Basin)	5.2 (to sediment)		4	90	Lein et al., 1982
Coral reef sediments					
The Great Barrier Reef Lagoon sediments	92[6] (benthic)	8[7]		20–32	Kinsey, 1985 Skyring, 1985

SRB : Sulphate-reducing bacteria.

Meadows and Campbell, 1978) in their distribution in the water column and this complicates estimates of primary productivity for large areas (e.g. Alvarez-Borrego, 1983). Estimates of primary productivity of seas and oceans are given by Koblentz-Mishke *et al.* (1970) on the basis of productivity zones which range from the highly productive neritic waters of coastal ecosystems and shallow seas to the lower production areas of the oligotrophic waters of the halistatic subtropics. Accession of organic matter by the sediments decreases exponentially as the depth of water increases and this is reflected in the flux of organic matter depositing on sediment surfaces (Billen, 1982; Suess, 1980; Suess and Muller, 1980). In a wide variety of lakes phytoplankton are consumed in the water column by grazing and metabolic processes. However, in some freshwater lakes, the situation may be different because Forsberg (1985) showed that in Lake Wingra (USA) 42% sank to the sediments.

6.1.4 Methods and variability: sulphate reduction rates

Several methods have been used for estimating sulphate reduction rates in sediments and this subject has been reviewed extensively by Skyring (1987). The following paragraphs summarize the methods briefly.

Mathematical modelling of the depth/concentration profiles of sulphate, sulphide, ammonia and sedimentation rates have been used to calculate sulphate reduction rates for estuarine and deep sea sediments (Aller and Yingst, 1980; Bender and Heggie, 1984; Berner, 1964; Goldhaber *et al.*, 1977; Goldhaber and Kaplan, 1980; Jørgensen, 1978; Murray *et al.*, 1978; Toth and Lerman, 1977; Westrich and Berner, 1984). These modelling techniques have some advantages over direct methods for calculating sulphate reduction rates in deep sea sediments. Release of the enormous pressures on raising the sediments to the surface may adversely affect some microorganisms; however, it appears that high temperatures are more detrimental (Jannasch, 1984). Whatever the environment, simple mathematical methods are only applicable where the sediment is not perturbed by fauna or water currents (Jørgensen, 1978; Westrich and Berner, 1984). Other indirect methods for calculating sulphate reduction rates have involved ammonia turnover rates (Blackburn, 1979) and electrical impedance of porewaters and sediment slurries (Oremland and Silverman, 1979).

Direct chemical methods involving measurements of sulphate depletion and increase of sulphide concentrations have also been used (e.g. Aller and Yingst, 1980; Bågander, 1977). Jørgensen (1978) found that long periods of incubation in a closed vessel resulted in an overestimation of net sulphate reduction rates because the natural rate of aerobic sulphide oxidation was lowered. He thus recommended short incubation periods with the natural orientation of the samples being maintained. In estimating *in situ* sulphate reduction rates there is a statistical problem in addition to these problems of chemical balances and fluxes. Observations have shown that variations in both sulphide and/or sulphate concentrations of replicate samples are generally too great to permit the recognition of relatively small changes due to sulphate reduction which occurs in short periods (Skyring et al., 1983). Radioactive sulphate (^{35}S) has been used by many investigators (first by Ivanov, 1956) as an extension of the direct methods, to trace the course of sulphate reduction in sediments because the production of very small amounts of radioactive sulphide may be detected. A sulphate reduction rate may be calculated from the specific activity of the sulphate, and the quantity of radioactive sulphide produced over a given period. However, while the principle is simple, there are some problems which may affect the reliability of the results (Howarth and Giblin, 1983; Howarth and Jørgensen, 1984; Jørgensen, 1978; King et al., 1985; Skyring, 1984, 1985, 1987, 1988; Skyring et al., 1979, 1983). Because of important recent developments in this method, some emphasis is given to this topic in the following paragraphs.

Pyrite (FeS$_2$) is the form in which most of the reduced sulphur exists in subsurface marine sediments (Berner, 1984) and its formation in sediments is complex (Berner, 1970; Hallberg, 1972; Krouse and McCready, 1979; Lindstrom, 1980; Luther et al., 1983; Morozov and Rosanov, 1981; Rickard, 1970; Sweeney and Kaplan, 1980; Trudinger, 1981; Volkov and Rosanov, 1983). However, pyrite is the most stable end-product of sulphate reduction and knowledge of the mechanism and rates of its formation are essential to understand its geochemical significance.

Pyrite is generally thought to form slowly in marine sediments (Berner, 1971) and for this reason, most investigators using ^{35}S-sulphate to determine sulphate reduction rates have considered that pyrite formation during short-term incubations (i.e. hours to days) could not significantly affect the determination if the rate of ^{35}S incorporation into pyrite was omitted. However, in a series of papers on this subject, (Howarth and Teal, 1979; Howarth and Giblin, 1983; Howarth and Marino, 1984; Howarth and Merkel, 1984), Howarth and his colleagues concluded that the formation of pyrite in marine sediments was rapid. They found that the ^{35}S in pyrite was 5 to 9.6 times that which appeared in the H$_2$S and FeS fractions in Sippiwissett (USA) *Spartina* marsh sediment. They used aqua regia oxidation and chromium reduction (Zhabina and Volkov, 1978) methods for the preparation of pyrite sulphur. Also, Ivanov et al. (1976, 1980) and Lein et al. (1982) found that most of the reduced ^{35}S-sulphate was in pyrite and organic sulphur when measuring sulphate reduction rates in sediments from the western Pacific Ocean and Baltic Sea. Howes et al. (1984) also measured the sulphate reduction rates in sediments from a *Spartina* marsh using similar methods and calculated that a lesser amount (60% and 45%) of the reduced ^{35}S appeared in a Cr-reducible sulphur from the 0–10 cm and 0–2 cm layers respectively. Subsequent to these studies, King et al. (1985) made a detailed study of O_2 uptake, CO_2 production and sulphate reduction in the 0–10 cm layer of Great Sippiwissett Marsh. They calculated that about 50% of the reduced ^{35}S-sulphate occurred in the acid-volatile fraction, 37% in the elemental S fraction, and 13% was presumed to be in the pyrite fraction. These results agree generally with those of Howes et al. (1984).

6.1.5 Quantitative relationships between organic carbon oxidation and sulphate reduction

Table 6.1 summarizes some quantitative relationships between primary productivity and sulphate reduction in various marine ecosystems.

Cyanobacterial mats
Cyanobacterial mats should be useful for quantifying the interaction of the carbon and sulphur cycles in marine environments because primary productivity is entirely local and several investigators have shown that most of the sulphate reduction in these ecosystems generally occurred at very high rates (up to 104 mmol m^{-2} d^{-1}) in close proximity or even within the cyanobacterial mat (Cohen, 1984; Jørgensen and Cohen, 1977; Howarth and Marino, 1984; Lyons et al., 1983; Nedwell and Abram, 1978; Skyring, 1984; Skyring et al., 1983). In the simplest systems, the complete set of reactions from the synthesis to the degeneration of organic matter occurs within a millimetre or less (Cohen, 1984; Skyring et al., 1983). Quantitative relationships between primary productivity and sulphate reduction were first calculated for Solar Lake cyanobacterial mats by Jørgensen and Cohen (1977). They found that 14–20% of the daily primary productivity (0.7 to 1.0 mol (organic-C) m^{-2} d^{-1}) was oxidized by the SRB. Skyring et al. (1983) investigated a much simpler system in the *Microcoleus* smooth mats of Spencer Gulf, Australia and concluded, from multivariate analysis, that possibly (correlation coefficient, r=0.7, probability p=0.015) 100% of the organic carbon synthesized by the cyanobacteria in the mat was eventually oxidized by the SRB. They also estimated that the molar ratio between photosynthetically fixed carbon and reduced sulphate was 2 : 1 (±20%). Estimates of quantitative relationships between primary productivity and sulphate reduction in cyanobacterial mats (and also in other sedimentary systems) may be complicated by the fact that most of the photosynthate is not immediately available for oxidation. For example, Skyring et al. (1983) showed that sulphate reduction rates correlated positively with primary productivity of the *Microcoleus* mat only when the productivity data were transformed by an equation which accounted for the rate of decomposition

of the photosynthate. They also derived a polynomial equation which described the quantitative relationships between the sulphate reduction rate, available organic carbon, temperature, porewater content and salinity in a cyanobacterial mat ecosystem. The mathematical model was independently consistent for the Spencer Gulf cyanobacterial ecosystem and it may have general applicability for those marine ecosystems where productivity is high and where the SRB populations are numerically similar.

Marshlands, sea grass beds and coastal environments
Quantitative relationships between the cycling of carbon and sulphur in the coastal *Spartina* salt marshes of the eastern USA have been calculated by several investigators (Howarth and Teal, 1979; and Howarth et al., 1983, 1984, 1985; Howes et al., 1984; King et al., 1985; Skyring et al., 1979). However, because of the difficult problems involved in the measurement of the accession of organic matter to the sediments and rates of sulphate reduction, some disparities occur in the data. Howes et al. (1984) estimated that the annual sulphate reduction rate accounted for the oxidation of about 40 mol (C) m^{-2} a^{-1} which was considerably lower than the 150 mol (C) m^{-2} a^{-1} calculated by Howarth et al. (1979, 1983). The sulphate reduction rates calculated for the sediments of the *Spartina* marshlands of Sippiwissett Marsh and Sapelo Island (USA) (despite some of the problems which are associated with the interpretation of the rate data) all demand an amount of organic matter which has been calculated to be from 30% to 150% of the primary productivity of the *Spartina*. Howes et al. (1984) calculated a below ground production rate of 58 to 75 mol (C) $m^{-2}a^{-1}$ for a *Spartina* marsh ecosystem and concluded that around 90% of the organic input to these sediments resulted from the below ground production of *Spartina*. It is clear from these studies on the salt marsh ecosystems of the Great Sippiwissett Marsh and Sapelo Island that calculations of the quantitative relationships between C and S diagenesis in higher plant ecosystems must include above and below ground productivity.

Quantitative relationships between C and S diagenesis in sea grass beds have not been as extensively investigated as those for salt marshes. However, it appears that the SRB are also responsible for the remineralization of significant amounts of photosynthate in these ecosystems. For example, Jørgensen and Fenchel (1974) and Jørgensen (1977) found that in a model system and in *Zostera*-colonized sediments of Limfjorden (Denmark), about 50% of *Zostera* substrate was oxidized by the SRB. Moriarty et al. (1985) calculated that the SRB oxidized around 80% of the unexported organic matter in *Zostera capricornia* beds in Moreton Bay, Australia. In both investigations, however, below ground productivity was not measured.

Planktonic ecosystems
Other marine systems may be much more complex because the relationships between the sites of primary productivity and the translocation and deposition of organic matter are difficult to establish both qualitatively and quantitatively. However, pioneering work was undertaken by B.B. Jørgensen and recently there have been estimates of the C/S relationships in several planktonic ecosystems. Aller and Yingst (1980) calculated that the SRB oxidized approximately 35% of organic matter reaching the sediment in Long Island Sound (USA) (11% of planktonic productivity). Lein *et al.* (1982) measured sulphate reduction rates in several Baltic Sea sediments and calculated that only 30% of planktonic primary productivity reached the sediment, around 90% of which was oxidized by the SRB.

6.1.6 Temporal relationships between organic synthesis, preservation and sulphate reduction

Equation (2) was generalized from the reaction for sulphate reduction in marine sediments given by Leventhal (1983) as:

$$2R(CH_2O) + SO_4^{2-} \rightarrow H_2S + 2HCO_3^- + 2R \qquad (2)$$

where R is the residue of organic matter not metabolized, but deposited and preserved in the sediment. The ratios between the reactants may depend on the varying nature of the organic matter (Lerman, 1982). For example, Westrich and Berner (1984) showed that about 35% of planktonic material (Long Island Sound) was decomposed over a much longer time scale than the rapidly metabolized (65%) fraction. Presumably their 35% fraction equates with Leventhal's R (residue) fraction. Volkov and Rosanov (1983) studied the relationship between the percentage of organic matter metabolized and the initial carbon content in deep-water sediments from the Pacific Ocean, the Bering Sea and the Sea of Okhotsk. They concluded that the rate of decomposition of organic carbon during reduction processes (e.g. sulphate reduction) is related to the initial organic concentration and suggested that there is a relatively constant proportion of fresh organic matter which is readily metabolizable. Westrich and Berner (1984) calculated that the kinetics of equation (2) were first order and that sulphate reduction in this planktonic system can be described by the sequential decomposition of two organic fractions with decay constants of $8\pm1a^{-1}$ and $0.94\pm0.25a^{-1}$. From the data of Jørgensen and Cohen (1977) and Skyring *et al.* (1983) it is evident that most cyanobacterial photosynthates have decay constants equivalent to the rapidly metabolized organic fraction described by Westrich and Berner (1984).

Leventhal (1983) and Berner and Raiswell (1983) found that there was a positive correlation between organic carbon and sulphide of the sediments from the Black Sea and other marine environments. The regression lines have the same slope, but that for the Black Sea data intercepts the S axis at 1.5%. Leventhal suggested that this was due to the quantity of sulphide which was not fixed as iron sulphide (thus lost from the systems) and that it was indicative of a euxinic sedimentary environment. He further suggested that such correlations between C and S may be useful for identifying analogous ancient environments. Contemporary and similar studies by Berner and Raiswell (1983) have led to similar conclusions. However, sediments which are very low in Fe, such as many marine carbonates, may not follow these trends (Berner and Raiswell, 1983; Gibson, 1985).

Skyring *et al.* (1983) calculated that most of the photosynthate particulate organic carbon (POC) synthesized by the cyanobacterial mat colonizing intertidal sediments in Spencer Gulf was oxidized by the SRB within the year and most oxidized within a month or two of synthesis. Further, Bauld (1984) reported that cyanobacterial mats from the same ecosystems produced 1–6% DOC during photosynthesis and subsequently Plumb (1985) showed, from molybdate inhibition studies, that this DOC was rapidly oxidized by the SRB. However, the contribution of this low molecular weight (LMW) component to the DOC pool would be too small on most occasions to be reflected in day- and night-time sulphate reduction rates. On the other hand, Howarth and Marino (1984) found a fivefold increase in sulphate reduction rates in cyanobacterial mats (in the Great Sippiwissett Marsh) on cold sunny days as opposed to warm cloudy days and they suggested that this indicated a short-term coupling of mat photosynthesis and sulphate reduction. In a study of several microbial processes occurring in the sediments of the sea grass beds of Moreton Bay, Moriarty *et al.* (1985) showed that there may have been a slight diurnal effect with sulphate reduction rates peaking from noon to 6 pm. This is more likely to have been due to light-induced excretion of DOC by the roots of the sea grass and the observation suggested that the SRB were active in the plant rhizospheres. Recently, Capone *et al.* (1983) isolated polyesters of β-hydroxybutyrate and β-hydroxypentanoate from the cyanobacterial mats and stromatolites from Shark Bay (Western Australia). It is not presently known if these compounds are oxidized by the SRB, but butyrate is a known substrate for several species, and Sørensen *et al.* (1981) calculated (from molybdate inhibition experiments) that around 10% of sulphate reduction could have been coupled to butyrate oxidation in coastal lagoon sediments (Denmark). Krom and Sholkovitz (1977) showed in sediments from Duich, a fjord type estuary, that the high molecular weight (HMW) fraction increased with depth down to 80 cm. They attributed this to a humification process. Once a compound enters this type of HMW reservoir, its residence time in the sediment would be much longer than the average residence time for LMW/DOC compounds. Krom and Sholkovitz (1977) also found that the LMW/DOC compounds decreased with depth and that the slow oxidation rate of the LMW compounds in the below surface sediments permitted condensation reactions which resulted in the formation of the HMW/DOC.

Studies of temporal relationships between photosynthesis and sulphate reduction have indicated that, in most marine environments, organic matter is turned over rapidly, resulting in little preservation. For example, in the cyanobacterial mats from Spencer Gulf there is very little preservation of organic matter in the sediment and Cohen *et al.* (1980) calculated that only 1% of mat photosynthate was preserved in the sediment in Solar Lake. However, preservation occurs in some marine (or near marine) situations which may be more complex than those in which high organic concentrations in the sediments result from high rates of organic sedimentation (Berner and Raiswell, 1983).

Figure 6.1 is a schematic diagram which relates the diagenetic relationships between various components of biogeochemical carbon reservoirs. High molecular weight (HMW) photosynthate (or particulate organic carbon, POC) is composed of HMW compounds which may be ultimately energy sources but are not directly metabolized by the SRB. The dissolved organic carbon (DOC) reservoir may be crudely divided into HMW and low molecular weight (LMW) fractions and the average residence time for C atoms in each may vary widely, depending on their suitability as substrates for microbial populations. The DOC reservoir in most sediments results from the anaerobic diagenesis of organic matter buried with sediment and in some instances may be a significant proportion of the total organic content. For example, an acid-soluble organic portion, presumably actual or potential DOC, was from 20 to 40% of the total organic carbon in an organic rich sediment from Hamelin Pool, Shark Bay, Western Australia (Skyring, 1985). Also, in cyanobacterial mats and other microbenthic ecosystems, significant amounts of DOC may be produced directly by photosynthesizers (Bauld, 1984; Chambers, 1985). The kerogen reservoir, which may originate in the R-fraction (Leventhal, 1983), is largely composed of water-insoluble HMW compounds and the residence time of compounds in this pool may be billions of years if the sedimentary formation survives crustal weathering and movement.

A detailed examination of the environmental factors which affect the preservation of organic matter is beyond the scope of this presentation. However, it is evident from the various data available that factors which retard sulphate reduction may also be important in retarding the complete oxidation of organic matter. It is possible that hypersalinity may be important in this respect. For example, in Spencer Gulf sediments, sulphate reduction rates correlated negatively with porewater salinity in cyanobacterial mats

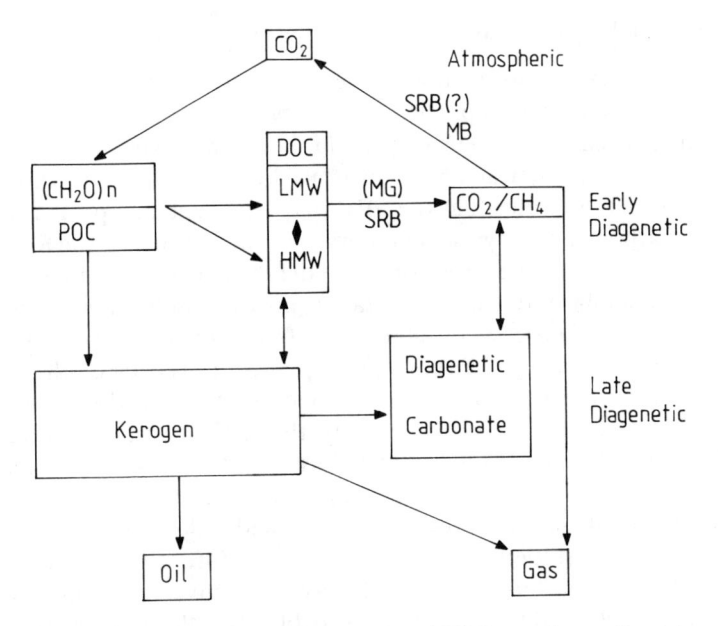

Sulphur and Carbon Cycles in Marine Sediments 141

Figure 6.1 Carbon reservoirs and the SRB in marine sediments.
DOC: dissolved organic carbon
POC: particulate organic carbon
LMW: low molecular weight organic compounds
HMW: high molecular weight organic compounds
MG: methanogens
MB: methanobacteria
SRB: sulphate-reducing bacteria

(Skyring *et al.*, 1983) and the cyanobacterial organic matter in the 2000 year old anoxic sediments of Solar Lake is a classic example of preservation in a hypersaline environment. Although the Great Salt Lake is not marine, it is hypersaline and Post (1980) observed that organic matter was best preserved in those sediments of the lake which were associated with saline inundations. Further, Lupton (personal communication) showed that the metabolism of both the SRB and the fatty acid producers in marine sediments from Shark Bay was retarded in hypersaline environments. On the other hand, there are other factors which may contribute to the preservation of organic matter in marine (or sulphate-rich) environments. For example, Aizenshtat *et al.* (1984) found that the secondary enrichment of organic matter with sulphur (from sulphate reduction) is a feature of the well-preserved cyanobacterial material in Solar Lake. They also observed sulphur enrichment of the organic matter in the very early stages of diagenesis pointing to the possibilty of the direct involvement of sulphate reduction in organic preservation.

6.1.7 Concluding remarks

Currently available correlations between primary productivity and sulphate reduction rates are too imprecise to identify exact quantitative relationships between the cycling of carbon and sulphur in marine environments. However, despite all the analytical and statistical problems, there is experimental evidence which indicates that the SRB may oxidize a high proportion of the organic carbon entering sedimentary marine systems. Clearly not all ecosystems are the same with respect to the dynamics of C and S diagenesis; however, cyanobacterial mats appear to provide useful model systems for investigating the process of and quantitative relationships between the biogeochemical cycling of carbon and sulphur. This is because the cyanobacteria are usually the major sources of photosynthate in the ecosystem and most of the organic C is oxidized by the SRB within or in very close proximity to the mat (Cohen, 1984; Skyring, 1984; Skyring et al., 1983). Also, cyanobacterial mats are not as complicated as other ecosystems where organic material may be produced above and below ground, transported long distances or settled through deep water before it enters the sediment. Generally, cyanobacterial mats are highly productive and support very high sulphate reduction rates and it is probable that the rate estimations are representative of the *in situ* condition even though (non-acid volatile sulphide) NAVS data have not been included in the initial calculations (Jørgensen and Cohen, 1977; Skyring et al., 1983). A more precise understanding of important diagenetic reactions may be obtained by a detailed examination of the biogeochemical dynamics of specific C, H and S cycling in an intact but isolated mat.

To date, investigators have relied on destructive methodology, intermittent sampling and insufficient data processing facilities. However, more precise calculations may be realized with the availability of high capacity data processors and the development of non-destructive physical and chemical probes which would permit the continuous, simultaneous monitoring of several key parameters. For example, recent developments in microelectrode (Revsbech and Ward, 1984) technology have resulted in the observation of unexpected and dramatic changes in the porewater chemistry of cyanobacterial mats over micron distances. Geochemical events (which may be irreversible) occur in these micro-layers and over a long period may result in the accumulation of significant quantities of diagenetic products. Also, microprobes employing radioactive tracers and gas chromatographic techniques (Cohen, 1984; Skyring, 1984; Skyring et al., 1983, 1987) have shown that strictly anaerobic processes such as sulphate reduction and methanogenesis occur within the cyanobacterial mat ecosystem. New developments in nuclear magnetic resonance (NMR) spectroscopy and Mössbauer analysis may also provide non-destructive methods for following the dynamics of C and S in intact ecosystems.

Sulphur and Carbon Cycles in Marine Sediments

It generally appears that organic matter and sulphide are best preserved in marine sediments made anoxic by the activities of the SRB. There are, however, still some gaps with respect to temporal relationships between the formation of various organic carbon reservoirs and sulphate reduction. In particular, more detailed information is required on the very slow carbon and sulphur metabolic rates in subsurface sediments, which may be important in pyrite (and other metal sulphide) formation. Also, the geochemical evidence for methane oxidation by the SRB needs more investigation to show whether or not the SRB can overcome thermodynamic barriers which appear to preclude this metabolic process.

Skyring (1987) drew attention to several important coastal and continental shelf/slope ecosystems where quantitative estimates of the relationships between carbon and sulphur were few or lacking completely. However, even though current knowledge is limited, it appears that 50–90% of global sulphate reduction occurs in coastal, shelf and slope sediments. Global cycles which are important in the burial of organic carbon and sulphur are also important for understanding global fluxes of oceanic and atmospheric constituents, and the formation of oil and some economic minerals.

6.2 INTERACTION OF THE SULPHUR AND CARBON CYCLES IN RECENT MARINE SEDIMENTS*

6.2.1 Introduction

Simulation of the recent sulphur cycle enables one to estimate some of the main fluxes of sulphur in marine sediments. These are: (a) sulphur produced at the early stages of sedimentary diagenesis, (b) sulphur oxidized and returned to the sulphate reservoir and (c) sulphur buried in sediments (Ivanov 1983). As seen in Section 6.1 the biogeochemical cycles of sulphur and carbon are intimately interrelated. Quantitative estimates of the sulphur fluxes made with the help of radioactive isotopes were used by the author to establish the balance of organic carbon in recent oceanic sediments at the stage of their early diagenesis.

6.2.2 Primary production and aerobic mineralization of organic matter in the oceans

Primary production is the initial step of the oceanic carbon cycle. Some estimates of the magnitude of primary production are presently available ranging from $20 \times 10^3 - 55 \times 10^3$ Tg (C) a^{-1} (Ryther, 1969; Romankevitch, 1977; Mopper and Degens, 1979) to 126×10^3 Tg (C)a^{-1}. The latter, long-

*A. Yu. Lein.

considered as overestimates, appear nowadays most reasonable. High estimates of primary production were favoured by the radioisotopic (^{14}C) techniques of measuring the rates of primary production in the oceans and by the researchers' due regard for biomass of *phytoplankton* (Shulenberger and Reid, 1981).

Living organisms are involved in mineralization of some 60–80% of the suspended organic matter in the euphotic layer of the oceans. The quantity of organic matter reaching the ocean floor is variable and depth-dependent. To depths of 100 m, some 40–80% of the organic matter from primary production reaches the ocean floor. The amount of C_{org} reaching the water–sediment interface decreases with depth to 2–10% or less (Jørgensen, 1983). At the sediment–water interface there is a major consumption of C_{org} in aerobic and anaerobic processes. Below, data are presented for oxygen consumption by aerobic respiration in recent sediments of different morphometric oceanic zones (Jørgensen, 1983):

Zone	Depth (m)	Oxygen consumption (mmol m^{-2} day^{-1})
Shelf	0–5	20
	50–200	10
Continental slope	200–1000	3
	1000–4000	0.3
Depressions	over 4000	0.05

Aerobic processes which occur in sediments of the world oceans, over the total surface area of 360×10^6 km^2, consume 527×10^9 mol (O_2) day^{-1} or 6155 Tg (O)a^{-1}.

In addition to the aerobic oxidation of organic matter described by Jørgensen (1983), these processes also embrace oxidation of reduced compounds which migrate from the anaerobic zone of bottom sediments. Most important are the oxidation of hydrogen sulphide, methane and ammonium. Regrettably, methane oxidation and nitrification in bottom sediments of the oceans are still poorly investigated (see Section 6.4). Therefore, our present knowledge is insufficient for estimating oxygen consumption in these processes.

Oxygen consumption for oxidation of biogenic hydrogen sulphide may, in the first approximation, be estimated on the following grounds. Total hydrogen sulphide production during microbial sulphate reduction makes

Sulphur and Carbon Cycles in Marine Sediments

up 492 Tg (S) a^{-1} (Lein, 1983) of which 111 Tg (S) are buried as sulphide in modern sediments (Volkov and Rosanov, 1983). The remaining 380 Tg (S) are oxidized to sulphate in the reaction:

$$S^{2-} + 2O_2 \rightarrow SO_4^{2-}$$

so it follows that this process requires about 760 Tg (O) a^{-1}.

Subtraction of this value from the total oxygen consumption by bottom sediments gives 5400 Tg (O) for the annual O_2 consumption in oxidation of organic matter.

Aerobic oxidation of organic matter schematized as:

$$(CH_2O) + O_2 \rightarrow CO_2 + H_2O$$

suggests that about 2000 Tg (C_{org}) are absorbed by aerobic processes in the upper horizons of marine mud sediments.

6.2.3 Estimation of organic matter consumption in anaerobic diagenetic processes

The assessment of C_{org} consumption in anaerobic processes which occur in the biogeochemically active sediments of the shelf and continental slope may be approached differently. The first approach is based on due regard for the variation in concentrations of C_{org} and N_{org} in the lower horizons of mud columns compared with their concentrations in the surface horizon which is assumed as the value of primary or initial C_{org} (Emery and Rittenberg, 1952). Firstly, such an approach does not account for possible changes in conditions of sediment formation which affect the magnitude of initial concentrations of C_{org}; secondly, primary production is assumed as the already residual C_{org}, as the most active processes of organic matter destruction take place at the water–sediment interface.

Another approach which is widely used by Soviet researchers estimates C_{org} consumption from the production of reduced compounds of manganese, iron and sulphur in compliance with the balance formulae of Uspensky-Strakhov (Bordovski, 1964; Strakhov and Zalmanzon 1955; Strakhov 1976). This second approach enables one to estimate the consumption of C_{org} on the basis of contents of solid-phase compounds. It essentially accounts for only C_{org} consumption during formation of stable disulphide of iron pyrite, since C_{org} uptake for the reduction of Fe^{3+}, Mn^{4+} and other oxidized components is extremely low and generally neglected. While substantiating this approach for estimating concentrations of primary C_{org}, Strakhov showed the approximate character of values thus obtained, since organic matter was depleted due to some other transformations, e.g. by splitting off $COOH^-$, CH_4, NH_3 etc. (Strakhov, 1976).

Recent biogeochemical studies suggest that the solid phase of sediments do not absorb all hydrogen sulphide produced *in situ* during bacterial sulphate reduction. The idea of migration and oxidation of a part of hydrogen sulphide in highly reduced sediments is supported by chemical analyses where the interstitial water shows a surface maximum for sulphate ion that is often enriched in the light isotope ^{32}S ($\delta^{34}S$ ranging from -18.8 to $-15.6‰$) compared to the isotopic composition of sulphate sulphur in overlying water (Lein, 1983). The latter fact may be considered as clear evidence of the emergence of a subsurface sulphate maximum from the oxidation of isotopically light metabolic hydrogen sulphide.

Thus the C_{org} consumption in anaerobic bacterial processes in reduced sediments, specifically those containing free H_2S, is significantly underestimated if assesssed on the basis of the pyrite sulphur content of the solid phase.

Taking into account the above arguments, yet another approach was proposed to estimate organic matter mineralized in the course of anaerobic destruction. This method is based on experimental measurements of the rate of biogeochemical processes using the radioisotopes of sulphur and carbon under conditions as close to natural as possible. (Lein, 1983).

Biogeochemical processes of diagenetic transformation of sediments involve microorganisms from various physiological groups: aerobic and anaerobic saprophytic and cellulose-degrading bacteria, sulphate reducers, methanogens, denitrifiers and methylotrophs. The total bacterial population in sediments, excluding the oligotrophic oceanic zones, amounts to several billion cells per gram of wet mud, with a biomass (C_{org}) of 1–100μg (C) per gram of wet mud (Lein and Namsarayev, 1986).

Under anaerobic conditions decomposition of organic matter is a multistage process: primary anaerobes decompose polymeric compounds to monomers which in turn serve as a substrate for fermentation agents and gas-producing bacteria.

Figure 6.2 illustrates anaerobic destruction of organic matter in the sediments of Aden Bay and carries quantitative estimates of specific reservoirs and fluxes (using ^{14}C and ^{35}S). Of particular interest are the so far unique, quantitative estimates of the carbon flux associated with the products of cellulose decomposition. These products are utilized as organic substrates by methanogens and sulphate reducers which bring about most of the consumption of organic matter during early diagenesis.

Sulphur and Carbon Cycles in Marine Sediments 147

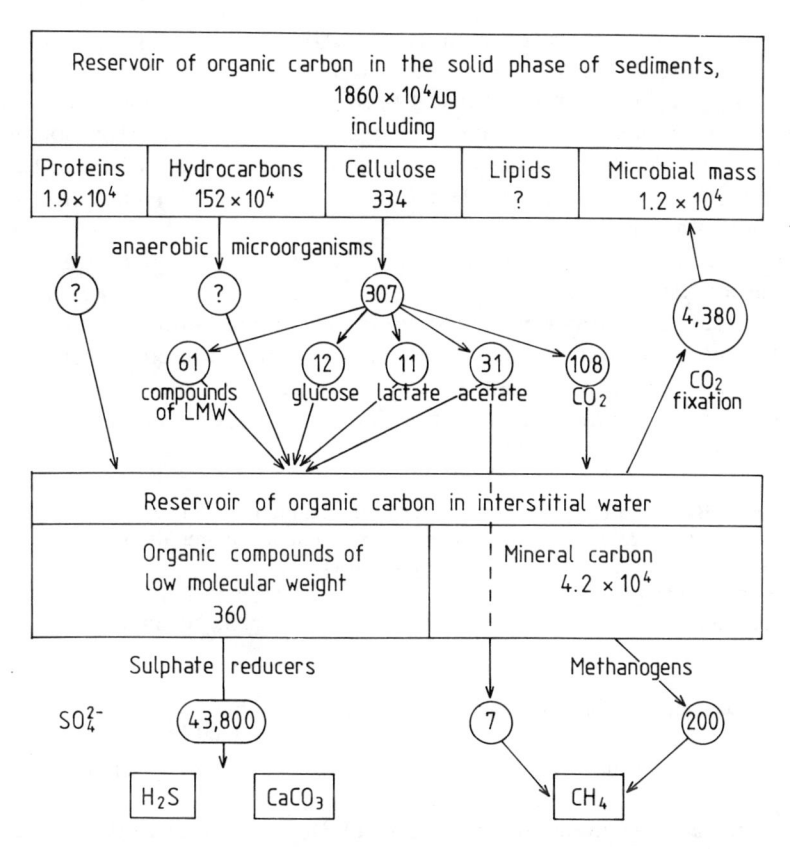

Figure 6.2 Diagram of anaerobic destruction of organic matter. Open boxes indicate reservoirs of C_{org}, μg kg^{-1} dry mud. Numbers within circles are fluxes, μg (C) kg^{-1} a^{-1} dry mud.

6.2.4 Quantitative estimation of C_{org} consumption during anaerobic diagenesis

Earlier studies established that 90–95% of C_{org} involved in anaerobic biogeochemical reactions is consumed by bacterial sulphate reduction (Lein and Ivanov, 1981; Belyaev *et al.*, 1980). Consequently, knowledge of the rate of bacterial sulphate reduction in sediments is sufficient for estimating the magnitude of C_{org} digestion under conditions of anaerobic diagenesis. Simple calculation using the equation of sulphate reduction reaction:

$$SO_4^{2-} + 2C \rightarrow S^{2-} + 2CO_2$$

enables one to obtain a quantitative expression of the rate of C_{org} mineralization during anaerobic destruction of organic matter.

Experimental data on the rates of bacterial processes in the top metre of recent marine sediments may be used for estimating C_{org} consumption under conditions of early diagenesis in reduced muds (Ivanov et al., 1976; Lein et al., 1981; Lein and Ivanov, 1981; Belyaev et al., 1980; Ivanov and Lein, 1980 and others).

In those rare cases where the rate of sediment accumulation is known, the carbon balance in muds may be assessed using measurements of the rates of reduction processes carried out in experiments with ^{35}S. An estimate of this kind is given in Table 6.2. Sediments of the top metre received 1400 kg(C_{org})m^{-2}. Twenty per cent of this was consumed in anaerobic processes, whereas all C_{org} which reached the floor was digested in the process of diagenesis.

Table 6.2 Balance of C_{org} in the top metre of sediments in the Mexican shelf (Site 668; depth 140 m; 23°23′N, 106°56′W), rate of sediment accumulation 100 mm/1000 a (Andel, 1964)

Balance item	C_{org} quantity
Mass of wet mud in the top metre with density of 1.5 g cm^{-3}	1500 kg m^{-2}
Primary production in the region	2.5 g (C) m^{-2} d^{-1}
Quantity of C_{org} reaching the floor, on the basis of 15% of the primary production (Jørgensen, 1983)	0.14 kg (C) m^{-2} a^{-1}
Quantity of C_{org} having reached the bottom over the past 10 000 a	1400 kg (C) m^{-2}
Consumption of C_{org} during bacterial sulphate reduction at the sulphate reduction rate of 55.7µg (S) kg^{-1}(wet mud) d^{-1} (Ivanov et al., 1976)	41.8µg (C) kg^{-1}d^{-1}
Consumption of C_{org} in bacterial sulphate reduction on the basis of wet mud mass in the uppermost metre of sediment per:	
24 hours	62.7mg (C) m^{-2}
1 year	22.9g (C) m^{-2}
10 000 years	229kg (C) m^{-2}
Quantity of C_{org} remaining over the total volume of wet mud in the top metre with the content of residual C_{org} = 3% dry mud, humidity 60%	18kg (C) m^{-2}
Consumption of C_{org} for aerobic respiration in the top metre over 10 000 a with oxygen absorption in sediments of the shelf of 10 mmol m^{-2} d^{-1} (Jørgensen, 1983)	1168 kg (C) m^{-2}
Total consumption of C_{org} in the top metre of muds (229 + 18 + 1168)	1415 kg m^{-2}

A large body of experimental material hitherto obtained is condensed in Table 6.3 which presents average values for the rates of reduction processes in the top metre of sediments in different morphometric oceanic zones.

In recent years researchers have increased studies of the rate of sulphate reduction in marshlands using ^{35}S (Howarth and Teal, 1979; Howarth and Giblin, 1983; Skyring *et al.*, 1983, and others). The rates of sulphate reduction in the well-studied saline marshes of Sapelo Island and Sippiwissett (USA) are very high at: 40 and 75 mol (S)m^{-2} a^{-1} (Howarth *et al.*, 1984). According to Skyring (see Section 6.1) all of the C_{org} from primary production may be consumed in processes of sulphate reduction in marshes.

The estimation of global C_{org} consumption in sulphate reduction in marshlands is difficult because their total surface area is not known accurately. Assuming a very rough value of 6×10^4 km^2 for the total marshland surface (*The World Ocean Atlas*, 1974, p.37, geomorphological map) and using the average of the two cited rates of sulphate reduction (57 mol (S) m^{-2} a^{-1} or 1.8 kg (S) m^{-2} a^{-1} one finds that the marshes produce up to 100 Tg (S) a^{-1}. This magnitude is of little importance to the sulphur cycle as nearly all of this sulphur is rapidly oxidized and channelled into the marine sulphate reservoir. The magnitude of C_{org} digestion during sulphate reduction in marshlands may, however, amount to 75 Tg (C) a^{-1}, i.e. comparable with C_{org} consumption in sediments of the shelf (Table 6.3). Hence, reliable measurements of total area of marshlands are much called for in view of new data on the rates of sulphate reduction in marshlands. Reverting to Table 6.2 let us underline an important geochemical inference: over 70% of the total mass of C_{org} digested by anaerobic processes, accounting specifically for marshes, is consumed in sediments of near-continental areas of the oceans at depth of 1000 m.

Annual production of metabolic hydrogen sulphide per unit area of sediment surface and consumption of C_{org} in this process were estimated from data on the rate of reduction processes (Table 6.4). The estimated magnitude of C_{org} consumption by anaerobic bacterial diagenesis accounts for about 15% of the total flux of C_{org} to the world ocean floor. The data presented also suggest that the minimum quantity of organic carbon reaching the water column–marine bottom sediment interface is about 2.5×10^3 Tg (C) a^{-1}.

6.3 COUPLING OF THE CARBON AND SULPHUR CYCLES THROUGH ANAEROBIC METHANE OXIDATION*

6.3.1 Introduction

Anaerobic methane oxidation is a process that appears to be important in controlling the flux of methane from anoxic marine sediments. There is

*W.S. Reeburgh.

Table 6.3 Productivity of bacterial sulphate reduction in the top metre of world ocean sediments

Geomorphological zone of the ocean floor	Depth (m)	Zone area (10^6 km^2)	Rate of sulphate reduction μg (S) kg^{-1} day^{-1}	Rate of sulphate reduction mg (S) kg^{-1} a^{-1}	Productivity of sulphate reduction g (S) m^{-2} a^{-1}	Flux of H$_2$S (Tg a^{-1})	Consumption of C$_{org}$ in sulphate reduction Tg (S) a^{-1}	%
Shelf	0–50	11(2)*	55.6	20.3	26.4	79.2	59.6	16
	50–200	16(3)	37.2	13.6	17.7	53.1	40.0	11
Continental slope	200–1000	(15)	28.2	10.3	13.4	201.0	151.0	41
	1000–3000	(61)	5.5	2.0	2.6	158.6	119.2	32
Slope's foot, depressions, ocean bed	over 3000	257	0.97	0.35	—	—	—	—
Total		360(81)				492.5	369.7	100

*In parentheses: the areas of biogeochemically active sediments where bacterial sulphate reduction takes place. For the shelf zone such areas are calculated on the basis of data on the distribution of recent sand–aleurite–pelitic muds (Lisitsyn, 1978).

Sulphur and Carbon Cycles in Marine Sediments

Table 6.4 Balance of C_{org} in sediments of the world oceans

Process	Tg (C) a^{-1}	%	Reference
Mineralization of C_{org}			
aerobic	2000	82	Jørgensen, 1983
anaerobic	370	15	see Table 6.2
Burial of C_{org}	85	3	Romankevitch, 1977
Total flux of C_{org} to the floor	2455	100	

general agreement that the process occurs, but the organisms responsible have not been isolated yet, and the mechanism remains unknown. There is a suggested, but unconfirmed, link with the sulphur cycle through sulphate reduction. This section briefly reviews the geochemical evidence for anaerobic methane oxidation and uses measured rates of anaerobic methane oxidation from several environments to make a preliminary estimate of the importance of the process to the global methane and sulphur cycles. Evidence for a link between anaerobic methane oxidation and sulphate reduction is summarized and results from some recent inhibition experiments that provide information on the possible mechanism of anaerobic methane oxidation are presented.

The earliest evidence for anaerobic methane oxidation was presented by Davis and Yarbrough (1966), who reported that methane was oxidized in small quantities by sulphate reducers growing on another substrate. Sorokin (1957) showed that sulphate reducers were unable to oxidize methane when it was the sole substrate. No further work was reported on the subject until the late 1970s, when several papers employing diagenetic models (Berner, 1971) were published.

6.3.2 Geochemical evidence for anaerobic methane oxidation

Anaerobic methane oxidation is a controversial subject among microbiologists, since the responsible organisms and mechanism are unknown. However, a growing body of geochemical evidence supports the occurrence of oxidation of methane in the absence of oxygen, and has been summarized and reviewed in Alperin and Reeburgh (1984). This geochemical evidence favouring anaerobic methane oxidation is derived from three independent approaches: diagenetic modelling, direct *quasi in situ* rate measurements using the radioactive tracer $^{14}CH_4$, and stable carbon isotope budgets. These three independent approaches were applied to samples from a single station in Skan Bay, Alaska (USA) (Alperin and Reeburgh, 1984), and gave consistent results. Figure 6.3 shows a schematic diagram with methane and sulphate depth distributions characteristic of anoxic marine sediments. The most

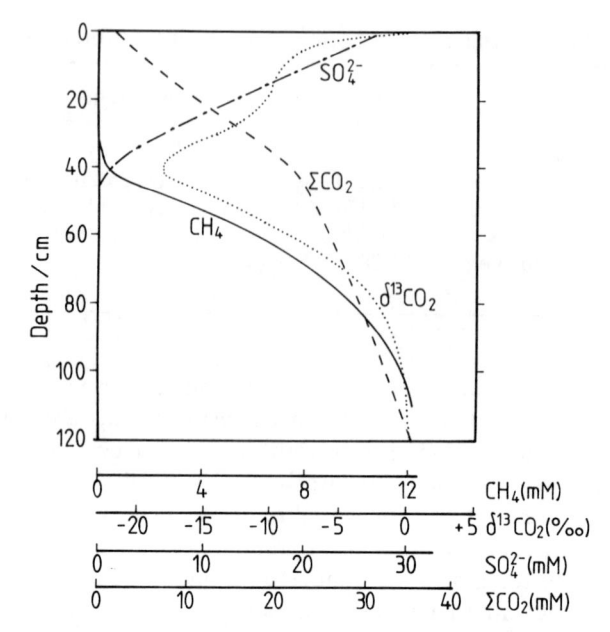

Figure 6.3 Depth distributions of methane, total carbon dioxide, sulphate and the stable isotope ratio of carbon dioxide in interstitial waters of a typical anoxic marine sediment. Note the low methane concentration surface zone, the curved methane distribution, the coincident slope changes and transitions in the distributions, and the minimum in the $\delta^{13}CO_2$ distribution. These distributions suggest that anaerobic methane oxidation occurs in a subsurface depth interval that coincides with the $\delta^{13}CO_2$ minimum. Adapted from Reeburgh (1982)

striking characteristics are the low methane concentration surface zone and the concave-up methane profile. The other distributions all undergo breaks or slope changes over the same depth interval, which is taken to be the zone of methane oxidation. This schematic diagram illustrates how methane diffusing upwards is consumed in the methane oxidizing zone, increasing the upward flux of CO_2 and producing a minimum in the $\delta^{13}CO_2$ distribution as a result of oxidizing isotopically light methane.

6.3.3 Diagenetic models

Diagenetic models and vertical advection–diffusion models (Craig, 1969) are conceptually identical; both balance sedimentation and diffusion (or in the case of advection–diffusion models, vertical advection and eddy diffusion) against reaction. Reeburgh (1976) treated methane data from the Cariaco Trench water column and sediments with vertical advection–diffusion and diagenetic models, respectively, and found that the modelled water column methane oxidation rates were 10^2–10^3-fold lower than those obtained for the sediments. Barnes and Goldberg (1976) and Martens and Berner (1977) considered methane in Santa Barbara Basin and Long Island Sound (both USA) sediments, respectively. Later model studies include those of Bernard (1979) on Gulf of Mexico sediments, and Whiticar (1982) on Baltic Sea sediments.

All of these model studies showed that a methane consumption process was required to explain the methane depth distributions in the sediment. These environments all contain sulphide and are anoxic, so the process is clearly anaerobic. These model studies all showed that anaerobic methane oxidation occurred in a subsurface zone in anoxic marine sediments and that there was net methane consumption. The model studies are based on actual field data and consider the result of production and consumption reactions, so they yield net rates.

6.3.4 Methane oxidation rate measurements

Tracer techniques using $^{14}CH_4$ following the $^{35}SO_4^{2-}$ sulphate reduction rate method of Jørgensen (1978) have been used in direct measurements of anaerobic methane oxidation rate (Reeburgh, 1980). These studies involve injecting $^{14}CH_4$ either into syringes containing sediment or into intact sediment cores, incubation at *in situ* temperatures and collection of the unreacted methane and CO_2, the product of oxidation. The diagenetic models predict that anaerobic methane oxidation will be restricted to a narrow subsurface zone. Measured anaerobic methane oxidation rate depth distributions from several studies and environments (Reeburgh, 1980; Alperin and Reeburgh, 1984; Devol, 1983; Devol et al., 1984; Iversen and Jørgensen, 1985) show this subsurface maximum in methane oxidation, which is illustrated in Figure 6.4. The locations and magnitudes of the rate maxima accord with model predictions. The results from several investigations, which are summarized in Table 6.5, agree well.

6.3.5 Stable carbon isotope distributions

The advantages of using stable isotopes as tracers in natural systems result from the fact that they require no additions, incubations or controls, and avoid the uncertainties introduced by each of these manipulations. Biogenic methane has a characteristic light (−60‰ to −100‰) stable carbon isotope signature, while the dissolved inorganic carbon (DIC) is much heavier isotopically (0‰ to −5‰); this large isotopic difference may be used to quantify additions of methane-derived CO_2. Figures 6.5 and 6.6 show characteristic features in the stable isotope distributions that result from anaerobic methane oxidation. Figure 6.5 shows a minimum of $\delta^{13}CO_2$,

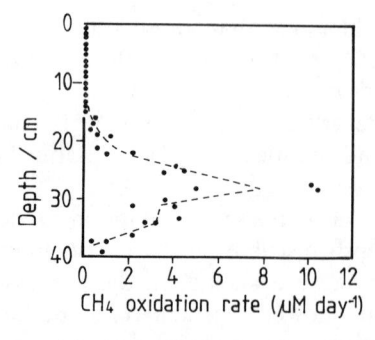

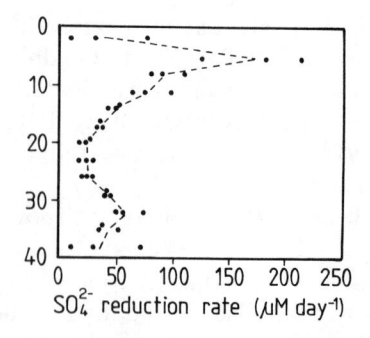

Figure 6.4 Depth distributions of measured anaerobic methane oxidation and sulphate reduction rates in Skan Bay sediments. Note the subsurface maximum in sulphate reduction rate as well as the maximum in methane oxidation coincident with the second subsurface maximum sulphate reduction rate. The coincident maxima in rates have been observed in other environments and are the strongest evidence to date for a link between anaerobic methane oxidation and sulphate reduction.
Adapted from Alperin and Reeburgh (1985)

Table 6.5 Integrated anaerobic methane oxidation rates

Location and depth	Integration depth of sediments	Oxidation rate (mol cm^{-2}a^{-1})	Study
Skan Bay (65 m)	25 cm	27.0	Alperin and Reeburg, 1984
Kattegat (65 m)	172 cm	30.3	Iversen and Jørgensen, 1985
Skagerrak (200 m)	112 cm	42.3	
Saanich Inlet (225 m) (measured)	27 cm	71.3 25.8 40.9	Devol, 1983
		Average 45.9	
Saanich Inlet (225 m) (modelled)	27 cm	95.5 90.2 139.3	Devol, 1983
		Average 108.4	

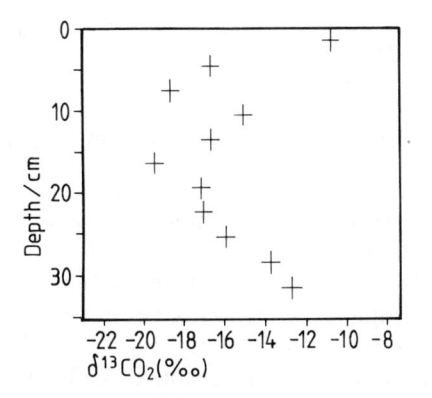

Sulphur and Carbon Cycles in Marine Sediments 155

Figure 6.5 Depth distribution of $\delta^{13}CO_2$ in Skan Bay sediment. Note the subsurface minimum, which corresponds to the subsurface methane oxidation rate maximum in Figure 6.4. Adapted from Alperin and Reeburgh (1984)

similar to the minimum in CH_4 concentration shown schematically in Figure 6.3. This minimum is caused by oxidation of isotopically light methane in the methane oxidizing zone, which results in a localized input of isotopically light, methane-derived CO_2. Figure 6.6 shows the depth distribution of $\delta^{13}CH_4$ in the same samples as used to obtain the data shown in Figure 6.5. This distribution shows the residual methane becoming progressively heavier in line with the preferential oxidation of lighter methane in the sediments. These data, fitted to a Rayleigh distillation model, yielded a fractionation factor of 1.004. A mixing model involving conservation of mass and stable isotopes was presented in Alperin and Reeburgh (1984) and gave results

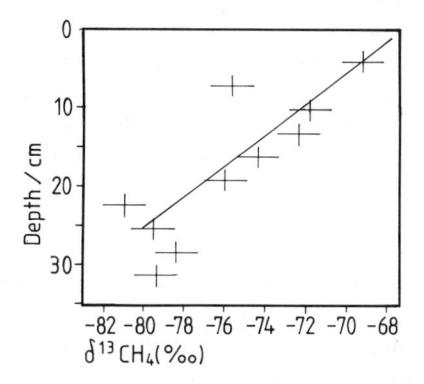

Figure 6.6 Depth distribution of $\delta^{13}CH_4$ in Skan Bay sediments showing sample depth intervals and standard error bars. This diagram shows residual methane becoming isotopically heavier as the lighter fraction is preferentially oxidized in surface sediments. A Rayleigh distillation model applied to the data yields a fractionation factor of 1.004. Adapted from Alperin and Reeburgh (1984)

that agreed with those from both diagenetic models and from direct rate measurements.

6.3.6 Importance of anaerobic methane oxidation in global methane and sulphur cycles

Anaerobic methane oxidation and sulphate reduction occur in the same environments—high organic carbon content coastal and shelf sediments. Attempts to determine what fraction of anaerobic methane oxidation can be accounted for by sulphate reduction have involved modelling as well as direct rate measurements of anaerobic methane oxidation and sulphate reduction. The integrated anaerobic methane oxidation during the process of sulphate reduction in the sediments, derived from model calculations, ranges from 30 to 50% (Reeburgh, 1982) to a high of 70% (Murray et al., 1978) of total consumption of organic carbon. Depth integrations of direct rate measurements give a somewhat different picture, with methane oxidation accounting for 23 to 40% of the total sulphate reduction in Saanich Inlet, USA and 12% in Skan Bay sediments (Devol et al., 1984). Iversen and Jørgensen (1985) showed this same ratio was 10% in Kattegat and Skagerrak sediments.

The measurements listed in Table 6.5 may be used to obtain a preliminary estimate of the importance of anaerobic methane oxidation in marine sediments to the global atmospheric methane budget. Depth-integrated anaerobic methane oxidation rate measurements are multiplied by the areas of estuaries and continental shelves; the two environments where anaerobic methane oxidation is known to be important. These areas have been tabulated by Sheppard et al. (1982) as part of a global compilation of methane sources and fluxes to the atmosphere. Table 6.6 shows the results of this estimate, which is probably high, since continental shelf sediments are not likely to produce methane over their entire area. Although there is no information on global distribution of methane in continental shelf sediments, these estimates may be checked by estimating the amount of primary production oxidized by sulphate reduction and using the methane oxidation: sulphate reduction ratio discussed above to estimate the importance of anaerobic methane oxidation.

Suess (1980) estimated that up to 50% of the primary production on continental shelves reaches the bottom in water depths of 100–200 m and Jørgensen (1983) estimated that 50% of the organic matter oxidation on shelves is carried out by sulphate reduction. Assuming a conservative continental shelf primary production rate of 100 g (C) m^{-2}a^{-1} and a methane oxidation: sulphate reduction ratio of 0.1, the anaerobic methane oxidation rate is 2.5 g (C) m^{-2} a^{-1}. Extended to the area of estuaries and shelves, 28×10^{12} m^2, this rate yields 70 Tg (C) a^{-1}, which agrees reasonably with

Table 6.6 Global importance of anaerobic methane oxidation

Measured Rates (Table 6.6) = 27 to 46 mol cm^{-2} a^{-1}

Environment areas	Methane consumption (Tg a^{-1})
Estuaries (1.4 × 10^{12}m^2)	6–10
Continental shelves (26.6 × 10^{12}m^2)	115–195
	121–205

the estimates derived from rate measurements.

The amount of reduced sulphur buried in marine sediments annually is 111.4 Tg (S) a^{-1} (Volkov and Rosanov, 1983; Ivanov, 1983). This number does not include the much larger amount of reduced sulphur that is re-oxidized at the sediments–water interface and made available for further sulphate reduction. The total sulphate reduction associated with primary production of 100 g (C) m^{-2} a^{-1} is 25 g (C) m^{-2} a^{-1}, or in terms of S (2 : 1 C : S stoichiometry), 33 g (S) m^{-2} a^{-1}. Extending this rate to the area of estuaries and shelves yields 900 Tg (S) a^{-1}. This estimate is somewhat higher than the 400–600 Tg (S) a^{-1} reported by Andreae and Galbally (1985). Methane oxidation would thus be responsible for 10% of the estimated total sulphate reduction, or 40–90 Tg (S) a^{-1}.

6.3.7 Evidence for a link between anaerobic methane oxidation and sulphate reduction

Evidence linking anaerobic methane oxidation and sulphate reduction comes from several sources. Coincident slope changes in the profiles of sulphate, methane, total carbon dioxide and $\delta^{13}CO_2$ (Figure 6.3) have been cited as evidence for a possible link (Reeburgh, 1982). Differences in methane distributions in freshwater and marine environments (Reeburgh and Heggie, 1977) also suggest the involvement of sulphate. The concave methane distribution observed in marine systems are not observed in low sulphate freshwater systems; linear methane profiles are observed in these systems, suggesting that methane produced in freshwater sediments diffuses into the water rather than being consumed within the sediments as in marine environments. Thermodynamic arguments also suggest that oxidation of methane by sulphate is possible at *in situ* concentrations (Martens and Berner, 1977). The quantitative dominance of sulphate reduction in marine sedimentary environments suggests that it is one of the few processes with sufficient oxidizing capacity to oxidize large quantities of methane (Reeburgh, 1983). The coincident maxima in measured methane oxidation and sulphate reduction rates that have been observed by Devol et al. (1984), Iversen and Jørgensen (1985) and Alperin and Reeburgh (1985) are perhaps the strongest evidence for a link between anaerobic methane oxidation and sulphate reduction.

6.3.8 Possible mechanisms for anaerobic methane oxidation and future work

It is not possible to determine from the above evidence whether the link between methane oxidation and sulphate reduction is direct or indirect. Experiments using specific inhibitors were conducted recently (Alperin and Reeburgh, 1985) to determine whether there is a direct couple between the two processes and also to narrow the range of possible organisms capable of mediating anaerobic methane oxidation. The organisms considered capable of mediating anaerobic methane oxidation are methanogens, sulphate reducers, an unknown organism or a consortium involving sulphate reducers. Methanogens have been observed to produce and consume methane (Zehnder and Brock, 1979; 1980), although net consumption has not been demonstrated. Sulphate reducers have been shown to oxidize methane; additional electron acceptors must be present (Iversen, 1984) and net methane consumption has not been demonstrated.

The Alperin and Reeburgh (1985) study was conducted on intact sediments and slurries from the anaerobic methane oxidation maximum in Skan Bay sediments. Sediments were injected with molybdate, an inhibitor of sulphate reduction, 2-bromoethanesulphonic acid (BES), an inhibitor of methanogenesis and methane oxidation (Zehnder and Brock, 1979) and fluoroacetate, an inhibitor of acetate oxidation; both sulphate reduction and methane oxidation rates were measured on sediments with each treatment. The results showed no inhibition of methane oxidation with any of the inhibitors; sulphate reduction was only inhibited by molybdate. The inhibition experiment results eliminate methanogens as responsible for anaerobic methane oxidation. The results also eliminate normal sulphate reducers as being directly responsible for anaerobic methane oxidation, although it is possible that the observed anaerobic methane oxidation could have been mediated by the 1–2% of sulphate reducers that were not inhibited in the experiment or by a small number of atypical organisms.

The results are consistent with two possibilities: anaerobic methane oxidation may be conducted by unknown organisms using iron or manganese oxides or reduced sulphur compounds (S (0) to S (IV)) as electron acceptors, or by a consortium involving an unknown organism and a sulphur-reducer which uses hydrogen as the coupling substrate. Simulation experiments will be necessary to confirm these possibilities.

6.4 QUANTITATIVE EVALUATION OF BIOGENIC METHANE GENERATION AND OXIDATION IN OCEANIC SEDIMENTS*

The problem of methane genesis and development of methods for determining the rates of its formation, accumulation and oxidation in modern sediments of the world ocean has recently acquired a more applied significance. This has been primarily connected with the utilization of geochemical means of prospecting for oil and gas deposits in sedimentary rocks beneath the oceans. These investigations are no less important for comprehending the conditions favouring the formation of crystallohydrate gas fields which presently occur in young oceanic sediments (Makagon et al., 1983; Claypool and Kaplan, 1974; Trofimuk et al., 1975; Geodakyan et al., 1979; Galimov and Kodina, 1982).

Rather abundant isotopic-geochemical evidence characterizing the $\delta^{13}C$ value of methane in the subsurface horizons of reduced bottom sediments of the Black, Bering and Baltic Seas and the Gulf of California (Table 6.7), suggests that the bulk of methane in these sediments is of a biogenic origin. The methane is characterized by an extremely light isotopic composition of carbon ($\delta^{13}C$ values vary from -60.0 to -82.0‰).

A similar isotopic composition of methane is peculiar also to most samples taken and analysed in the course of the work on the deep sea drilling project in sedimentary rocks to a depth of 1000–1600 m under the oceanic floor (Table 6.8). During microbial methanogenesis in bottom sediments of oceans, the bulk of methane is formed at the expense of carbon dioxide reduction (Belyaev and Finkelstein, 1976; Ivanov, 1979; Belyaev et al., 1980; Lein et al., 1981). This process not only enriches methane with isotopically light ^{12}C but also enriches the residual carbon dioxide with ^{13}C (Ivanov, 1979).

In the light of these data, of particular interest is the fact that, in many cases listed in Tables 6.7 and 6.8 the isotopically light methane is present together with carbon dioxide enriched in ^{13}C, as compared to the $\delta^{13}C$ value of inorganic carbon dissolved in oceanic water.

A principal scheme of biogeochemical processes resulting in methanogenesis under anaerobic conditions is given in Figure 6.7. It is seen that, in its geochemical essence, the microbial methanogenesis, alongside microbial sulphate reduction, is one of the final steps of the multi-stage process of organic matter decomposition under anaerobic conditions.

As seen from this scheme, there are two basic processes of microbial methanogenesis: methane formation from methyl groups of low molecular weight fatty acids, primarily acetate, and biogenic CO_2 reduction by hydrogen which is also one of the key intermediates in the anaerobic decomposition of organic matter.

*M.V. Ivanov and A.Yu. Lein.

Table 6.7 Isotopic composition ($\delta^{13}C$, ‰) of methane and carbon dioxide in upper horizons of reduced sediments of the Black, Baltic and Bering Seas and the Gulf of California

Sampling site, No. station	Depth of sea (m)	Horizon of silt sampling (cm)	$\delta^{13}C$-CH_4 (‰)	$\delta^{13}C$-CO_2 (‰)	Reference
Black Sea					
1. 113b	8	115–162	−64	−17	Alexeev and Lebedev, 1975
2. DSDP 380-1-6	9	115	−70	—	Hunt and Whelan, 1978
3. 114a	15	144–209	−60	−12	Alexeev and Lebedev, 1975
4. DSDP 379A-4-5	33	54	−66	—	Hunt and Whelan, 1978
5. DSDP 381-6-4	52	31	−72	—	Hunt and Whelan, 1978
6. 590	66	100–120	−82	—	Ivanov et al., 1983
7. 564	68	140–160	−78	—	Ivanov et al., 1983
8. DSDP 381-17-6	151	63	−63	—	Hunt and Whelan, 1978
9. DSDP 380A-8-5	406	116	−65	—	Hunt and Whelan, 1978
10. DSDP 379A-45-4	412	100	−66	—	Hunt and Whelan, 1978
11. 109	440	160–200	−62	+9	Alexeev and Lebedev, 1975
12. DSDP 381-51-3	469	50	−67	—	Hunt and Whelan, 1978
13. DSDP 379A-67-2	608	77	−65	—	Hunt and Whelan, 1978
14. DSDP 380A-46-4	756	76	−63	—	Hunt and Whelan, 1978
15. DSDP 380A-792	1066	0	−65	—	Hunt and Whelan, 1978
Baltic Sea					
16. 2679	78	40–75	−64	−23	Lein et al., 1982
Bering Sea					
17. Scan Bay	—	0–35	from −70 to −80 (10 tests)	from −11 to −20 (11 tests)	Reeburgh, 1983
Gulf of California					
18. South Guaymos depression	—	0–300	−76	from +2 to −16	Claypool and Kaplan, 1974

Table 6.8 Isotopic composition of methane and carbon dioxide ($\delta^{13}C$, ‰) in drilled wells of deep water oceanic sediments (DSDP)

Drilling site No. Drilled well (DSDP)	Depth of sea (m)	Thickness of drilled sediments, (m)	$\delta^{13}C$, ‰ CH$_4$	CO$_2$	Reference
The Atlantic Ocean					
Maroc depression					
1. 415	2817	220	−71	—	Galimov and Kodina, 1982
2. 415A	2817	1042	−61 to −71	—	
3. 416	4203	1624	−82	—	Galimov and Tchinenov, 1978
4. Blake-Outer-Ridge 533		400	−66 to −94	−4 to −25	Galimov and Kvenvolden, 1982
5. Blake-Babama Ridge		700	−70 to −90	+3 to −28	Claypool and Kaplan, 1974
6. Cariaco Trench	—	180	−60 to −80	+10 to −20	
The Pacific Ocean					
7. Axtoria Fan	—	800	−70 to −80	0 to −20	Claypool and Kaplan, 1974
8. Aleutian depression 180	—	450	−75 to −80	+5 to −20	
Gulf of California					
9. 474	3023	14	−75	—	Galimov and Kodina, 1982
10. 474A	3023	529	−40 to −60	—	
11. 477	2003	182	−40 to −44	−11 to −16	
12. 478	1889	309	−62 to −80	−6 to −9	
	747	435	−52 to −62	+1 to −10	
13. 479	—	119	−67	−4	
14. 480	1998	−55 to −77	−7 to −12		
15. 481A	159				

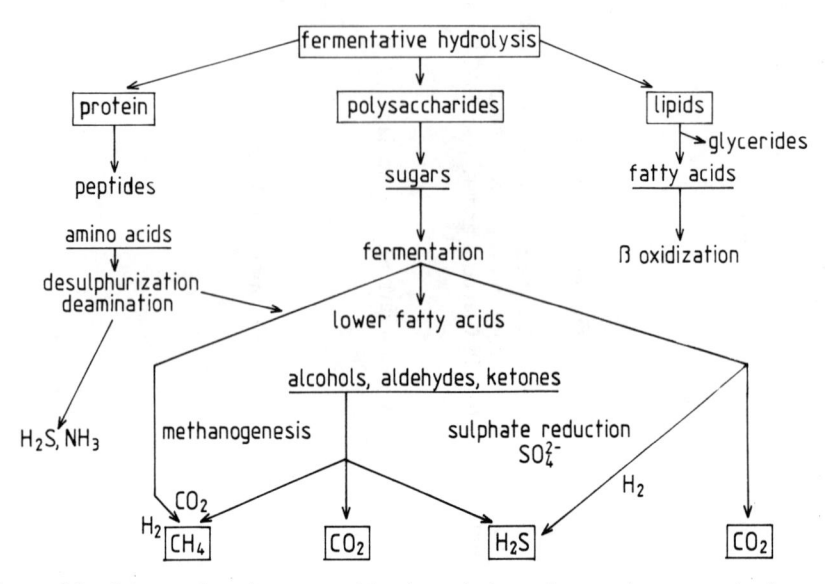

Figure 6.7 Scheme for the anaerobic degradation of organic matter in bottom sediments of seas and oceans

These two functions are found in the overwhelming majority of methanogenic bacteria studied hitherto, though the range of organic compounds known, i.e. substrates for microbial methanogenesis, has greatly broadened recently at the expense of methanol and compounds of a methylamine type.

The rest of this section reviews the quantitative data on microbial methanogenesis and oxidation of methane in oceanic sediments. The methods employed are described in detail in numerous reports (Belyaev and Ivanov, 1975; Ivanov et al., 1976; Lein, 1978; Belyaev et al., 1980; Lein et al., 1982).

Starting from 1973, these investigations included the following regions of the world ocean:

(a) The sub-equatorial part of the Pacific Ocean in the cross-section from Wake Atoll to Mexico touching the Gulf of California, the Southern Pacific including the Tasman Sea in the cross-section from New Zealand to Lima (Peru) including the Peruvian–Chilean trench.
(b) The South-China and Bering Seas.
(c) The Indian Ocean; during three expeditions basic investigations were carried out in the Arabian Sea, Gulf of Persia and Aden, and near the East Africa shores from the Somalia horn to Maputu.
(d) The Black, Baltic and Caspian Seas.

The results obtained showed rather a wide distribution of methanogenic bacteria in the upper horizons of reduced sediments of the world ocean. Their maximal quantities (up to 1000 cells per gram silt) were found in bottom sediments of continental seas surrounding the Soviet Union: Baltic, Black and Caspian. They are also present in the most samples of silt sediments of marginal seas (Bering and South-China) and sediments of the Gulf of California, Persia and Aden. Referring to oceanic sediments proper, methanogenic bacteria were found in samples of moderately and highly reduced sediments along the east coast of Africa and the sediments of the Peruvian–Chilean trench and the Pacific Ocean near the west coast of Mexico.

However, none of these bacteria were found in the red clay samples taken in the trans-Pacific cross-section along 19°N latitude and in the cross-section from New Zealand to Peru. The same results were obtained in the analysis of deep-water carbonate-clayish sediments from the northwest part of the Indian Ocean and oxidized sediments from the Tasman Sea.

We focused our attention on obtaining materials that quantitatively characterized the intensity of microbial methanogenesis using labelled $^{14}CO_2$ and $^{14}CH_3COOH$. Labelled compounds were injected in silt columns and the intensity of methanogenesis was calculated from the quantity of $^{14}CH_4$ generated in a sample.

A typical profile of the methanogenic activity in the sediment cores is given in Figure 6.8. As seen from these data, methanogenesis occurs throughout the reduced sediments under study from the subsurface horizons to a depth of more than 2 m. In many cases a clear-cut maximum of methanogenesis is observed in upper centimetres of reduced silts (Figure 6.8), i.e. methanogenesis occurs in horizons of sediments characterized by high activity of sulphate-reducing bacteria. No inhibition of the activity of methanogen by sulphates of porewaters or by hydrogen sulphide is observed.

The data on the intensity of methanogenesis in the upper horizons of sediments (to a depth of 2–3 m) are given in Table 6.9 which shows wide variations in sea and oceanic sediments—from dozens of milligrams to fractions of a milligram carbon per kilogram of wet silt per day. Maximal values are peculiar to sediments of the North and especially Baltic Seas, while minimal ones are characteristic of coastal sediments of the Indian (Arabian Sea and Somali depression) and Pacific Oceans (South-China Sea, Peruvian cross-section).

Rather intensive methanogenesis was observed in the bottom sediments in the west part of the Black Sea and deep-water sediments of the Bering Sea (Table 6.9). Comparatively low intensity of methanogenesis in sediments of the Gulfs of California, Persia and Oman, rich in organic matter, may be explained by the inadequate attention to sampling from the uppermost horizons of these sediments (Belyaev and Finkelstein, 1976; Ivanov et al.,

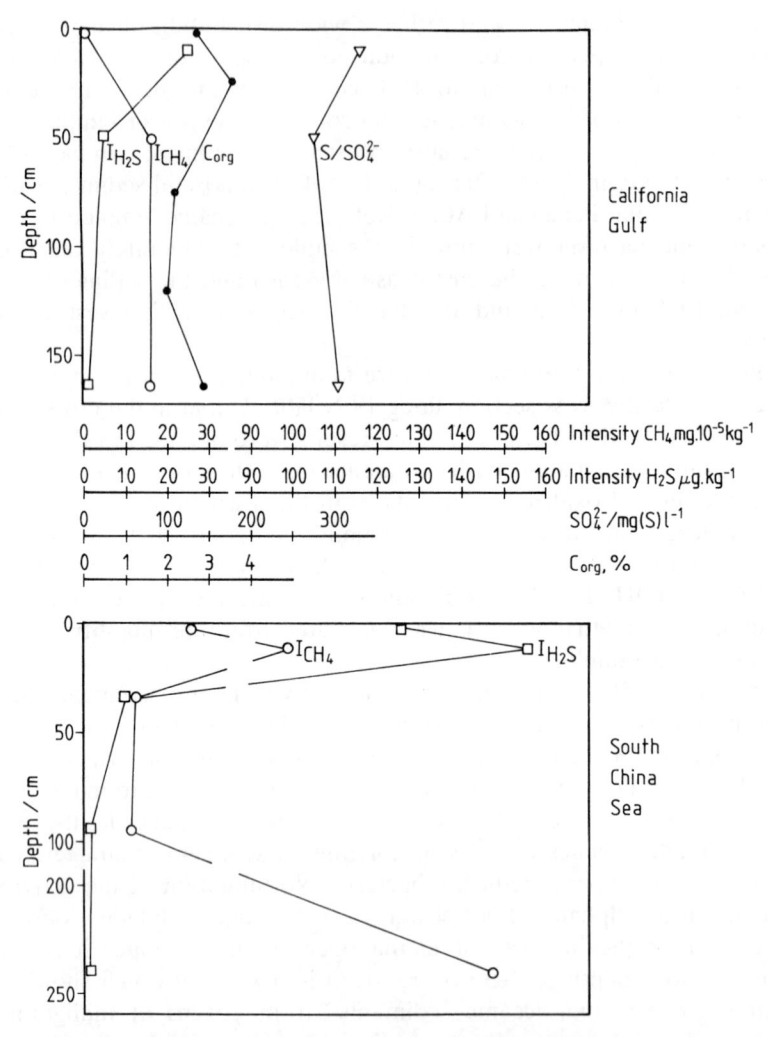

Figure 6.8 Distribution of the intensity of methanogenesis (I_{CH_4}), sulphate reduction (I_{H_2S}), content of sulphate sulphur (S/SO_4^{2-}) in porewater and C_{org} (% or dry weight) in cores in cores of reduced bottom sediments

1980). However, as became evident from further investigations, the highest methanogenic intensities were observed in the upper horizons (Figure 6.9).

Using two radioactively labelled compounds—carbon dioxide and acetate—we managed to show that in the overwhelming majority of sediments studied, the bulk of the methane was formed at the expense of carbon dioxide reduction (Table 6.9). This conclusion is important in interpreting pathways of anaerobic decomposition of organic matter in sediments (Figure 6.7),

Table 6.9 Intensity (mg 10^{-6} per kg wet silt per day) and productivity (mg m^{-3} (top metre) per day) of microbial methanogenesis in bottom sediments of seas and coeans, as well as the fraction of methane generated from carbon dioxide

Sampling site and number of analysed cores	Intensity of CH_4 formation	Productivity of CH_4 formation	Fraction of CH_4 from CO_2 %	References
I. Basin of the Atlantic Ocean				
Baltic Sea (7 cores)	200–11600	200–21000	56–99	Lein et al., 1982
North Sea	10–1200	—	the bulk	Senior et al., 1982
Black Sea, west part (8 cores)	0.5–220	7–68	70	Ivanov et al., 1983
II. Basin of the Pacific Ocean				
Bering Sea (10 cores)	6–680	2–195	the bulk	Gorlatov, 1984
Gulf of California (4 cores)	0.1–3.0	3–5	86–99	Belyaev and Finkelstein, 1976
Peruvian profile (6 cores)	0.1–34	2–25	the bulk	Laurinavitchus et al., 1981
South-China Sea (7 cores)	0.3–22	1–22	the bulk	Laurinavitchus et al., 1981
III. Basin of the Indian Ocean				
Arabian Sea (4 cores)	2–50	6–63	the bulk	Ivanov et al., 1980
Gulf of Oman (4 cores)	1–25	5–72	74–98	
Gulf of Persia (1 core)	4	—	83	
Somalian depression (1 core)	2–4	—	95–100	

and, as will be shown later, is in good agreement with isotope-geochemical data.

Table 6.9 also contains values of the biogenic methane production per square metre in a metre section of silt sediments calculated taking the results of all measurements of intensities down the sediment core from 0 to 100 cm into account. These data, corrected for microbial methane oxidation, can be employed to calculate the potential methanogenic activity of bottom sediments in various regions of the world ocean and thus forecast the possible formation of geologically young deposits of gaseous hydrocarbons in marine sediments.

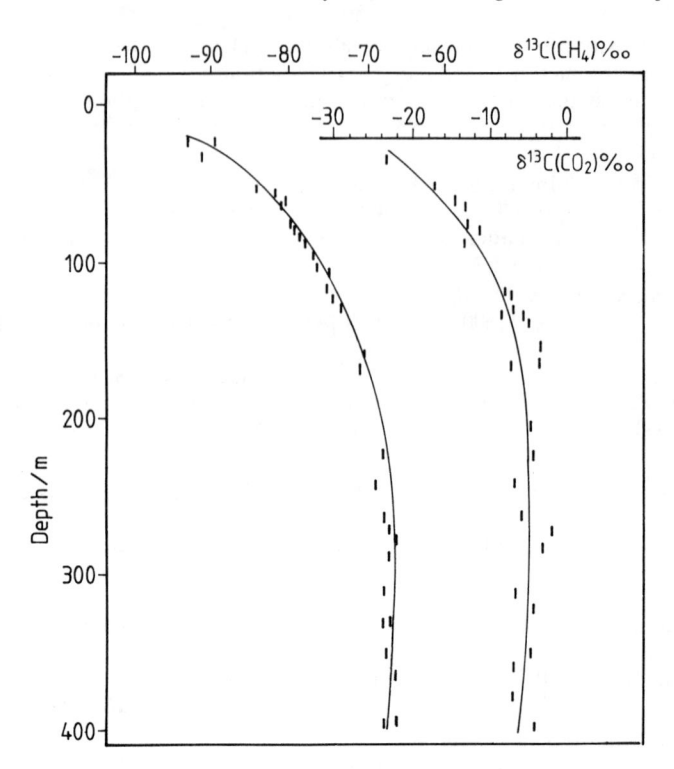

Figure 6.9 Simultaneous weighting of the isotopic composition ($\delta^{13}C$,) of methane and carbon dioxide along the seciton of well No. 533 drilled in Project DSDP in the Atlantic Ocean (Galimov and Kvenvolden, 1982)

It might be best to point out that, simultaneous with research into the intensity of methane formation and oxidation, the intensity of sulphate reduction in the same samples was also studied. This thus allowed us to show that sulphate reduction is the main mechanism of uptake of the low molecular weight organic compounds formed during anaerobic decomposition of organic matter in upper horizons of sea sediments (Ivanov, 1979; Belyaev et al., 1980; Lein et al., 1982). However, the intensity of sulphate reduction drops significantly in deeper parts of the silt mass (Figure 6.8), as do amounts of sulphate sulphur, while the relative contribution of methanogenesis to the transformation of organic matter increases markedly.

Based on isotope-geochemical evidence (changes in $\delta^{13}C$ values of methane and CO_2 in the mass of oceanic sediments (Figure 6.9)) one can assume that methanogenesis, at least that due to CO_2 reduction, continues down to a considerable depth.

Sulphur and Carbon Cycles in Marine Sediments

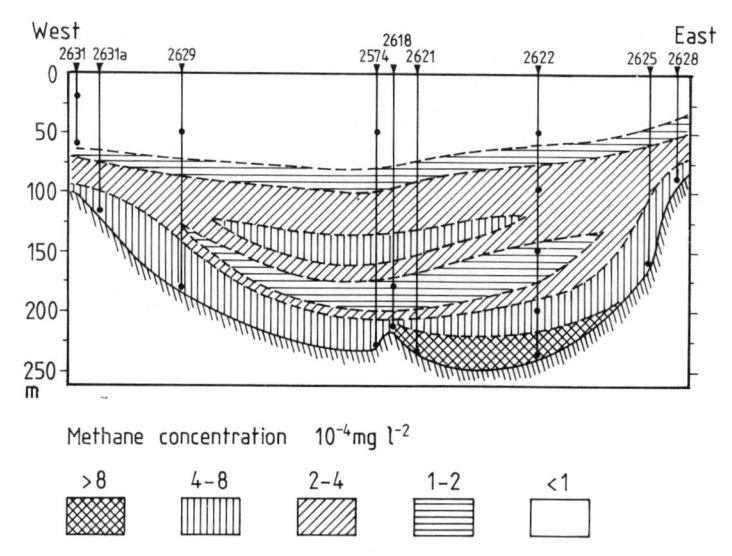

Figure 6.10 Content of dissolved methane in the Gotland depression (Trotzuk et al., 1984)

This is confirmed by simultaneously weighting the isotopic composition of both the substrate for methanogenesis (carbon dioxide) and the product (methane) (Figure 6.9).

The impact of anaerobic oxidation of methane on the character of its distribution in the upper horizons of marine sediments was considered in sufficient detail in the earlier section by Reeburgh (Section 6.3). Here we wish only to supplement his data with recent findings on the rate of methane oxidation in marine sediments.

Probably the most popular explanation for anaerobic oxidation of methane has been the one maintaining that, in the uppermost layers of the sediment, bacterial methanogenesis does not occur. This is because either conditions are insufficiently anaerobic or sulphate-reducing bacteria compete for organic matter and hydrogen more effectively than methanogenic bacteria. As shown above (Figure 6.8) this point of view has not been confirmed by new data: maximum intensity of methanogenesis is observed, in many cases, in zones of active sulphate reduction in sediments.

Two other hypotheses explain a sharp decrease in the methane content in subsurface horizons of bottom sediments either by the loss of methane from bottom sediments to the water mass or by its anaerobic oxidation in sediments. The geochemical data being reported here support both mechanisms. The validity of the migration theory is strengthened by numerous observations of the increased dissolved methane content in bottom water in contact with reduced sediments (Figure 6.10).

Table 6.10 Balance of methane formation and methane oxidation processes in sediments under 1m² for 1m layer

Place of sampling (station number)	Depth (m)	Intensity ($\mu g(C)\ m^{-2}\ day^{-1}$) methane formation	methane oxidation	Portion of oxidized CH_4, %
Black Sea				
555	22	11.2	6.22.9	55
559	26	6.9	2.7	42
568	86	13.4	6.5	20
580	340	28.0	1.0	23
546	1400	9.5	3.7	11
545	1700	68.2		5
Bering Sea				
Pp	89	14.1	2.4	17
26	300	21.2	4.6	22
9	870	23.2	4.9	21
22	1200	28.2	4.1	15
1	1320	18.3	1.4	8
27	3350	109.7	2.8	3
1p	3000	115.6	6.4	6
6	3650	256.6	10.3	4
28	3850	290.0	10.8	4
32	3850	301.5	4.3	2

Simultaneous with the accumulation of results from experiments with $^{14}CH_4$ added to water and silt samples from anaerobic zones of lakes (Laurinavitchus et al., 1981) and from reduced marine sediments, data also appear on the scale of methane oxidation in such ecosystems. Given the initial methane content of an analysed sample and the distribution of radioactivity between methane introduced to the biomass of methane-oxidizing bacteria, carbon dioxide released and the other products of $^{14}CH_4$ oxidation, it is possible to calculate the intensity of methane oxidation during the experiment.

Table 6.10, contains our own data characterizing the intensity of methane oxidation in upper horizons of the Black (Ivanov et al., 1983) and Bering Seas. As with the case of methanogenesis (Table 6.9) all figures have been scaled to a cubic metre of sediment in order to have comparable values. Table 6.10 also presents data on methanogenesis obtained at the same stations. The comparison between the activities of these two key processes of the methane cycle in bottom sediments of seas shows that a substantial part of newly formed microbial methane (up to 40–55%) may be oxidized within the sediments. In two regimes (sediments of Black and Bering Seas) the maximal part of biogenic methane is oxidized in shallow sediments of the shelf and the upper part of the continental slope.

In deep-water sediments (below 1000 m), characterized by a higher productivity of methanogenesis (Table 6.10), no more than 10% of newly formed methane is oxidized, making conditions more favourable for large accumulations.

In ten years of investigations carried out in the laboratory for Biogeochemistry (Institute of Biochemistry and Physiology of Microorganisms, USSR Academy of Sciences), we have developed quantitative methods for evaluating the intensities of microbial formation and oxidation of methane. Using these methods we have shown that such processes play an important role in final stages of mineralization of organic matter in sediments of different regions of the world ocean.

The data obtained show that the present methane generation and oxidation in sediments have considerable effect on sediment geochemistry and the isotopic composition of low molecular weight organic compounds. Therefore all the biogeochemical problems of methanogenesis in sediments can only be solved after accounting for a considerable microbial contribution.

REFERENCES

Aizenshtat, Z., Lipiner, G., and Cohen, Y. (1984). Biogeochemistry of carbon and sulfur cycle in the microbial mats Solar Lake (Sinai). In: Cohen, Y., Castenholz, R.W. and Halvorson, H.O., (Eds.) *Microbial Mats: Stromatolites*, Alan R. Liss., New York, pp.281–312.
Alexeev, F.A., and Lebedev, V.S. (1975). Carbon Isotopic composition of CO_2 and CH_4 of Black Sea bottom sediments. In: *Dissipated gases*, Moscow, VNii YaGG, pp.49–53.
Aller, R.C., and Yingst, J.Y. (1980). Relationships between microbial distributions and the anaerobic decomposition of organic matter in surface sediments of Long Island Sound, U.S.A. *Mar. Biol.*, **56**, 29–42.
Alperin, M.J., and Reeburgh, W.S. (1984). Geochemical observations supporting anaerobic methane oxidation. In: Crawford, R.L., and Hanson, R.S. (Eds.) *Microbial Growth on C-1 Compounds*, American Society for Microbiology, Washington, DC. pp. 282–289.
Alperin, M.J., and Reeburgh, W.S. (1985). Inhibition experiments on anaerobic methane oxidation. *Appl. Environ. Microbiol.*, **50**, 940–5.
Alvarez-Borrego, S. (1983). Gulf of California. In: Ketchum, B.H. (Ed.) *Ecosystems of the World. 26. Estuaries and Enclosed Seas*, Elsevier Scientific Publishing, Amsterdam, pp.427–49.
Andel, Van T.H. (1964). Recent marine sediments of Gulf of California. In: *Marine Geology of the Gulf of California*, Amer. Ass. Petrol. Geol., Tulsa, Oklahoma., p.216.
Andreae, M.O., and Galbally, I.E.(1985). The emission of sulfur and nitrogen to the remote atmosphere. In: Galloway, J.N., Andreae, M.O., Charlson, R.J., and

Rodhe H. (Eds.), *Biogeochemical Cycling of Sulfur and Nitrogen in Remote Atmosphere*, Dordrecht, Reidel.

Bågander, L.E. (1977). In situ studies of bacterial sulphate reduction at the sediment-water interface. *Ambio Special Report*, No. 5.

Balba, M.T., and Nedwell, D.B. (1982). Microbial metabolism of acetate, propionate and butyrate in anoxic sediment from Colne Point saltmarsh, Essex, U.K. *J. Gen. Microbiol.*, **128**, 1415–22.

Banat, I.M., Lindström, E.B., Nedwell, D.B., and Balba, M.T. (1981). Evidence for coexistence of two distinct functional groups of sulfate-reducing bacteria in salt marsh sediment. *Appl. Environ. Microbiol.*, **42**, 985–92.

Barcelona, M.J. (1980). Dissolved organic carbon and volatile fatty acids in marine sediment pore waters. *Geochim. Cosmochim. Acta*, **44**, 1977–84.

Barnes, R.O., and Goldberg, E.D. (1976). Methane production and consumption in anoxic marine sediments. *Geology*, **4**, 297–300.

Bauld, J. (1984). Microbial mats in marginal marine environments: Shark Bay, Western Australia, Spencer Gulf, South Australia. In: Cohen, Y., Castenholz, R.W., and Halvorson, H.O. (Eds.) *Microbial Mats: Stromatolites*, Alan R. Liss, New York, pp.39–58.

Bauld, J., Burne, R.V., Chambers, L.A., Ferguson, J., and Skyring, G.W. (1980). Sedimentological and geobiological studies of intertidal cyanobacterial mats in north-eastern Spencer Gulf, South Australia. Biogeochemistry of ancient and modern environments. In: Trudinger, P.A., Walter, M.R., and Ralph, B.J. (Eds.) *Proceedings of the Fourth International Symposium on Environmental Biogeochemistry*, Australian Academy of Science, Canberra pp.157–66.

Bauld, J., Chambers, L.A., and Skyring, G.W. (1979). Primary productivity, sulfate reduction and sulfur isotope fractionation in algal mats and sediments of Hamelin Pool, Shark Bay, W.A. *Aust. J. Mar. Freshwater Res.*, **30**, 753–64.

Belyaev, S.S., and Finkelstein, Z.I. (1976). Anaerobic gasforming bacteria in bottom sediments of Californian Gulf and Pacific Ocean. In: *Biogeochemistry of Diagenesis of Pacific Ocean Sediments*, Moscow, Nauka, pp.75–82.

Belyaev, S.S., and Ivanov, M.V. (1975). Radioisotopic method for determination of bacterial methanogenesis rate. *Microbiology*, **44**, (1), 166–8.

Belyaev, S.S., Ivanov, M.V., and Lein A.Y. (1981). Role of methane-producing and sulfate-reducing bacteria in the destruction of organic matter. In: *Biogeochemistry of Ancient and Modern Environments*, Canberra, Australian Academy of Science, pp.235–42.

Bernard, B.B. (1979). Methane in marine sediments. *Deep-Sea Res.*, **26**, 429–43.

Bender, M.L., and Heggie, D.T. (1984). Fate of organic carbon reaching the deep sea floor: a status report. *Geochim. Cosmochim. Acta*, **48**, 977–86.

Berner, R.A. (1964). An idealized model of dissolved sulfate distribution in recent sediments. *Geochim. Cosmochim. Acta*, **28**, 1497–503.

Berner, R.A. (1970). Sedimentary pyrite formation. *Am. J. Sci.*, **268**, 2–23.

Berner, R.A. (1971). *Principles of Chemical Sedimentology*, McGraw-Hill, New York.

Berner, R.A. (1984). Sedimentary pyrite formation: An update. *Geochim. Cosmochim. Acta*, **48**, 605–15.

Berner, R.A., and Raiswell, R. (1983). Burial of organic carbon and pyrite sulfur in sediments over Phanerozoic time: a new theory. *Geochim. Cosmochim. Acta*, **47**, 855–62.

Billen, G. (1982). Modelling and process of organic matter degradation and nutrient recycling in sedimentary systems. In: Nedwell, D.B., and Brown, C.M. (Eds.) *Sediment Microbiology*, Academic Press, New York. pp.15–52.

Blackburn, T.H. (1979). Nitrogen/carbon ratios and rates of ammonia turnover in anoxic sediments. In: Boroquin, A.W. and Pritchard, P.H. (Eds.) *Proceedings of the Workshop: Microbial Degradation of Pollutants in Marine Environments*, Paris, CNRS, pp. 148–53.

Bordovski, O.K. (1964). *Accumulation and Transformation of Organic Matter in Marine Sediments*, Moscow, Nedra, 128 pp.

Bunt, J.S. (1975). Primary productivity of marine ecosystems. In: Lieth, H. and Whittaker, R.H. (Eds.), *Primary Productivity of the Biosphere*, Springer-Verlag, New York. pp.177–83.

Burkholder, P.R., and Doheny, T.E. (1968). *The Biology of Eelgrass, with Special Reference to Hempstead and South Oyster Bays, Nassau County, Long Island, New York*, Contr. 3. Dep. Conserv. and Waterways, Town Hempstead, 120 pp.

Capone, D.G., Reese, D.G., and Kiene R.P. (1983) Effects of metals on methanogenesis. *Appl. Envir.*, **45**, 1586–91.

Castenholz, R.W. (1984). Composition of hot spring mats: A summary. In: Cohen, Y., Castenholz, R.W., and Halvorson, H.O. (Eds.) *Microbial Mats: Stromatolites*, Alan R. Liss, New York., pp.101–20.

Chambers, L.A. (1985). Biochemical aspects of the carbon metabolism of microbial mat communities In: Gabrie, C., Toffart, J.L., and Salvat, B. (Eds.) *Proc. Fifth Int. Coral Reef Cong.*, Antenne Museum–Ethe, Moorea, French Polynesia, pp.371–6.

Christensen, D. (1984). Determination of substrates oxidized by sulfate reduction in intact cores of marine sediments. *Limnol. Oceanogr.*, **29**, 189–92.

Claypool, G.E., and Kaplan, I.R., (1974). The origin and distribution of methane in marine sediments. In: *Natural Gases in Marine Sediments*, London, Plenum Press, pp.99–139.

Cohen, Y. (1984). The Solar Lake cyanobacterial mats: Strategies of photosynthetic life under sulfide, In: Cohen, Y., Castenholz, R.W., and Halvorson, H.O. (Eds.) *Microbial Mats: Stromatolites*. Alan R. Liss, New York, pp.133–48.

Cohen, Y., Aizenshtat, Z., Stoler, A. and Jørgensen, B.B (1980). The microbial geochemistry of Solar Lake, Sinai and modern environments. In: Trudinger, P.A., Walter, M.R., and Ralph, B.J. (Eds.) *Biogeochemistry of Ancient and Modern Environments*, Australian Academy of Science, Canberra, pp.167–72.

Cohen, Y., Castenholz, R.W., and Halvorson, H.O. (Eds.) (1984), *Microbial Mats: Stromatolites*, Alan R. Liss, New York.

Craig, H. (1969) Abyssal carbon and radiocarbon in the Pacific, *J. Geophys. Res.*, **74**, 5491–506.

Davis, J.B., and Yarbrough, H.F. (1966). Anaerobic oxidation of hydrocarbons by *Desulfovibrio desulfuricans*, *Chem. Geol.*, **1**, 137–44.

Devol, A.H. (1983). Methane oxidation rates in the anaerobic sediments of Saanich Inlet. *Limnol. Oceanogr.*, **28**, 738–42.

Devol, A.H., and Ahmed, S.I. (1981). Are high rates of sulfate reduction associated with anaerobic oxidation of methane? *Nature*, **291**, 407–8.

Devol, A.H., Anderson, J.J., Kuivila, K., and Murray, J.W. (1984). A model for coupled sulfate reduction and methane oxidation in the sediments of Saanich Inlet. *Geochim. Cosmochim. Acta*, **48**, 993–1004.

Emery, K.O., and Rittenberg, S.C. (1952). Early diagenesis of California basin sediments in relation to origin of oil. *Am. Assoc. Petrol. Geol. Bull.*, **36** (5), 735.

Forsberg, B.R. (1985). The fate of planktonic primary production. *Limnol. Oceanogr.*, **30** (4), 807–19.

Galimov, E.M., and Kodina, L.A. (1982). *Investigation of Organic Matter and Gases in Ocean Sediments*, Nauka, Moscow, 228 pp.

Galimov, E.M., and Tchinenov, V.A. (1978). Geochemistry of hydrocarbon gases and isotopic composition of methane in sediment of Marco trench. In: *YII Alljanio symposium: Stable Isotopes in Geochemistry*, Moscow, pp.186–8.

Geodakjan, A.A., Trotzjuk, V.Y., Verchovskaja, Z.I., and Avilov, V.I. (1979). Gases in modern sediments. In: *Chemistry of Ocean* Vol. 2, Nauka, Moscow, pp.291–311.

Gibson, D.L. (1985). Pyrite–organic matter relationships: Currant Bush Limestone, Georgina Basin, Australia. *Geochim. Cosmochim. Acta*, **14**, 989–992.

Goldhaber, M.B., Aller, R.C., Cochran, J.K., Rosenfeld, J.K., Martens, C.S., and Berner, R.A. (1977). Sulfate reduction, diffusion and bioturbation in Long Island Sound sediments: report of the FOAM group (Friends of Anoxic Mud). *Am. J. Sci.*, **277**, 193–237.

Goldhaber, M.B., and Kaplan, I.R. (1980). Mechanisms of sulfur incorporation and isotope fractionation during early diagenesis in sediments of the Gulf of California. *Mar. Chem.*, **9**, 95–143.

Gorlatov, S.N. (1984). Recent methane formation in the upper quarternary sediments of the Black Sea. *Geokhimiya*, **4**, 556–64.

Hall, C.A.S., and Moll, R. (1975). Methods of assessing aquatic primary productivity In: Leith, H., and Whittaken, R.H., (Eds.) *Primary Productivity of the Biosphere*, Springer Verlag, New York. pp.19–53.

Hallberg, R.O. (1970) An apparatus for the continuous cultivation of sulphate-reducing bacteria and its application to geomicrobiological purposes. *Antonie van Leeuwenhoek*, **36**, 241–54.

Hallberg, R.O. (1972). Iron and zinc sulfides formed in a continuous culture of sulfate-reducing bacteria. *N.J. Miner. Mh.*, **11**, 481–500.

Hines, M.E. (1981). Seasonal biogeochemistry in the sediments of the Great Bay estuarine complex, New Hampshire. PhD Dissertation. University of New Hampshire.

Honjo S., Manganini, S., and Cale, J. (1982). *Deep-Sea Res.*, **29**, 609–25.

Hood, D.W. (1983). The Bering Sea. In: Ketchum, B.H. (Ed.) *Ecosystems of the World. 26. Estuaries and Enclosed Seas.*, Elsevier Scientific Publishing, Amsterdam, pp.337–73.

Horner, S.M.J., and Smith, D.F. (1982). Measurements of total inorganic carbon in seawater by a substoichiometric assay using $NaH^{14}CO_3$. *Limnol. Oceanogr.*, **27**, 978–83.

Howarth, R.W., (1979). Pyrite: Its rapid formation in marshes and its importance in ecosystem metabolism. *Science*, **203**, 49–51.

Howarth, R.W. (1984). The ecological significance of sulphur in the energy dynamics of salt marsh and coastal marine sediments. *Biogeochemistry*, **1**, 5–27.

Howarth, R.W., and Giblin, A. (1983). Sulfate reduction in the salt marshes of Sapelo Island, Georgia. *Limnol. Oceanogr.*, **28**, 70–82.

Howarth, R.W., and Jørgensen, B.B. (1984). Formation of ^{35}S labelled elemental sulfur and pyrite in coastal marine sediments (Limfjorden and Kysing Fjord, Denmark) during short-term ^{35}S sulfate reduction measurements. *Geochim. Cosmochim. Acta*, **48**, 1807–18.

Howarth, R.W., and Marino, R. (1984). Sulfate reduction in salt marshes, with some comparisons to sulfate reduction in microbial mats. In: Cohen, Y., Castenholz, R.W., and Halvorson, H.O., (Eds.) *Microbial Mats: Stromatolites*, Alan R. Liss., New York, pp.254–63.

Howarth, R.W., and Merkel, S. (1984). Pyrite formation and the measurement of sulfate reduction in salt marsh sediments. *Limnol. Oceanogr.*, **29**, 598–608.

Howarth, R.W., and Teal, J.M. (1979). Sulfate reduction in a New England Saltmarsh. *Limnol. Oceanogr.*, **24**, 999–1013.
Howarth, R.W., Giblin, A., Gale, J., Peterson, B.J., and Luther, III G.W. (1983). Reduced sulfur compounds in the porewaters of a New England salt marsh. *Environ. Biogeochim. Ecol. Bull. (Stockholm)*, **35**, 135–52.
Howes, B.L., Dacey, W.J.H., and King, G.M. (1984). Carbon flow through oxygen and sulfate reduction pathways in salt marsh sediments. *Limnol. Oceanogr.*, **29**, 1037–51.
Howes, B.L., Dacey, W.J.H., and Teal J.M. (1985). Annual carbon mineralization and belowground production of Spartina alterniflora, *Ecology*, **66**, 595–605.
Hunt, J.M. and Whelan, J.K.(1978), Dissolve gases in Black Sea sediments, *Init. Rept. Deep Sea Drilling Project*, **42**, 661–5.
Ivanov, M.V. (1956). Isotopes in the determination of the sulfate-reduction rate in Lake Belovod. *Microbiologia*, **25**, 305–9.
Ivanov, M.V. (1978). Influence of microorganisms and microenvironment on the global sulfur cycle. In: *Envir. Biogeochem.* (Ann Arbor), **1**, 47–61.
Ivanov, M.V. (1979). Distribution and geochemical activity of microorganisms in ocean sediments. In: *Chemistry of ocean*, Vol. 2, Nauka, Moscow, pp.312–349.
Ivanov, M.V. (1983). Major fluxes of the global biogeochemical cycle of sulfur. In: Ivanov, M.V., and Freney, J.R. (Eds.) *The Global Biogeochemical Sulfur Cycle, Scope 19* Wiley, Chichester. pp.449–63.
Ivanov, M.V., and Lein, A. Yu. (1980). Distribution of microorganisms and their role in diagenetic mineralforming processes. In: *Geochemistry of Diagenesis of Pacific Ocean*, Nauka, Moscow., pp.117–37.
Ivanov, M.V., Lein, A. Yu., and Kashparova, E.V. (1976). Intensity of formation and diagenetic transformation of reduced sulfur compounds in sediments of the Pacific Ocean. In: *The Biogeochemistry of Diagenesis of Ocean Sediments*, Nauka, Moscow, pp.171–8 (in Russian).
Ivanov, M.V., Lein, A. Yu., Beylaev, S.S., Nesterov, A.I., Bonder, V.A., and Zhabina, N.N. (1980). Geochemical activity of sulfate-reducing bacteria in benthal sediments of the North-West Indian Ocean. *Geokhimiya*, **8**, 1238–49 (in Russian).
Ivanov, M.V., Vainstein, M.B., Galtohenko, V.F., Gorlatov, S.N. & Lein, A. Yu. (1983). Distribution and geochemical activity of bacteria in sediments of western part of Black Sea. In: *Geochemical Processes in Western Part of Black Sea*, Bulgarian Academy of Sciences, Sofia.
Iversen, N. (1984). Interaktioner mellen fermentings-processer og de terminate processer. *Thesis*, Åarhus University.
Iversen, N., and Blackburn, T.H. (1981). Seasonal rates of methane oxidation in anoxic marine sediments. *Appl. Environ. Microbiol.*, **41**, 1295–1300.
Iversen, N., and Jørgensen, B.B. (1985). Anaerobic methane oxidation rates at the sulfate–methane transition in marine sediments from Kattegat and Skagerrak (Denmark). *Limnol. Oceanogr.*, **30**, 944–55.
Jacobs, R.P.W. (1979). Distribution and aspects of the production and mass of eelgrass, Zoster marina L. at Rascott, France. *Aquatic Bot.*, **7**, 151–72.
Jannasch H.W. (1984). Microbes in the oceanic environment. In: Kelly, D.P., and Carr, N.G. (Eds.) *SGM Symposium 36 Part II*, The Society for General Microbiology Ltd, Pitman Press. Bath, pp.97–122.
Jannasch, H.W., and Wirsen, C.O. (1979). *Bioscience*, **29**, 592–8.
Javor, B.J., and Castenholz, R.W. (1984). Laminated microbial mats, Laguna Guerrero Negro. Mexico. *Geomicrobiol. J.*, **2**, 237–73.
Joint, I.R. (1978). Microbial production of an estuarine mudflat. *Estuar. Coast. Mar. Sci.*, **7**, 185–95.

Jørgensen, B.B. (1978). A comparison of methods for the quantification of bacterial sulfate reduction in coastal marine sediments. *Geomicrobial. J.*, **1**, 11–27; 29–47; 49–64.

Jørgensen, B.B. (1982) Mineralization of organic matter, *Nature*, **296**, 643–5.

Jørgensen, B.B. (1983). Processes at the sediment–water interface. In: Bolin, B., and Cook, R.B. (Eds.). *The Major Biogeochemical Cycles and their Interactions*, Wiley, Chichester, pp. 477–509.

Jørgensen, B.B., and Cohen, Y. (1977). Solar Lake (Sinai) 5. The sulfur cycle of the benthic cyanobacterial mats. *Limnol. Oceanogr.*, **22**, 657–66.

Jørgensen, B.B., and Fenchel, T. (1974). The sulfur cycle of a marine sediment model. *Mar. Biol.*, **24**, 189–201.

Jørgensen, B.B., Revsbech, N.P., and Cohen, Y. (1983). Photosynthesis and structure of benthic microbial mats: Microelectrode and SEM studies of four cyanobacterial communities. *Limnol. Oceanogr.*, **28**, 1057–93.

Jørgensen, N.O.G., Blackburn, T.H., Henriksen, K., and Bay, D. (1981). The importance of *Posidonia oceanica* and *Cymodocea nodosa* as contributors of free amino acids in water and sediment of sea grass beds. *Mar. Ecol.*, **2**, 97–112.

Karl, D.M., and Knauer, G.A. (1984). Vertical distribution, transport and exchange of carbon in the north east Pacific Ocean—evidence for multiple zones of biological activity, *Deep Sea Res.*, **31**, 221–43.

Khalil, M.A.K., and Rasmussen, R.A. (1983). Sources, sinks, and seasonal cycles of atmospheric methane. *J. Geophys. Res.*, **88**, 5131–44.

King, G.M., Howes, B.L., and Dacey, J.W.H. (1985). Short-term end-products of sulfate reduction in a salt marsh: formation of acid volatile sulfides, elemental sulfur and pyrite. *Geochim. Cosmochim. Acta*, **49**, 1561–6.

Kinsey, D.W. (1983). Short term indications of gross material flow in coral reefs–how far have we come and how much further can we go? In: Baker, J.T., Carter, R.M., Sammarco, P.W., and Stark, K.P., (Eds.) *Proc. Great Barrier Reef Conf.*, James Cook University, Townsville, pp.333–9.

Kinsey, D.W. (1985). The functional role of back-reef and lagoonal systems in central Great Barrier Reef. In: Gabrie, C., Toffart, J.L., and Salvart, B. (Eds.) *Proceedings of the Fifth Int. Cong. on Coral Reefs, Tahiti*, Antenne Museum–Ethe, Moorea, French Polynesia, pp.223–8.

Koblentz-Mishke, O.J., Volkovinsky, V.V., and Kabanova, J.G. (1970). Plankton primary production of the world ocean. In: Wooster, W.S. (Ed.). *Scientific Exploration of the South Pacific*, Nat. Sci., Washington DC, pp.183–93.

Kosiur, D.R., and Warford A.L. (1979). Methane production and oxidation in Santa Barbara basin sediments. *Estuar. Coast. Sci.* **8**, 379–85.

Krom, M.D., and Sholkovitz, E.R. (1977). Nature and reactions of dissolved organic matter in the interstitial waters of marine sediments. *Geochim. Cosmochim. Acta*, **41**, 1565–73.

Krouse, H.R., and McCready, R.G. (1979). Reductive reactions in the sulfur cycle. In: Trudinger, P.A., and Swaine, D.J. (Eds.) *Biogeochemical Cycling of Mineral-forming Elements*, Elsevier, Amsterdam, Oxford, New York, pp.315–68.

Krumbein, W.E., Cohen, Y., and Shilo, M. (1977). Stromatolitic cyanobacterial mats. *Limnol. Oeanogr.*, **22**, 535–65.

Laanbroek, H.J., and Veldkamp, H. (1982). Microbial interactions in sediment communities. *Phil. Trans. R. Soc. Lond. B.*, **297**, 533–50.

Laurinavitchus, K.K., Tokarev, V.G., and Mschenski, Ju. N. (1981). The rate of microbial methane forming processes in sediments of Pacific ocean and South China Sea. In: *Geochemical Activity of Microorganisms in Pacific Sediments*,

Pushchino, Academy of Science, pp.59–64.
Lein, A. Yu. (1978). Formation of carbonate and sulfide minerals during diagenesis of reduced sediments. *Environ. Biogeochem.* (Ann Arbor), **1**, 339–54.
Lein, A. Yu. (1983). C_{org} consumption by processes of organic matter mineralization in modern oceanic sediments. *Geochemistry*, **11**, 1634–9.
Lein, A. Yu., Grinenko, V.A., and Matrosov, A.G. (1981). Sulphur and carbon isotopes fractionation in modern oceanic sediments with different rates of bacterial sulphate reduction processes. In: *Geochemical Activity of Microorganisms in Pacific Sediments*, Pushchino, Academy of Science, pp.134–46.
Lein, A. Yu, and Ivanov, M.V., (1981). Dynamic biogeochemistry of anaerobic diagenesis of sediments. In: *Lithology and a New Stage of Geological Knowledge*, Nauka, Moscow, pp.62–76.
Lein, A. Yu., Ivanov, M.V., Masatov, A.G., and Zyakin A.M. (1976). Genesis of sulphur calcite ores *Geokhimiya*, **3**, 422–32.
Lein, A. Yu., Namsarayev, B.B., Trotsyuk, V.Ya., and Ivanov, M.V. (1981). Bacterial methanogenesis in Holocene sediments of the Baltic Sea. *Geomicrob. J.*, **2** (4), 299–317.
Lein, A. Yu., Vaynshteyn, M.B., Namsarayev, B.B., Kashparova E.V., Matrosov, A.G., Bondar, V.A., and Ivanov, M.V. (1982). Biogeochemistry of anaerobic diagenesis of recent Baltic Sea sediments. *Geochemistry International*, **19**, 90–103.
Lein, A. Yu., and Nansarayev, B.B. (1986). Biogeochemical processes of decomposition of organic matter in the rift sediments. In: *Geochemistry of carbon*, Academies of Science, Moscow, pp.173–5.
Lerman, A. (1982). Sedimentary balance through geological time. In: Holland, H.D., and Schidlowski, M. (Eds.) *Mineral Deposits and the Evolution of the Biosphere*, Springer-Verlag, Heidelberg, New York, pp.237–56.
Leventhal, J.S. (1983). An interpretation of carbon and sulfur relationships in Black Sea sediments as indicators of environments of deposition. *Geochim. Cosmochim. Acta*, **47**, 133–137.
Lindström, C.H. (1980). Transformation of iron constituents during early diagenesis—*in situ* studies of a Baltic Sea sediment–water interface. PhD. Thesis, Dept of Geology, University of Stockholm.
Lisitsyn, A.P. (1976). *The processes of ocean's sedimentation: Lithology and geochemistry*. Nauka, Moscow, p.358.
Livingstone, D.C., and Patriquin, D.G. (1981). Below ground growth of Spartina alterniflora Loisel.: Habitat, functional biomass and non-structural carbohydrates. *Estuar. Coast. Shelf. Sci.*, **12**, 579–87.
Luther, III, G.W., Giblin, A., Howarth, R.W., and Ryans, R.A. (1983). Pyrite and oxidised iron mineral phases formed from pyrite oxidation in salt marsh and estuarine sediments. *Geochim. Cosmochim. Acta*, **46**, 2665–9.
Lyons, W.B., Hines, M.E., and Gaudette, H.E. (1983). Major and minor element pore water geochemistry of modern marine sabkhas: the influence of cyanobacterial mats. In: Cohen, Y., Castenholz, R.W., and Halvorson, H.D. (Eds.) *Microbial Mats: Stromatolites*, Alan R. Liss, New York, pp.411–24.
Makagon, Yu, F., Trofimor, A.A., Tsarev, V.P., and Chersky, N.V. (1983). Possibilities of the formation of gas-hydrate deposits of natural gases in the bottom zones of seas and oceans. *Geology and Geofizika*, **4**, 37–77.
Martens, C.S., and Berner, R.A. (1977). Interstitial water chemistry of anoxic Long Island Sound sediments: Dissolved gases. *Limnol. Oceanogr.*, **22**, 10–25.
Meadows, P.S., and Campbell, J.I. (1978). *An Introduction to Marine Science*, Blackie, Glasgow, London.

Mopper, K., and Degens, E.T. (1979). Organic carbon in the ocean: nature and cycling. In: *Global Carbon Cycle, SCOPE 13*, Wiley, Chichester.

Moriarty, D.J.W., Boon, P., Hansen, J., Hunt, W.G., Poiner, I.R., Pollard, P.C., Skyring, G.W., and White, D.C. (1985). Microbial biomass and productivity in sea grass beds. *Geomicrobiol. J.*, **4**, 21–51.

Morozov, A.A., and Rozanov, A.G. (1981). On sulfide derivatives of iron (III), their analysis and possible formation in recent reduced sediments. *8th International Symposium on Environmental Biogeochemistry, Stockholm Sweden*. Abstracts, p.72.

Mountfort, D.O., and Asher, R.A. (1981). Role of sulfate reduction versus methanogenesis in terminal carbon flow in polluted intertidal sediment of Waimea Inlet, Nelson, New Zealand. *Appl. Environ. Microbiol.*, **42**, 252–8.

Murray, J.W., Grundmanis, V., and Smethie, Jr.W.M. (1978). Interstitial water chemistry in the sediments of Saanich Inlet. *Geochim. Cosmochim. Acta*, **42**, 1011–26.

Nedwell, D.B., and Abram, J.W. (1978). Bacterial reduction in relation to sulfur geochemistry in two contrasting areas of saltmarsh sediment. *Estuar. Coast. Mar. Sci.*, **6**, 341–51.

Oremland, R.S., and Silverman, M.P. (1979). Microbial sulfate reduction measured by an automated electrical impedance technique. *Geomicrobiol. J.*, **1**, 355–72.

Plumb, L.A., Bauld, J., Ho, D.T., and Reichstein, I.C. (1982). Production and fate of organic carbon in cyanobacterial mats, Baas Becking Geobiological Laboratory, Ann.Rep. 1982, pp.25–29.

Plumb, L.A., Bauld, J., Ho, D.T., and Reichstein, I.C. (1983). Production and fate of organic carbon in cyanobacterial mats, Baas Becking Geobiological Laboratory, Ann.Rep. 1983, pp.23–33.

Plumb, L.A., and Reichstein, I.C. (1984). Sedimentary organic carbon. In: Baas Becking Geobiological Laboratory, Ann.Rep. 1984, pp. 18–20.

Pomeroy, L.P., Darley, W.M., Dunn, E.L., Gallagher, J.L., Haines, E.B., and Whitney, D.M. (1981). Primary production. In: Pomeroy, L.R., and Wiegert, R.G. (Eds.) *The Ecology of a Salt Marsh*, Springer-Verlag, New York, pp.39–66.

Post, F.J. (1980). Biology of the North arm: In: Gwynn, J.W. (Ed.). *The Great Salt Lake: A Scientific, Historical and Economic Overview*. Utah Geological and Mineral Survey, Utah Department of Natural Resources Bulletin 116; pp.313–21.

Reeburgh, W.S. (1976). Methane consumption in Cariaco Trench waters and sediments. *Earth Planet. Sci. Lett.* **28**, 337–44.

Reeburgh, W.S. (1980). Anaerobic methane oxidation: rate depth distributions in Skan Bay sediments. *Earth Planet, Sci. Lett.*, **37**, 345–52.

Reeburgh, W.S. (1982). A major sink and flux control for methane in marine sediments: anaerobic methane oxidation. In: Fanning, K.A., and Manheim, F.T. (Eds.) *The Dynamic Environment of the Ocean Floor*, Lexington Books, Lexington, Mass. pp.203–17.

Reeburgh, W.S. (1983). Rates of biogeochemical processes in anoxic sediments. *Ann. Rev. Earth Planet. Sci.*, **11**, 269–98.

Reeburgh, W.S., and Heggie, D.T. (1977). Microbial methane consumption reactions and their effect on methane distributions in freshwater and marine environments. *Limnol. Oceanogr.* **22**, 1–9.

Revsbech, N.P., Jørgensen, B.B., and Blackburn, T.H. (1983). Microelectrode studies of the photosynthesis and O_2, H_2S and pH profiles of a microbial mat. *Limnol. Oceanogr.*, **28**, 1062–74.

Revsbech, N.P., Jørgensen, B.B., and Brix, O. (1981). Primary production of microalgae in sediments measured by oxygen microprofile, $H^{14}CO_3$-fixation and oxygen exchange methods. *Limnol. Oceanogr.*, **26**, 717–30.

Revsbech, N.P., and Ward, D.M. (1984). Microprofiles of dissolved substances and photosynthesis in microbial mats measured with microelectrodes. In: Cohen, Y., Castenholz, R.W., and Halvorson, H.O. (Eds.) *Microbial Mats: Stromatolites*, Alan R. Liss, New York, pp.171–88.

Rickard, D.T. (1970). The origin of framboids. *Lithos*, **3**, 269–93.

Roberts, M.J., Long, S.P., Tieszer, L.L., and Beadle, L.L. (1985). Measurement of plant biomass and net primary production. In: Coombs, J., Hall, D.O., Long, S.P., and Seurlock, J.M.W. (Eds.) *Techniques in Bioproductivity and Photosynthesis*, Pergamon Press, Oxford.

Romankevitch, V.A. (1977). *Geochemistry of Organic Matter in the Ocean*, Nauka, Moscow, 255 pp.

Rowe, G., and Howarth, R.W. (1985). Early diagenesis of organic matter in sediments off the coast of Peru. *Deep-Sea Res.*, **32** (1), 43–5.

Ryther, J.H. (1969). Photosynthesis and fish production in sea. *Science*, **166**, 72–6.

Sand-Jensen, K. (1975). Biomass, net production and growth dynamics in an eelgrass (Zostera marina L.) population in Vellerup Vig, Denmark. *Ophelia*, **14**, 185–201.

Sansone, F.J., and Martens, C.S. (1982). Volatile fatty acid cycling in organic-rich marine sediments. *Geochim. Cosmochim. Acta*, **46**, 1575–89.

Schubauer, J.P., and Hopkinson, C.S. (1984). Above- and below ground emergent macrophyte production and turnover in a coastal marsh ecosystem, Georgia. *Limnol. Oceanogr.*, **29**, 1052–65.

Shulenberger, E., and Reid, J.L. (1981). The Pacific Shallow oxygen maximum deep chlorophyll maximum and primary productivity reconsidered. *Deep-sea Res.*, **28A**, 901–19.

Senior, B., Lindström, E.B., Banat, I.M., Nedwell, D.B. (1982). Sulphate reduction and methanogenesis in the sediments of a salt marsh. *Appl. Envir.*, **43**, 987–96.

Sheppard, J.C., Westburg, H., Hopper, J.F., Ganesan, K., and Zimmerman, P. (1982). Inventory of global methane sources and their production rates. *J. Geophys. Res.*, **87**, 1303–12.

Skyring, G.W. (1984). Sulfate reduction in marine sediments associated with cyanobacterial mats in Australia. In: Cohen, Y., Castenholz, R.W., and Halvorson, H.O. (Eds.) *Microbial Mats. Stromatolites*, Alan R. Liss, New York pp.265–75.

Skyring, G.W. (1985). Anaerobic microbial processes in coral reef sediments. In: Delasalle, B., Gabrie, C., Galzin, R., Harmelin-Vivien, M., Toffart, J.L. and Salvat, B., (Eds.) *Proc. Fifth Int. Coral Reef Congress., Tahiti*, Antenne Museum–Ethe, Moorea, French Polynesia, pp.421–25.

Skyring, G.W. (1987). Sulfate reduction in coastal ecosystems *Geomicrobiology J.*, **5**, 295-374.

Skyring, G.W. (1988). Acetate as the main energy substrate for the sulfate-reducing bacteria in Lake Eliza, (SA) hypersaline sediments. *FEMS Microbiol. Ecol.* **53**, 87–94.

Skyring, G.W., and Chambers, L.A. (1980). Sulfate reduction in intertidal sediments. In: Freney, J.R., and Nicolson, A.J. (Eds.) *Sulfur in Australia*, Australian Academy of Science, Canberra, pp.88–94.

Skyring, G.W., Chambers, L.A., and Bauld, J. (1983). Sulfate reduction in sediments colonized by cyanobacteria, Spencer Gulf, South Australia. *Aust. J. Mar. Freshw. Res.*, **34**, 359–74.

Skyring, G.W., and Johns, I.A. (1980). Iron in cyanobacterial mats. *Micron*. **11**, 407–8.

Skyring, G.W., Jones, H.E., and Goodchild, D. (1977). The taxonomy of some new isolates of dissimilatory sulfate-reducing bacteria. *Can. J. Microbiol.*, **23**, 1415–25.

Skyring, G.W., and Lupton, F.S. (1984). Anaerobic microbial processes in peritidal and subtidal sediments of Hamelin Pool (Shark Bay, W.A.). Baas Becking Geobiological Laboratory Annual Reprt, 1984, pp.3–12.

Skyring, G.W., and Lupton, F.S. (1986). Anaerobic microbial activity in organic-rich sediments of a coastal lake. In: *Proceedings, 12th International Sedimentological Congress*. Bureau of Mineral Resources, Canberra, Australia.

Skyring, G.W., Oshrain, R.L., and Wiebe, W.J., (1979). Sulfate reduction rates in Georgia marshland soils. *Geomicrobiol. J.*, **1**, 389–400.

Skyring, G.W., Smith, G.D., and Lynch, R. (1987). Acetylene reduction and hydrogen metabolism by a cyanobacterial sulfate reducing bacterial mat ecosystem. *Geomicrobiol. J.*, **6**, 25–31.

Smith, R.L., and Klug, M.J. (1981). Electron donors utilized by sulfate-reducing bacteria in eutrophic lake sediments. *Appl. Environ. Microbiol.*, **42**, 116–21.

Sørensen, J., Christensen, D., and Jørgensen, B.B. (1981). Volatile fatty acids and hydrogen as substrates for sulfate-reducing bacteria in anaerobic marine sediment. *Appl. Environ. Microbiol.*, **42**, 5–11.

Sorokin, Y.I. (1957). Ability of sulfate-reducing bacteria to utilize methane for the reduction of sulfate to hydrogen sulfide. *Dokl. Akad. Nauk. SSSR*, **115**, 816–18.

Stams, A.J.M., Hansen, T.A., and Skyring, G,W, (1985). Utilization of amino acids as energy substrates by two marine Desulfovibrio strains. *Microbiol. Ecol. FEMS*, **31**, 11–15.

Stolz, J.F. (1984). Fine structure of stratified microbial community at Laguna Figueroa, Baja California, Mexico: II. Transmission electron microscopy as a diagnostic tool in studying microbial communities *in situ*. In: Cohen, Y., Castenholz, R.W. and Halvorson, H.O. (Eds.) *Microbial Mats: Stromatolites*, Alan R. Liss, New York, pp.23–38.

Steeman-Nielsen, E. (1953). The use of radioactive carbon (^{14}C) for measurement of organic production in the sea. *J. Cons. Perm. Int. Explor. Mer.*, **19**, 309–28.

Strakhov, N.M. (1976). *The Problems of Geochemistry of Modern Oceanic Lithogenesis*, Nauka, Moscow, 299 pp.

Strakhov, N.M., and Zalmanzon, E.S. (1955). Distribution of iron in sedimentary rocks and its significance for lithogenesis. *Izvestija AN SSSR, serija geol.*, **1**, 34.

Suess, E. (1980). Particulate organic carbon in the oceans surface productivity and oxygen utilization. *Nature*, **288**, 260–3.

Suess, E., and Muller, P.J. (1980). Productivity, sedimentation rate and sedimentary organic matter in the oceans II—elemental fractionation. *Colloq. Int. Du CNRS*, **293**, 17–26.

Sweeney, R.E,. and Kaplan, I.R. (1980). Diagenetic sulfate reduction in marine sediments. *Mar. Chem.*, **9**, 165–74.

Toth, D.J., and Lerman, A. (1977). Organic matter reactivity and sedimentation rates in the ocean. *Am. J. Sci.*, **277**, 465–85.

Trofimuk, A.A., Tschersky, N.B., and Tzarev, V.P. (1975). Resources of biogenic methane in the World Ocean. *Dokl. Acad. Nauk. SSSR*, **225**, No.4.

Trotzuk, V. Ya., Avilov, V.I., and Bolshakov (1984). The genetic composition of carbohydrates in water and sediments of Baltic Sea. In: *Geological History and Geochemistry of Baltic Sea*, Nauka, Moscow, pp.144–52.

Trudinger, P.A. (1981). Origins of sulfide in sediments. *BMR J. Australian Geology and Geophysics*, (Canberra, Australia) **6**, 279–85.

Trudinger, P.A., Swaine, D.J., and Skyring, G.W. (1979). Biogeochemical cycling of elements—general considerations. In: Trudinger, P.A., and Swaine, D.J. (Eds.). *Biogeochemical Cycling of Mineral-forming elements*, Elsevier, Amsterdam, pp.1–27.

Tshebotarev, E.N., and Ivanov, M.V. (1976). Distribution and activity of sulphate reducing bacteria in bottom sediments of Pacific ocean and Californian Bay. In: *Biogeochemistry of Oceanic Sediments Diagenesis*, Nauka, Moscow, pp.68–74.

Valiela, I., Howes, B., Howarth, R.W., Giblin, A., Forman, K., Teal, J., and Hobbie, J. (1982). Regulation of primary production in a salt marsh ecosystem. In: Gopal, B., Turner, R., Wetzel, R., and Whigman, D. (Eds.) *Wetlands: Ecology and Management*, National Institute of Ecology, Jaipur and International Scientific Publications, Jaipur, India, pp.261–90.

Volkov, I.I., and Rosanov, A.G. (1983). The sulfur cycle in oceans. Part I. Reservoirs and fluxes. In: Ivanov, M.V., and Freney, J.R. (Eds.) *The Global Biogeochemical Sulfur Cycle, SCOPE 19*, Wiley, Chichester, pp.357–423.

Westrich, J.T., and Berner, R.A. (1984). The role of sedimentary organic matter in bacterial sulfate reduction: The G model tested. *Limnol. Oceanogr.* **29**, 236–49.

Whiticar, M.J. (1982). The presence of methane bubbles in the acoustically turbid sediments of Eckernfordner Bay, Baltic Sea. In: Fanning, K.A., and Manheim, F.T. (Eds.) *The Dynamic Environment of the Ocean Floor*, Lexington Books, Lexington, Mass. pp.219–235.

Widdel, F. (1980). Anaerober Abbau von Fettsäuren und Benzoesäure durch neu isolierte Arten sulfat-reduzierender Bakterien. Dissertation, zur Erlangung des Doktorgrads der mathematisch-naturwissen-schaflichen Fakultät der Georg-August-Universität zur Göttingen.

Wiegert, R.G., Christian, R.R., and Wetzel, R.I. (1981). A model view of the marsh. In: Pomeroy, L.R. and Wiegert, R.G., (Eds.) *The Ecology of a Salt Marsh*, Springer-Verlag, New York, pp.183–218.

Wieder, R.K., Lang, G.E,. and Granus, V.A. (1985). An evaluation of wet chemical methods for quantifying sulfur. *Limnol. Oceanogr.* **30**, 1109–15.

Woodwell, G.M., Rich, P.H., and Hall, C.A.S (1973). Carbon in estuaries. In: Woodwell, G.M. and Pecan, E.V. (Eds.) *Carbon and the Biosphere*, United States Atomic Energy Commission, pp.221–40.

Zehnder, A.J.B., and Brock, T.D. (1979). Methane formation and methane oxidation by methanogenic bacteria. *J. Bacteriol.*, **137**, 420–32.

Zehnder, A.J.B., and Brock, T.D. (1980). Anaerobic methane oxidation: its occurrence and ecology. *Appl. Environ. Microbiol.*, **39**, 194–204.

Zhabina, N.N., and Volkov, I.I. (1978). A method of determining various sulfur compounds in sea sediments and rocks. In: Krumbein, W.E. (Ed.) *Environ. Biogeochem.*, Ann Arbor Science, Ann Arbor, Mich. **3**, 735–45.

Zieman, J.C., and Wetzel, R.G. (1980). Productivity in seagrasses: methods and rates. In: Phillips, R.C., and McRoy, C.P. (Eds.) *Handbook of Seagrass Biology: An Ecosystems Perspective*, Garland STPM Press, New York and London, pp.87–116.

Evolution of the Global Biogeochemical Sulphur Cycle
Edited by P. Brimblecombe and A. Yu. Lein
© 1989 SCOPE Published by John Wiley & Sons Ltd

CHAPTER 7
Sulphur Emission and Transformations at Deep Sea Hydrothermal Vents

HOLGER W. JANNASCH

7.1 INTRODUCTION

The search for submarine volcanism, based on the newly emerging concept of plate tectonics, led in 1977 to the discovery of sulphide-containing warm and hot springs or vents at ocean floor spreading zones. These areas of hydrothermal water circulation through the Earth's crust greatly affect the chemistry of seawater. Estimation of fluxes has been based on the present composition of seawater. The estimated rate of sulphide emission as compared to the seawater (sulphate) entrainment, as well as the various chemical and biological sulphur transformations determining its deposition as polymetal sulphides and anhydrites, are significant with respect to the global sulphur cycle. This report briefly presents the information so far available on deep sea hydrothermal conversions of sulphur.

7.2 CHARACTERISTICS OF DEEP SEA VENTS

After considerable theoretical work and exploratory studies with surface-deployed instrumentation, warm vents (<20°C) were first located by the research submersible ALVIN in 1977 at the Galapagos Rift spreading zone at a depth of 2550 m (Ballard, 1977; Lonsdale, 1977; Corliss et al., 1979; Edmond et al., 1979). The presence of hydrogen sulphide in the collected hydrothermal fluid and the observations of copious animal populations in the immediate vicinity of the vents led to studies on the primary production of organic carbon by sulphur-oxidizing bacteria (Jannasch and Wirsen, 1979, 1981; Karl et al., 1980). The fact that the emitted vent water contained between 10^5 and 10^8 bacterial cells per ml, and aerobic thiobacilli-type organisms could be isolated from greatly diluted samples, meant it could be concluded that subsurface mixing with oxygenated seawater had taken place prior to emission. This observation was corroborated by chemical analyses.

In 1979 hot vents, the so-called black smokers, were discovered at an active spreading zone of the East Pacific Rise at 21°N where hydrothermal fluid of about 350°C reaches the sea floor unmixed (Spiess et al., 1980). The heavy precipitation of polymetal sulphides in particulate form produces the smoke-like plume and part of the chimney-like structures. These chimneys may be up to 20 m high, and consist, in addition to metal sulphides, of calcium sulphate (as anhydrite). When a vent dies (see below) the anhydrite dissolves as a corollary of cooling, and the chimneys collapse forming mounds of varied mineral composition, largely polymetal sulphides covered by iron-manganese oxides and hydroxides. Life forms are less abundant in the vicinity of hot vents than at warm vents and appear to be clustered around warm water leakages near the bases of chimneys.

These deep sea vents and the entire cycling of seawater through the upper 3 km of the Earth's crust are powered by the consecutive cooling–contraction and warming–expansion processes. In comparison to the weathering cycle and mineral dispersion by runoff from the continents, the hydrothermal cycle is only about 0.5% of the total river flow but about 200 times higher in mineral content (Edmond and Von Damm, 1983).

Since basalt and seawater are the only two reactants, the chemistry of deep sea vents is simpler than that of terrestrial hot springs. Within the crust the penetrating seawater attains temperatures of more than 350°C (Figure 7.1). On reacting with basalt, the resulting hydrothermal fluid is enriched in hydrogen sulphide stemming both from the reduction of sulphate and additional sulphur leached from the rocks. Carbon occurs in its most oxidized and most reduced form. While magnesium is depleted, heavy metals and calcium are also enriched (Edmond et al., 1982).

The location and the rate at which the mixing of hydrothermal fluid with ambient seawater takes place has a strong effect on the dispersion of sulphur. In warm vents the mixing is estimated to occur in the extremely porous lava a few to 100 m below the sea floor. Subsurface precipitations may clog the flow of hydrothermal fluid, resulting in the formation of new conduits and contributing to the fact that individual vents have a relatively short life time. Hot vents are often surrounded by smaller warm leakages.

While the hot vents with flow rates of up to 2 m/s raise their plume several hundred metres above the sea floor, the plumes of warm vents (emission rates of 1 to 2 cm/s) can only be observed a few metres above the sea floor. The abundance of visible life forms (large clams (*Calyptogena magnifica*, Boss and Turner, 1980), vestimentiferan tube worms of the phylum Pogonophora (*Riftia pachyptila*, Jones, 1981), and clusters of a yet unnamed mussel) appears to correspond to the concentration of hydrogen sulphide.

A 'dead' vent results from the stop of flow or its diversion to a new outlet. The non-motile animals die rather quickly in the absence of hydrogen sulphide, and the soft parts are rapidly removed by scavengers. The rarity

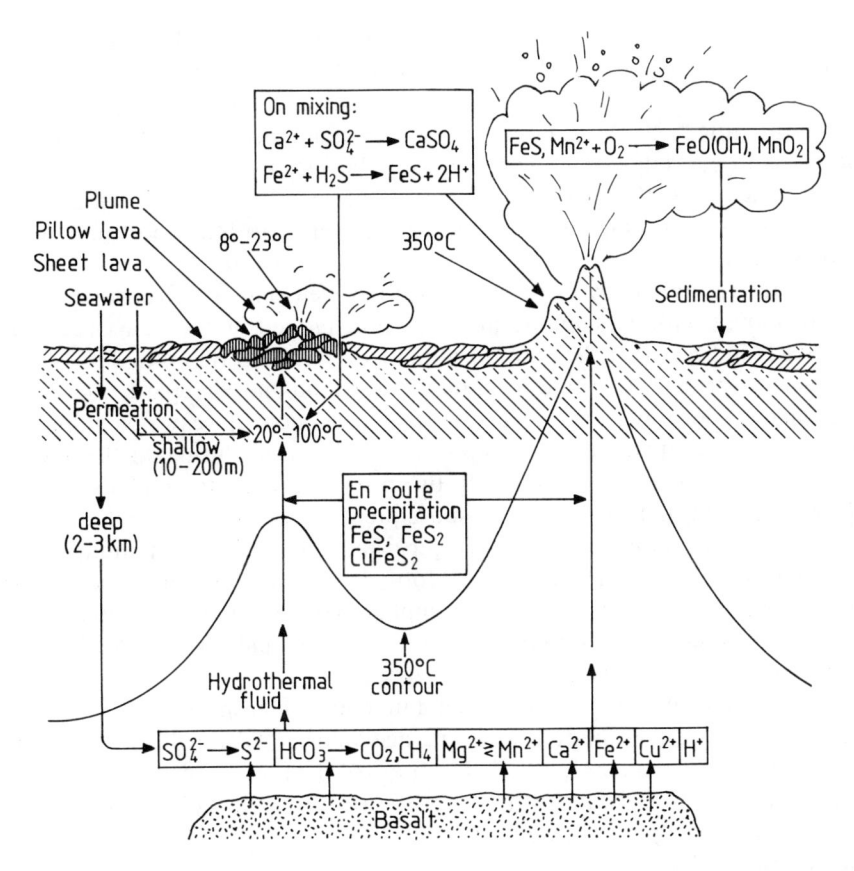

Figure 7.1 The major geochemical reactions occurring during the hydrothermal cycling of seawater through the Earth's crust at ocean floor spreading zones, indicating the two commonly observed types of vents (redrawn from Jannasch and Taylor, 1984)

of fossilized life forms in the metal-rich deposits of ancient spreading centres and presently mined ophiolites (Haymon et al., 1984) may be a result of the fast and efficient scavenging of such organic-rich 'islands' in an otherwise oligotrophic environment. The complete dissolution of clam shells has been estimated to last 20 to 23 years (R. A. Lutz, personal communication) and serves as an excellent indicator of short-term vent histories.

7.3 EMISSIONS OF SULPHUR

The unmixed hydrothermal fluid of the deep sea hot vents near the 21°N East Pacific Rise area contains 6.75 to 8.37 mM hydrogen sulphide at temperatures between 275 and 350°C (Von Damm, 1983). In several newly found hot vents of the 10°N to 13°N area, the concentrations range more widely between 0.5 and 10.0 mM hydrogen sulphide with pH values approximating 3.5 (J.M. Edmond, personal communication). The differences of the hydrogen sulphide concentrations in the various hot vents are assumed to be caused by the different quantities of metal sulphides and gypsum deposited prior to emission, and most likely relate to the age of individual vent systems.

Sulphur isotope analyses of dissolved hydrogen sulphide (Kerridge et al., 1983) and of various sulphide minerals precipitated at the vents (Styrt et al., 1981) indicate that the emitted sulphur derives largely (up to 90%) from leaching as primary sulphide from basalts (Arnold and Sheppard, 1981) and the smaller portion from the geothermal reduction of seawater sulphate. The latter is largely removed by deposition as anhydrite during the downward permeation into the crust and only the rest is reduced to sulphide. It has been suggested that the anhydrite subsequently redissolves in cooler waters when the crust moves off-axis (Wolery and Sleep, 1976; Humphris and Thompson, 1978; Mottl et al., 1979; Mottl, 1983). In laboratory experiments the inorganic reduction of seawater sulphate coupled with the oxidation of Fe^{2+} from basalt has been demonstrated at 300°C and above (Mottl et al., 1979; Shanks et al., 1981).

An estimate of the quantity of sulphur emission from an individual hot vent can be based on the following average values: (1) a hydrogen sulphide concentration of 5.0 mM, (2) a flow rate of 1 m/sec, and (3) a chimney orifice diameter of 4 cm (own observation). The calculation results in a value of 0.7 kg sulphur per hour or 6 tons per year emitted from an average-size black smoker.

Estimates on the global sulphur emission from deep sea vents can be made from calculations by Edmond et al. (1979, 1982) and McDuff and Edmond (1982). Based on the concentrations of several key minerals in seawater, they arrived at a figure of 7–8 million years for a total hydrothermal seawater cycling through the Earth's crust. Such estimates result in an upper limit of sulphate entrainment at 4×10^{12} mol per year, and an upper limit of hydrogen sulphide emission into the ocean of 1×10^{12} mol per year. In other words, there is a net loss of sulphur from seawater during hydrothermal cycling in the form of polymetal sulphide deposits.

The annual emission of 10^{12} moles or 32×10^6 tons of hydrogen sulphide corresponds to 5×10^6 average-size smokers as defined above. However, this total or global number of smokers must certainly be lower because the observed occurrence of extensive chemosynthetically supported animal populations (see below) at the spreading centres suggests that a considerable portion of the hydrogen sulphide is emitted via warm rather than hot vents (see also Mottl, 1983).

7.4 TRANSFORMATIONS OF SULPHUR

The conditions for the reaction of emitted hydrothermal fluid with oxygenated seawater are distinctly different for the hot and warm vents. The forceful dispersion of hydrothermal fluid from black smokers distributes particulate polymetal sulphides over a wide area. Most of the particles are in the μm-size range (M. Mottl, personal communication) and only a small fraction of this material is sedimented. The larger portion becomes part of deep- and midwater circulation. The sulphides will oxidize to elemental sulphur and ultimately to sulphate.

In the immediate surrounding of the warm vents little abiological or spontaneous-chemical oxidation of emitted hydrogen sulphide is evident from deposits. If Kevex X-ray spectra of microbial mats sampled from an active vent plume (*ca* 20°C) are compared with those of a mat further removed from the vent orifice (*ca* 2°C), a change in the iron/manganese ratio becomes apparent (Figure 7.2). It results from the different dissolution products of the two metal compounds leading to a successively changing deposition during the outward flow of vent water. In contrast, the amount of sulphur within the mat deposits is hardly affected by the distance from the vent orifice. The relatively low and non-toxic concentrations of hydrogen sulphide in the warm vent plumes are, in contrast to hot vents, caused by the higher subsurface metal sulphide depositions (Figure 7.1).

A subsurface biological oxidation of hydrogen sulphide is evident in some vents where a turbidity and whitish-bluish iridescence of the emitted fluid was microscopically and biochemically resolved to represent dense bacterial cell suspensions (Jannasch and Wirsen, 1979; Karl *et al.*, 1980; Jannasch, 1984a). A high physiological activity was indicated by considerable adenosine 5'-triphosphate (ATP) concentrations and guanosine 5'-triphosphate/ATP ratios (Karl *et al.*, 1980). Measurements of CO_2 uptake rates conducted *in situ* with the appropriate controls (at 1 atm, 2°C, 23°C, and in the presence of a thiosulphate enrichment) indicated that the microbial chemosynthetic activity was (a) highly barotolerant, (b) not psychrophilic (low temperature adapted), and (c) readily stimulated by an additional source of reduced sulphur (Jannasch, 1984b).

Although there are a number of non-sulphur energy sources for autochemolithotrophic bacteria present in vent water (Table 7.1), the relative abundance of sulphur compounds is assumed to lead to the predominance of sulphur-oxidizing bacteria (Jannasch, 1984a). In turn, by far the largest part of the chemosynthetic production of organic carbon proceeds via sulphur oxidation. The existence of whole ecosystems dependent on terrestrial (geothermal) rather than solar (light) energy is unique for the warm deep sea vents. It represents the biologically most remarkable finding of their discovery.

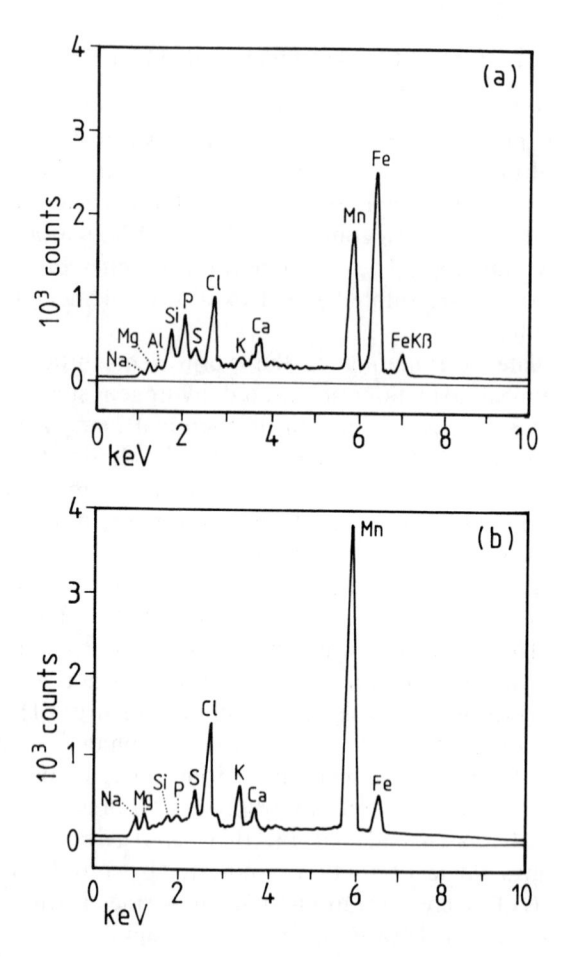

Figure 7.2 Spectra taken with a Kevex non-dispersive X-ray detector:
(a) from a microbial mat grown on a glass slide incubated for 10.5 months in an active vent fissure;
(b) from a microbial mat covering a mussel shell collected several metres away from the vent orifice (from Jannasch and Wirsen, 1981)

The biochemical versatility observed within the sulphur-oxidizing bacteria appears to be the key for the extraordinary efficiency of this process at the warm vents (Jannasch and Wirsen, 1985). The two most prominent characteristics are (1) the diversity of metabolic types, including the mixotrophic and facultative chemoautotrophic sulphur bacteria, and (2) the most unexpected symbiotic associations between certain invertebrates and chemolithoautotrophic bacteria. This symbiosis may account for up to 90% of the organic carbon production at the vents (Jannasch and Nelson, 1984).

Table 7.1 Types of bacterial chemotrophy*

	Electron donor	Electron acceptor	Carbon source	Organisms
Autolithotrophic	H_2	O_2	CO_2	Hydrogen-oxidizing bacteria
	S^{-2}, $S°$, $S_2O_3^{2-}$	O_2	CO_2	Sulphur-oxidizing bacteria
	Fe^{2+} (Mn^{2+})	O_2	CO_2	Iron and manganese-oxidizing bacteria
	NH_4^+, NO_2^-	O_2	CO_2	Nitrifying bacteria
	S^{2-}, $S°$, $S_2O_3^{2-}$	NO_3^-	CO_2	Denitrifying/sulphur-oxidizing bacteria
	H_2	$S°$, SO_4^{2-}	CO_2	Sulphur- and sulphate-reducing bacteria
	H_2	CO_2	CO_2	Methanogenic and acetogenic bacteria
Hetero-organotrophic	Organic substrate	O_2	Organic substrate	Aerobic heterotrophic bacteria
	Organic substrate	NO_3^-	Organic substrate	Denitrifying bacteria
	Organic substrate	$S°$, SO_4^{2-}	Organic substrate	Sulphur- and sulphate-reducing bacteria
	Organic substrate	Organic substrate	Organic substrate	Anaerobic heterotrophic and fermenting bacteria

*Modified after Jannasch and Wirsen (1985).

The discovery of symbioses between vent invertebrates and sulphur bacteria has subsequently led to similar findings with certain bivalves in shallow marine waters (Cavanaugh et al., 1981; Felbeck, 1981; Cavanaugh, 1983). In this environment, however, the generation of hydrogen sulphide is a result of bacterial sulphate reduction dependent on organic energy sources. The comparison between primary and secondary production of organic carbon with reduced sulphur functioning as the source of energy has been discussed at an earlier SCOPE Workshop (Jannasch, 1983).

Recently similar deep sea animal communities were discovered in the absence of hydrothermal seawater cycling (Paull et al., 1984). Saline seepages occur at the bottom of the escarpment of the western Florida continental platform at a depth of 3000 m. The extruding waters contain hydrogen sulphide and produce an invertebrate-dominated ecosystem of less spectacular productivity than that at the vents, but possibly stretched along the entire escarpment area. The source of sulphur might be the nearby sulphur domes, and its reduction is presently hypothesized to depend on the microbial oxidation of fossilized organic materials. For the discussion of global sulphur cycling, too little information is available on this new type of natural sulphur containing deep water vents at this time.

The role of high temperatures in the biological transformations of sulphur at deep sea vents is not yet understood. The isolation of a highly thermophilic methanogenic bacterium from the base of a black smoker (Jones et al., 1983) implies the possibility of reduction of elemental sulphur or possibly sulphate at high temperatures and pressures. Like many methanogenic bacteria (Stetter and Gaag, 1983), the above isolate is capable of reducing elemental sulphur anaerobically in the presence of hydrogen (W.J. Jones, personal communication). This chemolithoautotrophic sulphur and possibly sulphate reduction (included in Table 7.1) may be found to be an important process in deep sea vent sulphur transformations. The recent claim of bacterial growth at temperatures of 250°C (Baross et al., 1984) has not yet been substantiated by cultural or physiological studies.

ACKNOWLEDGEMENTS

The account of sulphur emissions was prepared with the kind help of M. J. Mottl. Our work on the microbiology of hydrothermal vents is supported by the National Science Foundation Grants OCE80-24253 and OCE83-08631. Contribution No. 5726 of the Woods Hole Oceanographic Institution.

ADDENDUM IN PROOF

Since this paper was submitted in December 1984, the 'unnamed' vent mussel has been described as *Bathymodiolus thermophilus* (Kenk, D.M. and Wilson, B.R. 1985. A new mussel (Bivalvia, Mytilidae) from hydrothermal vents in the Galapagos Rift zone. *Malacologia*, **26**, 253–71). Furthermore, the new observations on cold and slightly saline seepages in the deep sea have also been published (Paull, C.K., Hecker, B., Commeau, R., Freeman-Lynde, R.P., Neumann, C., Golubic, W.P., Hook, J.E., Sikes, E., and Curray, J., 1984. Biological communities at the Florida Escarpment resemble hydrothermal vent taza. *Science* (Washington), **226**, 911–33). It appears that the growth of mussels at this particular deep sea site is based on the symbiotic microbial oxidation of methane rather than sulphide.

REFERENCES

Arnold, M., and Sheppard, S.M.F. (1981). East Pacific Rise at latitude 21°N: isotopic composition and origin of the hydrothermal sulphur. *Earth Planet Sci. Lett.*, **56**, 148–56.

Ballard, R.D. (1977). Notes on a major oceanographic find. *Oceanus*, **20**, 25–44.

Baross, J.A., Deming, J.W., and Becker, R.R. (1984). Evidence for microbial growth in high-pressure, high-temperature environments. In: Klug, J., and Reddy,

C.A. (Eds.), *Current Perspectives in Microbial Ecology*, Amer. Soc. Microbiol Publ., Washington, pp. 186–95.
Boss, K.J., and Turner, R.D. (1980). The giant white clam from the Galapagos Rift *Calyptogena magnifica* species novum. *Malacologia*, **20**, 161–94.
Cavanaugh, C.M. (1983). Symbiotic chemotrophic bacteria in marine invertebrates from sulphide-rich habitats. *Nature*, **302**, 58–61.
Cavanaugh, C.M., Gardiner, S.L., Jones, M.L., Jannasch, H.W., and Waterbury, J.B. (1981). Procaryotic cells in the hydrothermal vent tube worm *Riftia pachyptila* Jones: possible chemautotrophic symbionts. *Science*, **213**, 340–1.
Corliss, J.B., Dymond, J., Gordon, L.I., Edmond, J.M., von Herzen, R.P., Ballard, R.D., Green, K., Williams, D., Bainbridge, A., Crane, K., and van Andel, T.H. (1979). Submarine thermal springs on the Galapagos Rift. *Science*, **203**, 1073–83.
Edmond, J.M., and Von Damm, K.L. (1983). Hot springs on the ocean floor. *Sci. Amer.*, **248**, 78–93.
Edmond, J.M., Measures, C., McDuff, R.E., Chan, L.H., Collier, R., Grant, B., Gordon, L.I., and Corliss, J.B. (1979). Ridge crest hydrothermal activity and the balances of the major and minor elements in the ocean: The Galapagos data. *Earth Planet. Sci. Lett.*, **46**, 1–18.
Edmond, J.M., Von Damm, K.L., McDuff, R.E., and Measures, C.I. (1982). Chemistry of hot springs on the East Pacific Rise and their effluent dispersal. *Nature*, **297**, 187–91.
Felbeck, H. (1981). Chemoautotrophic potentials of the hydrothermal vent tube worm, *Riftia pachyptila* (Ventimentifera). *Science*, **213**, 336–8.
Haymon, R.M., Koski, R.A., and Sinclair, C. (1984). Fossils of hydrothermal vent worms from Cretaceous sulphide ores of the Samail ophiolite, Oman. *Science*, **223**, 1407–9.
Humphris, S.E., and Thompson, G. (1978). Hydrothermal alteration of oceanic basalts by seawater. *Geochim. Cosmochim. Acta*, **42**, 107–25.
Jannasch, H.W. (1983). Interactions between the carbon and sulphur cycles in the marine environment. In: Bolin, B. and Cook, R.B. (Eds.) *The Major Biogeochemical Cycles and their Interactions*, Wiley, Chichester, pp. 517–25.
Jannasch, H.W. (1984a). Microbial processes at deep-sea hydrothermal vents. In: Rona, P.A. (Ed.) *Hydrothermal Processes at Seafloor Spreading Centers*, Plenum Publishing, New York, pp. 677–710.
Jannasch, H.W. (1984b). Microbes in the oceanic environment. In: Kelly, D.P., and Carr, N.G. (Eds.) *The Microbe 1984, Part III. Prokaryotes and Eukaryotes*, Soc. Gen. Microbiol. Symp., Cambridge University Press, pp. 97–122.
Jannasch, H.W., and Nelson, D.C. (1984). Recent progress in the microbiology of hydrothermal vents. In: Klug, M.J., and Reddy, C.A. (Eds.) *Current Perspectives in Microbial Ecology*, Amer. Soc. Microbiol. Publ., Washington, pp. 170–6.
Jannasch, H.W., and Taylor, C.D. (1984). Deep sea microbiology. *Ann. Rev. Microbiol.*, **38**, 487–514.
Jannasch, H.W., and Wirsen, C.O. (1979). Chemosynthetic primary production at East Pacific sea floor spreading centers. *BioScience*, **29**, 592–8.
Jannasch, H.W., and Wirsen, C.O. (1981). Morphological survey of microbial mats near deep sea thermal vents. *Appl. Environ. Microbiol.*, **41**, 528–38.
Jannasch, H.W., and Wirsen, C.O. (1985). The biochemical versatility of chemosynthetic bacteria at deep-sea hydrothermal vents. *Proc. Biol. Soc., Washington*, D.C. **6**, 325–34.
Jones, M.L. (1981). *Riftia pachyptila* Jones: observations on the vestimentiferan worm from the Galapagos Rift. *Science*, **213**, 333–6.
Jones, W.J., Leigh, J.A., Mayer, F., Woese, C.R., and Wolfe, R.S. (1983).

Methanococcus jannaschii sp. nov., an extremely thermophilic methanogen from a submarine hydrothermal vent. *Arch. Microbiol.*, **136**, 254–61.

Karl, D.M., Wirsen, C.O., and Jannasch, H.W. (1980). Deep-sea primary production at the Galapagos hydrothermal vents. *Science*, **207**, 1345–7.

Kerridge, J.F., Haymon, R.M., and Kastner, M. (1983). Sulphur isotope systematics at the 21°N site, East Pacific Rise. *Earth Planet. Sci. Lett.*, **66**, 91–100.

Lonsdale, P.F. (1977). Clustering of suspension-feeding macrobenthos near abyssal hydrothermal vents at oceanic spreading centers. *Deep-Sea Res.*, **24**, 857–63.

McDuff, R.E., and Edmond, J.M. (1982). On the fate of sulphate during hydrothermal circulation at mid-ocean ridges. *Earth Planet. Sci. Lett.*, **57**, 117–32.

Mottl, M.J. (1983). Metabasalts, axial hot springs, and the structure of hydrothermal systems at mid-ocean ridges. *Geol. Soc. Am. Bull.*, **94**, 161–80.

Mottl, M.J., Holland, H.D., and Corr, R.F. (1979). Chemical exchange during hydrothermal alteration of basalt by seawater—II. Experimental results for Fe, Mn, and sulphur species. *Geochim. Cosmochim. Acta*, **43**, 869–84.

Paull, C.K., Hecker, B., Commeau, R., Freeman-Lynde, R.B., Neumann, R.P., Corso, W.P., Golubic, S., Hook, J.E., Sikes, E. and Curray, J. (1984). Biological communities at the Florida escarpment resemble hydrothermal vent taxa. *Science*, **226**, 965–9.

Shanks, W.C. III, Bischoff, J.L., and Rosenbauer, R.J. (1981). Seawater sulphate reduction and sulphur isotope fractionation in basaltic systems: interaction of seawater with fayalite and magnetite at 200–350°C. *Geochim. Cosmochim. Acta*, **45**, 1977–95.

Spiess, F.N., MacDonald, K.C., Atwater, T., Ballard, R., Carranza, A., Cordoba, D., Cox, C., Diaz Garcia, V.M., Francheteau, J., Guerrero, J., Hawkins, J., Haymon, R., Hessler, R., Juteau, J., Kastner, M., Larson, R., Luyendyk, B., Macdougall, J.D., Miller, S., Normark, W., Orcutt, J., and Rangin, C. (1980). East Pacific Rise: Hot springs and geophysical experiments. *Science*, **207**, 1421–33.

Stetter, K.O., and Gaag, G. (1983). Reduction of molecular sulphur by methanogenic bacteria. *Nature*, **305**, 309–11.

Styrt, M.M., Brackmann, A.J., Holland, H.D., Clark, B.C., Pisutha-Arnond, V., Eldridge, C.S., and Ohmoto, H. (1981). The mineralogy and the isotopic composition of sulphur in hydrothermal sulphide/sulphate deposits on the East Pacific Rise, 21°N latitude. *Earth Planet Sci. Lett.*, **53**, 382–90.

Von Damm, K.L. (1983). Chemistry of submarine hydrothermal solutions at 21°N, East Pacific Rise and Guaymas Basin, Gulf of California. Ph.D. Thesis, WHOI/MIT, Woods Hole—Cambridge, Mass.

Wolery, T.J., and Sleep, N.H. (1976). Hydrothermal circulation and geochemical flux at mid-ocean ridges. *J. Geol.*, **84**, 249–75.

Evolution of the Global Biogeochemical Sulphur Cycle
Edited by P. Brimblecombe and A. Yu. Lein
© 1989 SCOPE Published by John Wiley & Sons Ltd

CHAPTER 8
Interaction of Sulphur and Carbon Cycles in Microbial Mats

Y. Cohen, V.M. Gorlenko and E.A. Bonch-Osmolovskaya

8.1 PHOTOSYNTHESIS IN CYANOBACTERIAL MATS AND ITS RELATION TO THE SULPHUR CYCLE

Cyanobacterial mats are organosedimentary structures composed primarily of benthic cyanobacteria together with diverse communities of microorganisms which trap, bind and precipitate sediment particles. Often cyanobacterial mats produce laminated sediments—biogenic stromatolites due to seasonal changes in the environment of deposition.

Cyanobacterial mats may serve as an ideal model for the study of basic mechanisms of the microbial sulphur cycle and the evolution of the sulphur cycle for several reasons:

(a) Cyanobacterial mats develop under environmental conditions excluding, or at least limiting, eukaryotic organisms such as grazing metazoans. This allows the study of microbial interactions with limited bioturbation.
(b) In most microbial mats the organic matter produced is autochthonous, allowing the study of *in situ* primary production and the coupled processes of mineralization of primary organic carbon of a known source. In contrast, in most other systems, part or all of the primary organic carbon is transported from other, often unknown, sources and is partially decomposed during transport under unknown conditions. This aspect is particularly important in the study of the decomposition and maturation of microbial mats as a model for oil shale formation (Aizenshtat *et al.*, 1984).
(c) Cyanobacterial mats are the oldest known biogenic sedimentary structures found as stromatolites dating back to 3.5×10^9 years ago. As such

*Yehuda Cohen.

stromatolites are the only fossil record which can be found throughout the entire geological era, even though they were most abundant during the Precambrian (Shopf, 1983).

Presently, cyanobacterial mats are confined to restricted habitats including hypersaline coastal marine environments, hot springs and alkaline lakes. Major marine cyanobacterial mats studied include: Solar Lake, Gulf of Aquaba (Krumbein and Cohen, 1974, 1977; Krumbein et al., 1977; Cohen et al., 1980; Jørgensen et al., 1979, 1983; Cohen 1984); Shark Bay, Western Australia (Logan et al., 1970; Playford and Cockbain, 1969; Bauld et al., 1979; Bauld, 1984); Spencer Gulf, South Australia (Bauld et al., 1979; Bauld, 1984); Laguna Figuerea, Baja California, Mexico (Stolz, 1984); Guerrero Negro, Baja California, Mexico (Javor and Castenholz, 1984a, 1984b) and many other mats in Bermuda, Persian Gulf; Bonair, Caribbean Sea; the Bahamas, and the USSR (Gorlenko et al., 1984). Hot spring cyanobacterial mats have been primarily studied in Yellowstone National Park, Wyoming, USA (Brock, 1978; Castenholz, 1984), Hunter's Spring, Oregon, USA (Castenholz, 1973) and other hot springs in the USA, New Zealand (Castenholz, 1976), Iceland (Schwabe, 1960; Castenholz, 1976), in the Far East Kamchatka (Gorlenko and Bonch-Osmolovskaya, 1986a) and in various hot springs in Europe. Cyanobacterial mats of alkaline lakes have been described in Ethiopia, USSR and various locations in the USA.

8.1.1 Oxygenic and anoxygenic photosynthesis

Cyanobacteria carry out oxygenic photosynthesis with two photosystems in series, in a manner identical to eukaryotic phototrophs. Water is the electron donor in oxygenic photosynthesis, and oxygen is the ultimate oxidation product. Cyanobacteria are similar to eukaryotic algae and plants in that they have chlorophyll a. In addition, the main light-harvesting pigments are phycobiliproteins, which are also found in eukaryotic red algae together with β-carotene and zeaxanthin as the most common carotenoids (Stanier, 1974).

The phototrophic green bacteria and the purple bacteria, in contrast, carry out anoxygenic photosynthesis using only one photosystem and then require electron donors of redox potential lower than that of water. The suitable donors are reduced sulphur compounds and CO_2. The photosynthetic pigments are various bacteriochlorophylls—a, b, c, d or e, together with a great variety of accessory carotenoids. Though many phototrophic bacteria can grow aerobically in a chemoheterotrophic model of growth, their long-

---

*The term 'sulphide' will be used to designate the total dissolved sulphide including H_2S, HS^-, and S^{2-}. The population of these species is pH-dependent.

term photosynthesis is generally restricted to anaerobic reduced environments where a suitable electron donor can be found. The physiological and ecological consequences of the differences between oxygenic photosynthesis of the cyanobacteria and the anoxygenic photosynthesis of the other phototrophs are profound. However, cyanobacteria often share the same ecological niche together with other phototrophs.

8.1.2 The sulphide-rich environment and toxicity

Cyanobacteria are sometimes found in environments with continuous exposure to sulphide, such as hot springs with a constant sulphide supply or in sulfureta, where sulphide is produced biogenically and where fluctuations of anaerobic and aerobic conditions are common. These oscillations may occur seasonally (Kuznetsov, 1970), or diurnally (Ganf and Viner, 1973) or even during shorter time periods (Jørgensen et al., 1979).

Life in sulphide-rich environments has both advantages and disadvantages. Sulphide is an effective reducing agent and provides the required low redox potential for growth under anaerobic conditions. Sulphide toxicity is quenched by chemical or biochemical combination with oxygen and oxidized compounds. This may reduce or even eliminate the oxygen effects of photoinhibition, photo-oxidation and photorespiration. Furthermore, sulphide serves as an electron donor to anoxygenic photosynthesis. It may also serve as an assimilatory sulphur source, especially in organisms lacking the capability of assimilating sulphate.

Sulphide is highly toxic for most microorganisms. It reacts with various cytochromes, haemoproteins and other compounds. It inhibits the electron transport chain, blocking respiration as well as oxygenic and anoxygenic photosynthesis. Sulphide toxicity is also apparent in organisms which are well adapted to sulphide-rich environments, but the sensitivity varies markedly among the different groups of microorganisms. The green phototrophic sulphur bacterium *Chlorobium* is the most tolerant. This organism may grow in the presence of as much as 8 mM sulphide (Pfennig, 1975). *Chloroflexus* is found in sulphide concentrations up to 1.5 mM (Castenholz, 1976). Purple sulphur bacteria (*Chromatium, Thiocystis,* and *Thiocapsa*) can tolerate 0.8–4 mM sulphide, whereas purple non-sulphur bacteria are less tolerant (0.4–2 mM).

8.1.3 Anoxygenic photosynthesis in *Oscillatoria limnetica* and other cyanobacteria

A flocculent mat from the bottom of the hypolimnion of Solar Lake, Sinai (Cohen et al., 1977b) is composed mostly of *Oscillatoria limnetica* together with the green sulphur bacterium *Prosthecochloris* sp. and the purple sulphur bacteria *Lamprocystis* sp. and *Chromatium* sp. Sulphide concentrations of up to 5 mM accumulate in this layer and persist for about ten months before holomixis occurs. Sulphide is oxidized during holomixis, and the sulphur-dependent phototrophic community disappears while *Oscillatoria limnetica* thrives under the oxic conditions which persist for two months before stratification sets in again (Cohen *et al.*, 1977a).

High primary production of up to 8 g (C) cm^{-2}d^{-1} was measured at Solar Lake under stratification (Cohen *et al.*, 1977b), of which about 60% may be attributed to the activity of *Oscillatoria limnetica* in the presence of sulphide.

This organism, though capable of oxygenic photosynthesis whenever sulphide is absent, switches off photosynthesis II (PS II) when sulphide is present and conducts anoxygenic photosynthesis (Cohen *et al.*, 1975a, 1975b). Very low concentrations (0.1–0.2 mM) of sulphide immediately inhibit the oxygenic system (Oren *et al.*, 1979). However, after two hours of exposure to light in the presence of high sulphide concentration (3 mM), photoassimilation reappears and is insensitive to the photosystem II inhibitor DCMU 3(3,4,dichlorophenyl) 1,1 dimethylurea (Cohen *et al.*, 1975a). The possible participation of PS II in anoxygenic CO$_2$ photoassimilation has been further excluded (Cohen *et al.*, 1975a) by the results of experiments in which PS II was simply not activated (rather than being inhibited) by the use of infrared light. Oxygenic photosynthesis requiring the operation in both photosystems decreased drastically in the infrared (red drop), whereas anoxygenic photosynthesis with sulphide was fully operative under such conditions. Furthermore, if both photosystems could contribute to the reaction, the enhancement in quantum yield would be predicted with respect to that obtained with only PS I in operation. However, the 'enhancement phenomenon' was observed only with oxygenic photosynthesis (Oren *et al.*, 1977). The new photosynthesis type is therefore anoxygenic, independent of PS II and driven by PS I with sulphide as electron donor.

Two hours' preincubation in the presence of sulphide and light is required for anoxygenic photosynthesis, indicating that induction may be involved. Indeed, the protein synthesis inhibitor chloramphenicol inhibits the initiation of anoxygenic photosynthesis (Oren and Padan, 1978). In order to survive sulphide inhibition of photosystem II an adaptation which allows the use of sulphide with photosystem I is necessary. We suggest that low redox leads to a sulphide resistant modification which enables sulphide electrons to be used. Protein synthesis occurs and electron carriers may be reduced. Addition of sodium dithionate eliminates the lag period observed when *Oscillatoria limnetica* is transferred to sulphide, possibly by providing the low redox potential needed for the reduction and modification of electron carriers.

Sulphide is oxidized to elemental S° by *Oscillatoria limnetica* according to the following stoichiometric relationship (Cohen *et al.*, 1975a):

$$2H_2S + CO_2 \rightarrow CH_2O + H_2O + 2S$$

Elemental sulphur was observed as refractile granules either free in the medium or adhering to the cyanobacterial filament. Otherwise the appearance of the cells under the electron microscope was the same under both anoxygenic and oxygenic conditions.

If sulphide is removed from sulphide-adapted cells, *Oscillatoria limnetica* immediately returns to oxygenic photosynthesis (Oren and Padan, 1978). This capacity must therefore be present in the cells. The photosynthetic system of *Oscillatoria limnetica* thus operates facultatively both oxygenically and anoxygenically.

When CO_2 is eliminated from the reaction system in *Oscillatoria limnetica*, sulphide donates electrons to hydrogen evolution (Belkin and Padan, 1979). This reaction occurs only in cells originally adapted to sulphide in the presence of CO_2. It is dependent on light and sulphide and is also insensitive to DCMU. Like the CO_2 photoassimilation reaction, it requires an induction period of two hours and is driven by PS I. Whereas CO_2 photoassimilation is fully induced after two hours of incubation in the presence of sulphide, the capacity of hydrogen evolution is very low. An additional 46 hours of incubation in the presence of sulphide is needed for a full induction of hydrogen evolution. Sulphide-dependent hydrogen evolution must require an additional step (S) to that required for the sulphide-dependent photoassimilation. Yet, addition of sodium dithionate eliminates the required incubation period, as in the case of anoxygenic photosynthesis (Belkin and Padan, 1984).

The concentration of light harvesting pigments (chlorophyll a, phycobilins and carotenoids) in the cell are identical under both oxygenic and anoxygenic photosynthesis (Oren *et al.*, 1977). Quantum yield spectra of the two photosynthetic modes show that, whereas only a narrow light band is efficiently utilized in the oxygen mode, the entire absorbed spectrum is used at high quantum efficiency in the anoxygenic mode. The drop in the quantum efficiency of oxygenic photosynthesis at both the blue and red ends of the visible spectrum is marked. This limited range of utilization of the visible light spectrum in oxygenic photosynthesis of *Oscillatoria limnetica* is similar to that of other cyanobacteria (Lemasson *et al.*, 1973). It is markedly different, however, from that of eukaryotic algae and plants which contain in addition chlorophyll b in their light harvesting system. In these organisms, almost the entire absorbed spectrum is utilized in oxygenic photosynthesis, with the exception of the far 'red drop' (Govindjee *et al.*, 1968).

A comparison of rates of anoxygenic photosynthesis as a function of sulphide concentration was made for 11 strains of cyanobacteria including the mat-forming *Lynogbia* 7104, *Aphanotheca halophilica* and *Oscillatoria limnetica* (Garlick *et al.*, 1977). In these organisms, the dependence on sulphide concentration was similar, generating an optimum curve rather than a saturation curve. The drop in the photosynthetic rates at higher sulphide concentrations is caused by sulphide toxicity effects on photosystem I. The maximal rates of oxygenic and anoxygenic photosynthesis are similar in both *Osc. limnetica* (1–2 μm (C) mg protein^{-1} h^{-1}) and *Aphanotheca halophilica* (0.5–1 μm (C) mg protein^{-1} h^{-1}). While the dependencies on sulphide are similar, both the affinities to sulphide and tolerances are different. Each strain exhibits a different range of sulphide concentrations at which anoxygenic photosynthesis can be performed. The sulphide ranges at pH 6.8 are 0.1–0.3 mM for *Lynogbia 7104*, 0.1–1.5 mM for *Aphanotheca halophilica* and 0.7–9.5 mM for *Osc. limnetica*. Furthermore, in each of these cases the range is markedly affected by the pH of the medium (Howsley and Pearson, 1979), which governs the dissociation of H$_2$S. Differences in pH-dependent sulphide ranges are known to constitute a determining factor in the ecology of photosynthetic sulphur bacteria (Baas-Becking and Wood, 1955; Pfennig, 1975).

The pattern of sulphide oxidation during anoxygenic photosynthesis of cyanobacteria seems to be thermodynamically less efficient than oxygenic photosynthesis. Photosynthetic bacteria oxidize sulphide to sulphate (Pfennig, 1975, 1977) gaining eight electrons for each H$_2$S molecule, while the cyanobacteria *Oscillatoria limnetica* and *Aphanotheca halophilica* remove only two electrons per molecule of sulphur. Elemental S° is the only end-product of anoxygenic photosynthesis in these cyanobacteria (Cohen *et al.*, 1975b). Elemental sulphur may also be intermediate in the sulphide oxidation process of several sulphur phototrophic bacteria (Trüper, 1973). Finally, other sulphur-containing electron donors which are utilized by photosynthetic bacteria do not seem to serve the photosynthetic system of sulphide-utilizing cyanobacteria (Gromet-Elhanan, 1977; Pfennig, 1975, 1977). *Anacystis nidulans*, however, can oxidize S$_2$O$_3^{2-}$ to SO$_4^{2-}$, yet S$_2$O$_3^{2-}$ cannot serve as the sole or even the major electron donor (Utkilen, 1976).

As in many photosynthetic bacteria and heterocystous cyanobacteria, H$_2$ has been shown to serve as an efficient electron donor for CO$_2$ photoassimilation in an anoxygenic reaction driven by photosystem I in both *Oscillatoria limnetica* and *Aphanotheca halophilica*. Hydrogen evolution dependent on sulphide oxidation has been demonstrated in both these strains (Belkin and Padan, 1979) and in heterocystous species (Bothe *et al.*, 1977; Tel-Or *et al.*, 1977; Weisman and Benemann, 1977).

In cyanobacteria, sulphide appears to be more toxic to oxygenic than to anoxygenic photosynthesis. In *Oscillatoria limnetica*, the former was inhibited by 0.1 mM sulphide (Oren *et al.*, 1979) while the latter was partially inhibited by 4 mM (at pH 6.8) (Garlick *et al.*, 1977). These differences have also been observed in other facultative anoxygenic strains. This difference in toxicity implies that at the higher range of sulphide concentrations, oxygenic photosynthesis will make only a minor contribution, if any, to photoassimilation of CO_2. However, at lower ranges, oxygenic photosynthesis may occur simultaneously with anoxygenic photosynthesis as demonstrated in *Oscillatoroa amphigranulata* (Utkilen and Castenholz, 1979) and in the chemocline of Solar Lake (Jørgensen *et al.*, 1979). Low sulphide concentrations (0.2 mM) are toxic to eukaryotic phototrophs and cyanobacterial strains that do not carry out anoxygenic photosynthesis (Castenholz, 1976; Knobloch, 1969; Howsley and Pearson, 1979).

The greater sulphide tolerance of anoxygenic photosynthesis operative with PS I is a selective advantage in sulphide-rich habitats. Some cyanobacteria cannot grow continuously under strictly anaerobic conditions but they can temporarily maintain themselves in the presence of sulphide. Other cyanobacteria are capable of permanent growth under these conditions. *Oscillatoria limnetica* thrives at the hypolimnion of Solar Lake under a 3cm column of sulphide-rich water and grew well in the laboratory with 3.5 mM sulphide and at pH 6.8 for a period of several months (Oren *et al.*, 1977). Facultative anoxygenic photosynthesis is clearly an advantage for life under prolonged periods of exposure to sulphide.

8.1.4 Sulphide–oxygen fluctuation in cyanobacterial mats

The *in situ* use of microelectrodes (O_2, sulphide, and pH) in various microbial mats demonstrated the establishment of a sharp redoxcline of less than 1 mm as a result of oxygenic photosynthesis of cyanobacteria at the proximity of the sulphide-rich microzone. The first measurements of microgradients of O_2, sulphide and Eh in microbial mats were carried out in the flat shallow cyanobacterial mats at Solar Lake, Sinai (Jørgensen *et al.*, 1979; Padan and Cohen 1982). This mat, dominated by the cosmopolitan mat-forming cyanobacterium *Microcoleus chthonoplastes*, is situated under 30 cm of oxic water. Extreme diurnal fluctuations were found, in the position of the O_2 maximum (0.5 mM at 1 to 2 mm depth at night). The O_2–H_2S interface migrates diurnally from the mat surface at night to 3 mm below surface at noon. The photic zone extends down to 2.5 mm below the surface and H_2S peaked right below the photic zone. H_2S and O_2 were found to coexist at 2.5 mm depth over a depth interval of 0.2–1 mm, with a turnover rate of less than a minute.

A more detailed study of the microstructures of the various microbial mats in Solar Lake and their O_2, H_2S and pH microprofiles was carried out (Jørgensen *et al.*, 1983). The dominant cyanobacterium in the flat and blister mats was *Microcoleus chthonoplastes* in association with filamentous flexibacteria, tentatively identified as *Chloroflexus*-like organisms. Similar associations were observed in cyanobacterial mats at Sabhat Gavish, southern Sinai (Gerdes and Krumbein, 1984), microbial mats in Laguna Figueroa,

Baja California, Mexico (Stolz, 1984), and in various microbial mats in Shark Bay, Western Australia, and Spencer Gulf, South Australia (Bauld, 1984). The flat mat from the most shallow parts of Solar Lake has a photic zone of merely 0.8 mm with a maximum photosynthetic activity of 50 mM O_2 cm^{-3} h^{-1} at 0.3–0.4 mm depth. In deeper waters, the mats are less compacted and the photic zones extend from 2.5 to over 10 mm with increasing depth of overlying water. In all the microbial mats examined, a pH shift of up to two units developed at the maximal photosynthetically active zone causing rapid deposition of $CaCO_3$ in this microzone (Revsbech et al., 1983).

In the hypolimnic flocculent mat at Solar Lake, Sinai, where *Oscillatoria limnetica* is the dominant cyanobacterium at 3 mM sulphide (Krumbein and Cohen, 1977; Krumbein et al., 1977), no O_2 could be detected at the photosynthetically active layer. Yet, this organism can produce O_2 well at higher light intensities, even at 3 mM sulphide (Cohen et al., 1985).

All cyanobacterial mats examined so far exhibit sharp microgradients of sulphide, O_2 and pH which fluctuate diurnally, thus exposing both the cyanobacteria and sulphate-reducing bacteria to alternating conditions of highly oxygenated water and elevated sulphide concentrations.

8.1.5 Strategies of oxygenic and anoxygenic photosynthesis in mat-forming cyanobacterial isolates

Axenic cultures of cyanobacteria isolated from various biotopes exposed to varying sulphide concentrations were examined for their capacity to carry out oxygenic and/or anoxygenic photosynthesis under a range of sulphide concentrations. Four different strategies of photosynthetic life under varying degrees of sulphide exposure were detected.

1. *Irreversible cessation of CO_2 photoassimilation upon brief exposure to sulphide*

Anacystis nidulans isolated from planktonic blooms is highly sensitive to sulphide. Exposure to 100 μm sulphide causes 50% inhibition of CO_2 photoassimilation, and at 200 μm sulphide it is blocked completely. When sulphide was removed after 2 hours of incubation in the presence of sulphide in the light, no regenerated photoassimilation could be detected. A similar sensitivity to sulphide was shown by Schwabe (1960) for *Mastigocladus*, the thermophilic cyanobacterium found under very low sulphide concentrations (up to 2 μm) in various hot springs in Iceland, New Zealand, and the USA (Schwabe, 1960; Castenholz, 1976, 1977; Brock, 1978).

2. *Enhancement of oxygenic photosynthesis upon exposure to sulphide and inability to utilize sulphide as an electron donor for anoxygenic photosynthesis*

This type of photosynthesis is represented by *Oscillatoria* sp. isolated from Wilbor Springs, California. This organism, which grows at 1 mM sulphide at neutral pH, shows 45 × enhancement of oxygenic photosynthesis upon exposure to 800 μm sulphide at neutral pH. With increasing sulphide concentrations CO_2 photoassimilation was gradually inhibited, yet no anoxygenic, DCMU-insensitive photosynthesis could be detected. Similar activities were reported for *Phormidium* sp. isolated from a hot spring at Yellowstone National Park by Weller et al., (1975).

3. *Enhancement of oxygenic photosynthesis at low sulphide concentrations, inhibition of photosystem II at higher sulphide concentrations and a concomitant induction of anoxygenic photosynthesis operating in concert with the partially inhibited oxygenic photosynthesis at higher sulphide concentration Microcoleus chthonoplastes*, representing this type of photosynthesis, is a cosmopolitan mat-forming cyanobacterium in hypersaline coastal lagoons. Isolates from Solar Lake and Sabhat Gavish, Sinai; Laguna Figueroa and Guerrero Negro salt pans, Baja California, Mexico; Spencer Gulf, South Australia; and Shark Bay, Western Australia all show virtually the same photosynthetic activity. The ultrastructures of the cyanobacterial mats dominated by *Microcoleus chthonoplastes* from Solar Lake (Cohen, 1984); Laguna Figueroa (Stolz, 1984), Shark Bay (Bauld et al., 1979; Bauld, 1984) and the Persian Gulf (Stolz, personal communication) are all extremely similar. The same type of photosynthetic activity was described by Utkilen and Castenholz (1979) for *Oscillatoria amphigranulata* isolated from alkaline hot springs in New Zealand at 2.2 mM sulphide.

4. *Complete reversible inhibition of photosystem II at low sulphide concentrations and induction of efficient anoxygenic photosynthesis at higher sulphide levels.*
This type of photosynthesis was initially described in *Oscillatoria limnetica* from Solar Lake by Cohen et al. (1975a), and later for other cyanobacteria by Garlick et al. (1977). Oren et al. (1979) demonstrated that, unlike types 2 and 3, photosystem II is completely blocked at 100 μM sulphide. When the cyanobacterium is exposed to high sulphide levels an induction period of 2 hours is needed for anoxygenic, photosystem I-dependent photosynthesis to be fully induced.

Recently this type of activity was also found in a hot spring isolate, *Oscillatoria* sp., growing in Stinky Spring, Utah at 1.1 mM sulphide and pH of 6.3.

Cyanobacteria exhibit a great degree of variability in their photosynthesis as a function of varying exposures to sulphide. These four strategies represent increasing degrees of adaptation to photosynthetic life under sulphide. Prolonged exposures to increasing sulphide concentrations result in the dominance of cyanobacterial strains which can better utilize sulphide, and hence one finds a gradual shift from type 1 to type 4.

Generally, type 1 are either not exposed at all to sulphide or exposed to negligibly low concentrations of it. Type 2 are found among cyanobacteria exposed to up to 1 mM sulphide at neutral or basic pH. Anoxygenic photosynthesis could not be demonstrated in these organisms, yet their oxygenic activity is enhanced at low redox potentials which allow protection from photoinhibition and increased efficiency of CO_2 photoassimilation. Types 3 and 4 develop at relatively high sulphide concentrations at neutral or acidic pH. Sulphide toxicity is pH-dependent; it is more toxic at lower pH. Presumably, H_2S penetrates passively through the cell membrane obeying diffusion laws, whereas the ionized forms, HS^- and S^{2-}, need active transport mechanisms. Thus, *Oscillatoria amphigranulata* occurring at 2.2 mM sulphide at alkaline pH is a type 3 organism, whereas *Oscillatoria* sp. from Stinky Spring, Utah, developing under 1.1 mM sulphide but at a pH of 6.3, is a type 4 cyanobacterium.

Photosystem II of the various types of cyanobacteria show different sensitivities to sulphide inhibition. Exposure to 50 μM sulphide at pH 7.5 induces maximal variable fluorescence of PS II in both *Anacystis nidulans* (type 1) and *Oscillatoria limnetica* (type 4). The addition of 10^{-4} M DCMU does not further enhance PS II fluorescence. Yet, in *Oscillatoria* sp. from Wilbor Springs, California (type 2), PS II variable fluorescence is only partially affected even at 1 mM sulphide and pH 7.5. Further addition of 10^{-4} M DCMU induces the maximal variable fluorescence of PS II. Similar results were obtained in cultures of *Synechococcus lividus* from Yellowstone National Park, *Oscillatoria sp.* from Stinky Spring, Utah, and the various *Microcoleus chthonoplastes* isolates from Solar Lake, Sabhat Gavish, Baja California and Spencer Gulf.

The ability of various cyanobacterial mat communities to produce and accumulate oxygen under increasing sulphide concentrations as well as the efficiency of recovery of oxygenic photosynthesis upon gradual removal of sulphide was measured by the introduction of O_2, S^{2-} and pH microelectrodes into small blocks of various cyanobacterial mats suspended in sulphide-containing media. Given sufficiently high light intensities, all cyanobacteria of the types 2, 3 and 4 are capable of producing and accumulating oxygen under sulphide. Yet, the level of sulphide at which oxygen production is detected changes significantly from one type to another. *Oscillatoria limnetica*—type 4—produces O_2 at only low sulphide concentrations of 50 μM, whereas *Microcoleus chthonoplastes*—type 3—undergoes efficient oxygenic photosynthesis at higher sulphide levels of 250 μM at neutral pH, and *Oscillatoria* sp. from Wilbor Springs—type 2—produce O_2 at much higher sulphide (i.e. of several mM) concentrations (Jørgensen et al,. 1983).

Various types of photosynthesis are found both in hot sulphur springs and various sulfureta. The degree of exposure to sulphide in sulfureta depends on the coupling of primary production and sulphate reduction—the only source of sulphide in these biotopes. The importance of sulphate reduction as the major process in the mineralization of the produced organic matter was demonstrated by Jørgensen and Cohen (1977) and by Skyring (1984).

8.1.6 Coupling of primary production and sulphate reduction in cyanobacterial mats

Since the coupling of the two processes in cyanobacterial mats must occur in close proximity to the photic zone, which is highly oxygenated during most of the day, the usual technique for the measurement of sulphate reduction cannot be accurately applied in these systems. Sulphide undergoes fast turnover due to efficient oxidation photosynthetically, chemolithotrophically, heterotrophically or even chemically in the presence of O_2. Hence, H_2S produced by sulphate-reducing bacteria cannot be quantitatively measured by injecting $Na_2^{35}SO_4$ into sediment cores. A new method has been developed in order to assess the degree of coupling of sulphate reduction to primary production in these systems (Cohen, 1985a). Silver wires, 0.15 mm in diameter and 1 cm in length, are coated with $Na_2^{35}SO_4$ of high activity and are introduced into the mat by means of a micromanipulator alongside the microelectrodes for O_2, sulphide and pH. The mat is then incubated either in the light or in the dark for periods of 10 minutes to 2 hours, after a period of preincubation under the same conditions. A series of silver wires are pulled out of the mat at different time intervals. They are autoradiographed for the activity of the $Ag^{35}S$, after washing to remove all remaining $^{35}SO_4^{2-}$. The vertical microprofiles of sulphate reduction are then compared to the microprofiles of O_2, sulphide and pH as well as to the profiles of oxygenic primary production. The silver wires are later cut into 1 mm segments and each is counted in a scintillation counter. This new method has several major advantages, but also presents some problems compared to the conventional method for the determination of sulphate reduction. The new method allows the determination of SO_4^{2-} reduction rate in close proximity to oxic microzones since the sulphide produced binds to the silver wire at high affinity and resists oxidation. Although there may not be an overall sulphide accumulation, as it is immediately oxidized, it can still be measured using this technique while this was impossible to determine in the previous method. The other advantage is the understanding of the microenvironmental conditions of sulphate reduction and a better insight of its coupling to primary production. The disadvantage of this technique is the difficulties in the quantification of the activities since diffusion of the labile SO_4^{2-} causes a decrease of specific activity with time. Therefore, the time dependence of the sulphate reduction activity is necessary to quantify its activity.

The new technique reveals an extremely tight coupling of sulphate reduction and primary production in the Solar Lake cyanobacterial mat. Not only can sulphate reduction be detected under the highly oxygenated conditions at the photic microzone of the cyanobacterial mat, but this activity is enhanced in the light where O_2 concentrations may increase to 4.5 × the saturation value at 1 atm O_2. Furthermore, the light spectrum that is most efficient in oxygenic photosynthesis is responsible for the induction of sulphate reduction. Specifically, wavelengths of 590–660 nm which are absorbed by the major light-harvesting pigment, phycocyanine, induce both primary production and sulphate reduction. The sulphate reduction activity under these conditions is fuelled directly by photosynthates excreted by the cyanobacterial oxygenic photosynthesis under high O_2 concentrations and high light intensities. The nature of these excretions is currently being studied.

Many sulphate-reducing bacteria have been shown to be strict anaerobes and highly sensitive to O_2 toxicity (LeGall and Postgate, 1973; Pfennig *et al.*, 1981). Yet, in the Solar Lake cyanobacterial mats, and probably in many other mat systems, these organisms operate well under periodic exposures to high O_2 concentrations. The mechanisms allowing sulphate reduction under these conditions are not fully understood. Preliminary results indicate the involvement of H_2 in the coupling of primary production and sulphate reduction. Several mat-forming cyanobacteria have been shown to produce H_2 under CO_2 limitations. Temporal CO_2 limitations may well occur at the photic microzone of the cyanobacterial mats due to the high specific rate of CO_2 photoassimilation creating high pH values (>9.5) and precipitation of $CaCO_3$ in this Ca-rich system.

8.1.7 Fe^{2+}-dependent photosynthesis in benthic cyanobacteria

The diurnal migration of the redoxcline through the photosynthetic layer assures the release of Fe^{2+} from the pool of monosulphide. High concentrations of dissolved Fe^{2+} were observed in the interstitial water in the upper 2 mm of the Solar Lake cyanobacterial mat in the morning hours. The ferrous ion is a good potential electron donor for photosynthesis in cyanobacteria and may thermodynamically operate well in the range of redox potential values of −50 mV to +50 mV, which is typical for the photic microzone during the diurnal transitions from fully reducing conditions at night to the high O_2 concentrations in the daytime. Banded microlayers of iron oxides are not very common in recent cyanobacterial mats, yet they are found in several mats in Spencer Gulf, South Australia and in Shark Bay, Western Australia as well as in several coastal lagoons such as the Sippiwissett Marsh at Cape Cod, USA. Iron-dependent photosynthesis of mat-forming cyanobacteria has long been speculated to be responsible, at least in part, for the extended deposition of Banded Iron Formations during the Precambrian (Knoll, 1979; Cloud and Gibor, 1983). This type of photosynthesis was speculated to be an important step in the evolution of oxygenic photosynthesis (Hartman, 1983).

Several cyanobacteria were examined for their capacity to use Fe^{2+} as an electron donor in photosynthesis. *Oscillatoria* sp. from Wiebor Springs and Stinky Hot Spring as well as all *Microcoleus chthonoplastes* isolates showed Fe^{2+}-dependent CO_2 photoassimilation (Cohen and Gack, 1985). Ferrous ion donates electrons primarily to photosystem II and this activity is thus DCMU sensitive. Ferrous ion initially blocks photosynthesis as indicated by a sharp temporal decrease in the PS II variable fluorescence in the presence of 10^{-5} M DCMU. Yet, after a short incubation in the presence of Fe^{2+}, the fluorescence reappears and with it an efficient rate of CO_2 photoassimilation and iron oxidation. Iron is oxidized through a ferritine intermediate. The end-product of the oxidation, iron oxide, is excreted outside the cell, similar to the excretion of S° in sulphide-dependent anoxygenic photosynthesis.

Preliminary results of the diurnal fluctuation of Fe^{2+} and $\delta^{13}C$ values for ΣCO_2 in the interstitial water in the upper 10 mm of the sediment core from the San Francisco Marsh, showed loss of Fe^{2+} from the photic zone during the day. There were two peaks of ^{13}C enrichment, one at the deepest part of the photic zone at 4–6 mm, and one at the surface. This indicated two microzones of autotrophic activities. The upper zone is clearly the result of oxygenic photosynthesis, whereas at the lower layer a Fe^{2+}-dependent CO_2 photoassimilation is inferred.

The importance of Fe^{2+}-dependent photosynthesis as an intermediate between sulphide-dependent activity and oxygenic photosynthesis is not yet understood.

8.1.8 Regulation of the sulphur cycle in recent cyanobacterial mats

Over 99% of CO_2 photoassimilation and SO_4^{2-} reduction as well as all sulphide oxidation processes in cyanobacterial mats take place in the upper 5–10 mm of the sediment. Any attempt to estimate the rates of processes in this system must be carried out on a microscale. Analysis of bulk samples must be cautiously treated.

In these microdimensions, very sharp gradients of O_2, sulphide, pH, Eh, Fe^{3+} and other parameters are established. These sharp gradients within a few millimetres are the result of highly specific photosynthetic activities, sulphate reduction and sulphide oxidation and heterotrophic microbial activity. The chemical microgradients fluctuate drastically daily and expose the microbial communities to 1 mM O_2, pH of 9–10, Eh of +200 mV, and over 10 mM Fe^{3+} at noon, in contrast to 5 mM sulphide, pH 6.5–7.0, and

Eh of -100 mV at night. Microorganisms must adapt to these drastic microenvironmental changes by developing elaborate regulatory mechanisms.

The regulation of photosynthetic mat-forming cyanobacteria has been studied in detail since 1974. Anoxygenic, sulphide-dependent photosynthesis and oxygenic photosynthesis under sulphide both play a major role in sulphide oxidation processes in these environments. Using our present knowledge we can predict which type of cyanobacteria will dominate a given biotope by measuring the diurnal microgradients of oxygen and sulphide (Cohen et al., in press).

Since iron serves as a major trap of sulphide, the regulation of Fe^{2+}, Fe^{3+} in sediments is highly interrelated to the sulphur cycle. The information on Fe^{2+} photosynthesis in cyanobacteria is presently inadequate to properly assess its ecological importance.

8.1.9 Precambrian cyanobacteria and stromatolites

Cyanobacterial mats are an extremely ancient phenomenon, documented in the oldest fossils known, dating 3.5×10^9 years ago. For the remaining period of the Archaean and throughout the Proterozoic era, up to 0.57×10^9 years ago, stromatolitic communities of cyanobacterial mats are the most abundant fossils. In the Palaeozoic era, the fossil record of stromatolites is limited and restricted to intertidal and supratidal hypersaline marine environments, thermal springs and alkaline lakes (Awramik, 1984), where they were protected from the newly evolved grazing activity.

Another major sedimentary record of the Proterozoic is widely spread deposition of finely laminated ferro–ferric oxides known as the Banded Iron Formations. The understanding of microbial activities in recent cyanobacterial mats may throw light on our understanding of evolution processes in the Precambrian. The recent mat-forming cyanobacteria present a protocyanobacterial group which differs markedly from modern planktonic cyanobacteria.

8.2 MICROORGANISMS OF THE SULPHUR CYCLE AND THEIR ACTIVITY IN MICROBIAL MATS OF HOT SPRINGS

8.2.1 Introduction

Certain aquatic microorganisms find a favourable ecological niche at solid–liquid interfaces where particular physicochemical conditions have been established (Marshall, 1980; Gorlenko et al., 1983). In waters lacking nutrients, such as springs, inorganic ions and organic ions may be concentrated

*V.M. Gorlenko and E.A. Bonch-Osmolovskaya.

on the sediment surface. This promotes the colonization of the spring bed by microorganisms which form benthic associations referred to as microbial mats. The composition of microbial mats depends upon the characteristics of the spring.

Microbial mats often contain a phototrophic component, mainly cyanobacteria, which are primary producers of organic matter (Bauld, 1981a; Krumbein, 1983). However, in some cases, phototrophic microorganisms are not found in microbial mats. Examples are sulphur mats of submarine vents (see Chapter 7) and regions near extremely hot sulphur spring orifices. In these systems, the primary producers are chemolithotrophic sulphur bacteria (Brock, 1978; Caldwell et al., 1976; Gorlenko et al., 1988).

According to Bauld (1981b) microbial mats are associations of microorganisms which colonize benthic surfaces and form adhesive, extended and often layered structures. The terms cyanobacterial and algabacterial mats should be used carefully, since they characterize only certain types of mats.

Emphasis has been placed on benthic microbial associations inhabiting shallow marine systems (Krumbein et al., 1979; Cohen et al., 1984), but information about mats in thermal springs is rather scanty (Brock, 1978; Cohen et al., 1984). These environments (dominant sulphide, CH_4, CO_2 and N_2) are of particular interest because they are believed to be similar to Precambrian systems (Awramic, 1984; Walter, 1976; Zavarzin, 1984).

Due to the diversity of types, springs are convenient natural model systems. It is possible to study individual factors which determine populations of species and regulate microbial processes, in different springs.

Hot springs are particularly interesting because high temperatures restrict the number of species (Brock, 1978; Gorlenko et al., 1985). Downstream, at lower temperatures more diverse microbial associations develop. Hence, one spring can provide considerable information about the structure of microbial communities.

The present paper considers microbial mats in thermal springs with high concentrations of sulphur compounds. The sulphide concentration determines species and regulates carbon and sulphur metabolism in these systems (Brock, 1978; Castenholz, 1984a).

8.2.2 Types of hot springs

The mineral composition of a given spring depends on the composition of the rocks through which the hydrothermal solution migrates. The total mineralization of springs varies over the range of 10^{-2} to 10^2 g(salt) l^{-1}. Therefore both fresh water and halophilic microorganisms can develop in different springs. In some mineral springs, the cation concentration sequence is $Na^+ > Ca^{2+} > Mg^{2+}$ (Miyaki, 1965), and the anion sequence may be $SO_4^{2-} > Cl^- > HCO_3^-$: $Cl^- > SO_4^{2-} > HCO_3^-$: or $Cl^- > HCO_3^- > SO_4^{2-}$. The second anion sequence resembles seawater in composition. The water of some hot springs contains high concentrations of heavy metals and toxic compounds such as Sb and As salts. These substances might influence the microbial population.

Hot springs can be divided into two main groups: alkaline to neutral and acid. Acid springs have unique microflora which have adapted to extreme conditions: pH (1–5), high temperature and H_2S, the latter being especially toxic at low pH values. Acid springs are found in zones of solfataric and fumarole activity. Here, the water may be acidified by bacterial oxidation of fumarole elemental sulphur to sulphate (Zavarzin, 1984).

Boiling acid sulphur springs are especially interesting since they are inhabited by archaebacteria including the sulphur oxidizers Sulpholobales and the elemental sulphur reducers, Thermoproteales. In acid hydrothermal springs, neither mats nor prokaryotic phototrophs have been found. However, the eukaryotic alga *Cyanidium acidocaldarium* may be found at 45–55°C (Brock, 1978). In contrast to acid springs, microbial mats may develop to a considerable thickness near the orifices of neutral or alkaline hot springs.

Hydrotherms are subdivided according to their gas composition (Zavarzin, 1984). Nitrogen–CO_2 thermal springs have highest temperatures and are formed by mantle emanations. Examples that have been studied microbiologically include springs in Iceland, North America (Yellowstone National Park), Kamchatka (caldera of the Uzon volcano) and Kunashir Island (the southernmost Kuril island). Nitrogen-rich hydrothermal springs are formed outside the regions of active volcanism and have lower temperatures. There are also CO_2 springs which are relatively cold. Rather rare hydrogen springs are found in Iceland.

Hydrogen sulphide, usually a product of mantle exhalations, is found in many types of springs: high concentrations are present in association with CO_2 in thermal springs in craters and calderas of volcanoes. Secondary biogenic H_2S is generated in springs flowing through sedimentary sulphate rocks with organic inclusions. Exogenic gases from deep crustal or upper mantle sources may serve as energy and carbon sources for microflora. Potential hydrogen donors are H_2, CH_4 and H_2S whereas CO_2 is an electron acceptor. Carbon dioxide (bicarbonate) and dissolved organic substances may be assimilated by microorganisms. Mantle gases usually contain no more than 0.05% oxygen, but downstream water may derive O_2 from the atmosphere and photosynthesis by cyanobacteria.

The microbial communities are relatively stable in neutral-alkaline springs because of the steady water inflow. Seasonal fluctuations of temperature, mineralization and other physicochemical factors are minimal (Castenholz, 1984a). Photo- and chemolithotrophic microorganisms dominate these waters because the small quantities of nutrients, especially organic matter, do not meet the requirements of most heterotroph species (Campbell, 1983).

However, excreted organic matter and decaying cells create conditions suitable for the development of some heterotrophic microorganisms (Ward et al., 1984). Therefore, hot springs are characterized by a great diversity of microorganisms dependent upon factors such as concentrations of H_2S, pH and salinity.

8.2.3 Microorganisms of the sulphur cycle

8.2.3.1 Cyanobacteria

Cyanobacteria are primary producers which form the basis of most hot spring microbial mats at temperatures below 74°C and pH higher than 5.5 (Castenholz, 1984a). The unicellular bacterium *Synechococcus lividus* was found at 74°C in alkaline springs of Yellowstone National Park. One of the isolated strains was cultured at 63–67°C (Meeks and Castenholz, 1971). In nature, other species of *Synechococcus* are found at lower temperatures such as *S. elongatus* (66–70°C) and the widely spread *S. minervae* (60°C) which prefers hot springs with low sulphide content (Castenholz, 1969; Gerasimenko et al., 1983). Usually these species develop jointly with the filamentous *Chloroflexus* which imparts rigidity to the mat. Other species of cyanobacteria colonize low-temperature (45–55°C) zones of thermal springs. Filamentous cyanobacteria of the genera *Oscillatoria, Spirulina* and *Phormidium* are also widespread. Some species tolerate temperatures as high as 60°C, e.g. *Oscillatoria okenii, Spirulina* sp., *Phormidium laminosus*. Highly differentiated cyanobacteria which are capable of N_2 fixation (e.g. *Mastigocladus luminosus*) occur in many low sulphide springs of the USA, New Zealand, Iceland and the USSR (Castenholz, 1969).

In 1975, it was shown for the first time that *Oscillatoria limnetica* may completely switch to anoxic photosynthesis consuming dissolved sulphide as a hydrogen donor (Cohen et al., 1975a, 1975b) and oxidizing it to extracellular S^0. It was further shown that cyanobacteria can be divided into four groups, on the basis of their tolerance to sulphide and ability to use sulphide as an H-donor (Castenholz, 1973, 1976, 1977; Garlick et al., 1977; Howsley and Pearson, 1979; Cohen et al., 1984). Various cyanobacteria from hot spring environments are listed in their respective groups in Table 8.1. It is seen that a large variety of bacteria inhabiting hot sulphur springs participate actively in the sulphur cycle. In some species, sulphide oxidation occurs by anoxic photosynthesis similar to bacterial photosynthesis, whereas in others it occurs with the help of O_2 released by O_2 photosynthesis.

Cyanobacteria performing anoxic photosynthesis are able to compete for H-donors with green and purple bacteria which also inhabit microbial mats of sulphur springs. The role of thiophilic cyanobacteria in the sulphur cycle is not confined to oxidative processes. Some, belonging to the fourth group,

Table 8.1 Cyanobacteria–sulphide relation in hot springs

Species	[HS$^-$] in nature	Response* to HS$^-$	Reference
Mastigocladus laminosus	2 µM	I	Castenholz, 1976; 1977; Gerasimenko *et al.*, 1983
Synechococcus lividas	trace	II	Castenholz, 1977; Brock, 1978
Oscillatoria sp.	6 µM	II	Cohen (this volume, Section 8.1)
Phormidium sp.	—	II	Weller *et al.*, 1975
Oscillatoria amphigranulata	2.2 µM	III	Cohen (this volume, Section 8.1)
Phormidium amphigranulata	0.1 µM	III	Bildushkinov and Gerasimenko, 1985
Oscillatoria sp.	1.1 µM	IV	Cohen (this volume, Section 8.1)

* I—HS$^-$ is not used as H-donor by cyanobacteria: they are highly sensitive to HS$^-$.
II—HS$^-$ is not used as H-donor but it stimulates oxygenic photosynthesis.
III—HS$^-$ is used as H-donor; photosystem II is tolerant to [HS$^-$].
IV—HS$^-$ is used as H-donor; photosystem II is sensitive to low [HS$^-$].

are able to consume S^0 as an electron acceptor for oxidation of endogenic substrates under dark anaerobic conditions (Oren and Shilo, 1979). As a result of 'sulphur respiration', elemental sulphur is reduced to sulphide. Sulphur respiration has also been found in *Chloroflexus* and in purple sulphur bacteria (Van Gemerden, 1968).

Organisms in microbial mats can also participate in the sulphur cycle by assimilatory reactions. Like other prokaryotes, cyanobacteria are able to assimilate sulphate. Some of them prefer to utilize more reduced products for assimilatory purposes. It has been shown that *Oscillatoria amphigranulata* can utilize S^0 as well as sulphate during phototrophic growth (Castenholz and Utkilen, 1984). Sulphur accumulated by such microorganisms undergoes further transformation during the decomposition of dead cells and diagenesis.

8.2.3.2 Anoxic phototrophs

The green filamentous bacterium *Chloroflexus* is a common component of microbial mats in hot springs with low mineralization (Pierson and Castenholz, 1974; Bauld and Brock, 1973; Castenholz, 1984a, 1984b; Gorlenko *et al.*, 1985). This organism is capable of phototrophic growth when consuming sulphide or organic compounds as electron donors and chemotrophic growth in the dark (Madigan and Brock, 1973, 1975; Madigan *et al.*, 1974). The most rapid growth for all isolated strains occurs photoheterotrophically

(Castenholz, 1984a, 1984b). *Chloroflexus*, considered to be the most active consumer, uses both excretion products of oxygenic phototrophs, i.e. cyanobacteria (Bauld and Brock, 1974) and decomposition products of dead microorganisms (Ward et al., 1984). Bildushkinov and Gerasimenko (1985) are of the opinion that *Chloroflexus* plays an important role in the oxidation of organic matter with O_2, since it inhabits the interface between the anaerobic and aerobic zones. *Chloroflexus* occurs both in high sulphide springs and in hydrothermal springs containing trace sulphide. Evidently the role of *Chloroflexus* in different mats varies depending upon redox conditions, availability of sulphide, and light intensity (Doemel and Brock, 1977). At high sulphide concentrations in mats which contain *Chloroflexus* as the sole phototrophic component (Castenholz, 1976, 1984b) sulphide inhibits many species of cyanobacteria. Therefore, in a number of sulphur springs, only *Chloroflexus* develops, and its upper temperature limit (65–66°C) is lower than that for low sulphide springs (72°C). Anaerobic mats are common in springs of Iceland, New Zealand and some regions of North America. Primary production in anaerobic mats is maintained exclusively at the expense of anoxic phototrophic green bacteria. Elemental sulphur produced by photo-oxidation serves as terminal electron acceptor for anaerobic destructors, such as sulphur-reducing bacteria. Since S^0 does not accumulate in macrobial mats, one can assume that sulphur reduction is a basic removal process.

Anaerobic sulphur mats of hot springs are characterized by relatively simple interdependent cycles of carbon and sulphur. They can be regarded as the most probable analogues of ancient mats or stromatolites which existed before the appearance of O_2 photosynthesis (Awramic, 1984).

It is noteworthy that strains isolated from the upper part of anaerobic mats cannot develop under anaerobic conditions in the dark (Castenholz, 1984b). It is still unknown whether these strains are ecotypes of *Chlorofexus aurantiacus* or an independent species.

In warm sulphur springs (28–40°C), there are filamentous green bacteria, e.g. *Oscillochloris trichoides*, containing gas vacuoles (Gorlenko and Korotkov, 1979). This green bacterium occurs in springs of Kamchatka, Dagestan, and Lenkoran (Gorlenko et al., 1985) and has been studied in a monoculture (Gorlenko and Korotkov, 1979). It has been shown that the physiology of *O. trichoides* is similar to *Chloroflexus*. However, *C. aurantiacus* and *O. trichoides* have ecologically different temperature regimes. *Oscillochloris* tends towards anaerobiosis and photoautotrophy and exhibits a greater tolerance to sulphide (204 mM). During photosynthesis, sulphide is oxidized by *Oscillochloris* to S^0. Field studies show that *O. trichoides* inhabits low temperature sulphur mats often beneath purple sulphur bacteria. From these observations one can assume that photosynthesis by *O. trichoides* depends upon sulphide ascending from lower layers of the mat.

Purple sulphur bacteria have often been observed in hot springs (Miyoshi, 1897; Castenholz, 1969, 1977). Madigan (1984) isolated the thermophilic strain *Chromatius* strain MC from a reddish bacterial mat embedded in carbonate sinter of a sulphide thermal spring (45°C) in the Upper Terrace area of Mammoth hot springs, Yellowstone National Park, Wyoming, USA. The optimal temperature range of this organism in laboratory cultures was found to be 48–50°C and the upper limit, 58°C. *Chromatium* strain MC morphologically resembles the smaller species *C. vinosum*. It differs from the other mesophilic species by its inability to grow on organic substrates in the absence of sulphide and CO_2. The thermophilic culture also differs from *C. vinosum* in carotenoid composition, its basic carotenoid being rhodovibrin.

Other thermophilic strains of *Chromatium* were obtained from thermal springs in New Mexico (47–48°C) (Madigan, 1984). Evidently, purple sulphur bacteria are widely spread in springs enriched in sulphide at temperatures of 40–58°C. Mesophilic *Chromatium* are often found in mats of hot springs in the temperature range 28–45°C (Gorlenko et al., 1985).

Hot springs of high salinity in Dagestan in the Caucasus (Baftugai springs; mineralization, 70–100 g salt l^{-1}; pH, 6.5–7.0; sulphide, 30–40 mg l^{-1}; 45–52°C) contain sulphur mats with a red-brown layer consisting of *Ectothiorhodospira* sp. (authors, unpublished data). Some species of this genus (*E. halophila* and *E. abdelmalekii*) develop at elevated temperatures (45–47°C) in shallow saline water basins and soda lakes (Raymond and Sistrom, 1969; Imhoff et al., 1978).

Purple sulphur bacteria in microbial mats utilize dissolved sulphide during anoxic photosynthesis. Since they occupy the upper boundary of the anaerobic zone, it is evident that they primarily consume biogenic sulphide arriving from lower layers of the mat. It is well known that many small forms of Chromatiaceae can grow aerobically in the dark (Bogorov, 1974; Gorlenko, 1974; Kämpf and Pfennig, 1980). Therefore, one can assume that during the dark and perhaps at low light intensities during the day, purple bacteria oxidize sulphide by chemosynthesis. Thus they perform the same function as colourless chemosynthetic bacteria in the community. In summary, the role of purple bacteria as producers of organic matter is very important since they function both photosynthetically and chemosynthetically.

Non-sulphur purple bacteria are widely spread in thermal springs at moderate temperatures (28–55°C) (Gorlenko et al., 1985). *Rhodopseudomonas palustris*, *Rhodobacter capsulatus*, and *Rhodocyclus gelatinosus* are frequently encountered (Table 8.2). The first two are able to consume sulphide at low concentrations both in light and the dark (Gorlenko, 1981).

Hence, some species of non-sulphur purple bacteria may participate in sulphur metabolism in microbial mats, performing the function of sulphur bacteria. The physiology of non-sulphur purple bacteria resembles that of

Table 8.2 Temperature limits for anoxic phototrophic bacteria in some springs

Species	\	\	Temperature (°C)	\	\	\	\
	10	20	30	40	50	60	70
Phodopseudomonas viridis		*	*	*			
Phodopseudomonas palustris	*	*	*	*	*	*	
Rhodobacter capsulatus			*	*	*	*	
Rhodocyclus gelatinosus		*	*	*	*	*	
Thiocapsa roseopersicina	*	*	*				
Chromatium vinosum		*	*	*			
Chromatium minutissimus			*	*			
Chlorobium limicola			*	*			
Chloroflexus aurantiacus			*	*	*	*	*
Oscillochloris trichoides	*	*	*	*			
Rhodospirillum mediosalinum				*	*		
Rhodobacter sp.			*	*			
Ectothiorhodospira sp.			*	*	*		
Prosthecochloris sp.			*	*	*		

Chloroflexus aurantiacus. The temperature ranges for *Chloroflexus* and non-sulphur purple bacteria barely overlap. So it is reasonable to conclude that as the temperature decreases downstream, *Chloroflexus* is replaced functionally by mesophilic species of non-sulphur purple bacteria (Gorlenko et al., 1985).

In the Dagestan saline springs, the sulphide-tolerant halophilic non-sulphur bacterium *Rhodobacter* sp. (similar to *R. euruhaline*) coexists with *Ectothiorhodospira* sp. (Kompantseva, 1985). *Rhodobacter* sp. easily switches from photoheterotrophic to photoautotrophic metabolism, consuming sulphide as an electron donor in the latter state. Extracellular S^0 is the main oxidation product. This purple bacterium functions like *Chloroflexus* which has never been found in saline springs.

All known species of green sulphur bacteria (Chlorobiaceae) develop only in stable anaerobic ecosystems (Pfenning, 1967; Gorlenko et al., 1984). These obligate photolithoautotrophic bacteria utilize sulphide as H-donor. They can only use certain organic acids together with CO_2, and in the presence of sulphide they utilize H_2. These highly specialized organisms are restricted ecologically. They must compete for sulphide with the more universal purple bacteria and some non-sulphur purple bacteria. However, their chlorosomes, containing particularly bacteriochlorophylls c, d and e which absorb light with wave lengths from 730 to 760 nm, allowing them to occupy an independent niche. Green sulphur bacteria are widely distributed in cold, high sulphide springs. A high sulphide tolerance facilitates their development at sulphide concentrations of 200–300 mg l^{-1} without extra specific competition (van Neil, 1931; Pfennig, 1967, 1975, 1978). Such sulphide concentrations at pH 6.5–7.0 inhibit the growth of other phototrophs including purple bacteria. It is also remarkable that green bacteria can develop in association with sulphate- and sulphur-reducing bacteria at very low sulphide concentrations (Pfennig, 1978).

Gorlenko et al., (1985) isolated *Chlorobium* at temperatures below 45°C in thermal sulphide springs at Lenkoran, Dagestan, and Kamchatka in line with Kaplan (1956) who concluded that they grew above 42°C. In the hot saline springs of Dagestan, the temperature range for green sulphur bacteria is wider, the upper limit being 52–55°C. Here only the species *Prosthecochloris* sp. has been found. It is morphologically identical to *P. aestuarii*, which is widespread in seas and shallow saline basins. Green sulphur bacteria may occupy a narrow microzone adjacent to anaerobic destructors in highly developed multi-component microbial mats in warm springs.

8.2.3.3 Non-photosynthetic aerobic bacteria of the sulphur cycle

In microbial mats of sulphide-containing hot springs, aeration and O$_2$ photosynthesis may create aerobic conditions, especially during the day (Revsbech and Ward, 1984). The availability of both sulphide and O$_2$ is favourable for the development of chemolithotrophic sulphur bacteria.

The thermophilic facultative autotrophic thiobacterium *Thiobacillus thermophila* was isolated from a hot spring in Yellowstone National Park (Golovacheva, 1984). The growth of this bacterium occurs at 35–55°C, the optimum being 50°C (at pH 5.6). Of particular interest are sulphur bacteria such as *Thermothrix thiopara* which can develop at temperatures above 55°C (Caldwell et al., 1976). In nature, *Thermothrix* exists at 72°C, pH 7.0 and a sulphide concentration of 17.4 µM (0.57 mg l^{-1}). In the laboratory, it grows between 67 and 77°C, 70–73°C being the optimum at pH 7.0. This facultative autotroph, neutralophile and facultative anaerobe capable of denitrification utilizes sulphide or thiosulphate substrates. In addition, it may develop heterotrophically on amino acids and simple media aerobically or anaerobically, reducing nitrate (Brannan and Caldwell, 1980). *Thermothrix thiopara* exhibits polymorphism. In hot springs, it grows as slimy interwoven filaments containing sulphur particles, *Thermothrix* communities may consist of several morphologically different bacteria: filamentous, small rods and cocci (Balashova, 1986).

Some authors believe that *Thermothrix thiopara* was described earlier as *Thiospirillum pistiense*. These microorganisms were observed in thermal sulphide springs of Japan, and in the Ozerkovo and Yuzhno-Koshelevo hot springs in Kamchatka (Kuznetsov, 1955).

Microorganisms similar to *Thermothrix* are found in hot sulphur springs of caldera of the Uzon volcano in Kamchatka (Gerasimenko *et al.*, 1983; Gorlenko *et al.*, 1987a). In the Thermophilny spring, sulphur bacteria dominate and develop as interwoven filaments covered by amorphous sulphur (66–70°C; pH, 6.3–6.7; sulphide, 10–16 mg l^{-1}; trace O$_2$). There are also abundant large curved bacteria belonging to the genus *Macromonas*. Thermophilic (65°C) microorganisms morphologically similar to *Thermothrix thiopara* and a thermophilic variant of *Thiobacterium bovista* were isolated from thermal sulphur mats. Small rod-shaped cells of these sulphur bacteria were submerged in globular or dentritic gel-like masses. Cells grew on a mineral medium containing 1% Na$_2$S$_2$O$_3$ at 65°C and a pH of 4.5 (Balashova 1985, 1986).

8.2.3.4 Anaerobic microorganisms of the sulphur cycle

Thermodesulfobacterium commune is at present the only sulphate-reducing organism isolated from very hot springs (Zeikus *et al.*, 1983). It was found in different springs (50–72°C) and in cyanobacterial mats at 59°C. *Th. commune* is a small gram-negative rod able to grow on lactate and pyruvate (more slowly on acetate and hydrogen) in the presence of sulphate. Its temperature optimum is around 70°C. However, it grows between 45 and 85°C. *Th. commune* is not an archaebacterium but it possesses some unique properties, e.g. lipids in this organism are branched diesters of glycerol. It contains cytochrome C$_3$, no desulphoviridine, and differs from other sulphate-reducing bacteria by its low content of G+C nucleotide pairs in DNA (34.4 mol %).

Microbial mats are probable sites for sulphur-reducing bacteria but information about these organisms is scanty. An organism identified as a thermophilic variant of *Pseudomonas mendocina* (Balashova, 1986) was isolated from a *Thermothrix* association. This facultative aerobe can utilize different inorganic compounds, including S^0 as an electron acceptor. In springs with a high colloidal sulphur content near the Uzon volcano, there is a predominance of obligate anaerobic microflora in cyanobacterial mats (Bonch-Osmolovskaya, 1986). The number of bacteria reducing sulphur during caseine hydrolysate oxidation was found to be very high (10^{10} cell cm^{-3}). The dominant forms are small obligate anaerobic rods. They reduce sulphur and are resistant to streptomycin.

Organisms utilizing other organic substrates to reduce sulphur in thermophilic cyanobacterial associations have not been found as yet. Undoubtedly, they would be of much interest since the known sulphur-reducing bacteria are either mesophiles (Pfennig and Biebl, 1976) or extreme thermophiles (optimal temperatures for growth >80°C) (Stetter, 1985). The latter are Thermoproteales belonging to archaebacteria with their distinct structural and molecular organization as Thermoproteales (Zillig *et al.*, 1981a). Extreme thermophilic sulphur-reducing archaebacteria belonging to the genera *Thermoproteus* (Zillig et al., 1981b). *Thermofilum* (Zillig et al., 1983) and *Desulfurococcus* (Zillig et al., 1982) were isolated from thermal springs in Iceland. The first two are obligate sulphur reducers. The latter can grow by sulphur respiration and anaerobic decomposition of organic matter in the absence of sulphur. Thermoproteales are capable of both lithoautotrophic growth by hydrogen oxidation (Fisher et al., 1983) and organotrophic growth. The range of organic substances consumed is very wide (Zillig et al., 1981).

A new extreme thermophilic archaebacteria was isolated from Kamchatka hot springs—*Desulfurococcus amylolyticus*—with a temperature optimum at about 90°C (Bonch-Osmolovskaya et al., 1988). The lower temperature limit for Thermoproteales is 70°C, but this temperature is too high for their participation in cyanobacterial associations. Perhaps Thermoproteales associates with *Thermothrix thiopara* but this has not been demonstrated. A high number of extreme thermophilic bacteria (10^5–10^7 cells cm^{-3}) capable of utilizing casein hydrolysate for sulphur reduction were found in the caldera of the Uzon volcano (Bonch-Osmolovskaya and Svetlichny, 1987). Evidently their substrate is dissolved organic matter arriving from low temperature zones by thermal convection. Therefore, extreme thermophilic sulphur reducers can interact indirectly with cyanobacterial associations and couple the anaerobic cycles of carbon and sulphur in very high temperature ecosystems.

8.2.4 Extreme factors in the formation of microbial mats in hot sulphur springs

In the following sections we discuss how the structure and function of sulphur microbial mats in hot springs are controlled by extremes of temperatures, sulphide concentrations, pH and salinity.

8.2.4.1 Temperature

Temperature greatly affects the diversity of microbial associations in hot springs (Brock, 1978). In alkaline and neutral low mineralized springs (temperature over 72°C), phototrophic microorganisms are absent. In high temperature sulphur springs, sulphur oxidizers and reducers develop. The primary producers of organic matter in sulphur mats are colourless sulphur bacteria which develop in microaerophilic conditions. At extremely high temperatures, the most abundant sulphur bacteria are *Thermothrix thiopara*, *Thiobacterium bovista* and some thionic bacteria (Balashova, 1985; Brock, 1978).

Under natural conditions (e.g. Thermophilny spring), sulphur bacteria associate with sulphur-reducing microorganisms. The conditions for the development of strictly anaerobic bacteria are created inside slimy structures during degradation of sulphur bacterial biomass (Gorlenko et al., 1987a).

Therefore, in high temperature regimes of hot sulphur springs, unique mats develop which consist of microorganisms highly dependent on sulphur compounds. Biomass accumulates due to chemosynthesis by sulphur bacteria. Sulphur, and probably sulphate-reducers, are the dominant destructors. Oxygen plays an important role as an electron acceptor during bacterial oxidation of sulphur compounds. However, in the Thermophilny spring *Thermothrix* grows abundantly at an extremely low O_2 content (less than 1 mg l^{-1}). There are always phototrophic microorganisms in microbial mats in sulphur springs at temperatures below 72–74°C. These include unicellular and filamentous thermophilic cyanobacteria, as well as the green filamentous *Chloroflexus surantiacus*.

As mentioned previously, the distribution of some species and groups of cyanobacteria is regulated by their temperature tolerance and strategy in relation to sulphide (Table 8.1). Among anoxic phototrophs, the most thermophilic is *Chloroflexus aurantiacus* (Table 8.2). This microorganism dominates in all springs with low mineralization.

Purple non-sulphur and sulphur bacteria as well as green sulphur bacteria are found in microbial mats below 60–65°C. Their density can be very high (10^2–10^7 cells l^{-1}) (Gorlenko et al., 1985). The diversity of phototrophs increases gradually as the temperature decreases to 45°C (Figure 8.1, Gorlenko et al., 1985). If the index of diversity (Odum, 1983) is zero at temperatures higher than 60°C, then it gradually decreases from 0.27 to 0.5 over the range of 60 to 45°C. At lower temperatures, the variability index

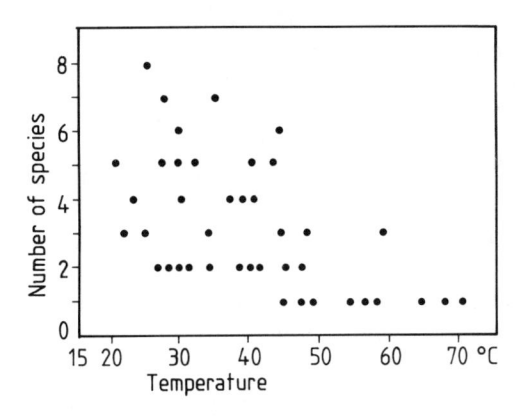

Figure 8.1 Number of species of phototrophic bacteria in 41 different microbial mats at different temperatures

fluctuates irregularly from 0.52 to 0.83. In the temperature range of 20 to 45°C, the phototrophic community is most diverse.

Gorlenko et al. (1985) identified a number of species of phototrophic bacteria in hot springs of the USSR (Table 8.2). It is noteworthy that at elevated temperatures (45–60°C), non-sulphur purple bacteria were found (*Rhodopseudomonas palustris, Rhodobacter capsulatus, Rhodocyclus gelatinosus, Chromatium vinosum, Rhodospirilum mediosalinium*). Below 45°C, the purple *Rhodopseudomonas viridis*, a number of species of purple and green sulphur bacteria, and the green filamentous *Oscillochloris trichoides* were found.

Thus, temperature exerts a significant effect on the composition of microflora in hot spring microbial mats. It is evident, however, that below 45°C, its influence is minimal and the structure of microbial associations is determined by other parameters.

8.2.4.2 Sulphide concentration

Even trace concentrations of sulphide completely exclude certain cyanobacteria in microbial mats (Castenholz, 1973). The effect of increasing sulphide concentrations on other cyanobacteria is highly species-dependent (see Section 8.1). *Chloroflexus aurantiacus* exhibits rather a low tolerance of sulphide (1–2 mM) which is comparable to that of some non-sulphur purple bacteria (Hansen and Van Gemerden, 1972). Since sulphide concentrations are relatively low in hot springs, conditions are favourable for *Chloroflexus aurantiacus*. In contrast, extremely high sulphide contents may be observed in warm and cold springs. For instance, in Talgi spring (Dagestan, Caucasus) (28°C; pH, 6.3), the sulphide concentration may be as high as 218 mg l^{-1} (authors, unpublished data). Sulphide in this spring is the main factor which determines the phototrophic population.

Near the spring orifices, there are only green sulphur bacteria *Chlorobium limicola* which form thin films on stones. At decreasing sulphide concentrations (less than 190 mg l^{-1}) and pH greater than 6.7, a sulphur mat consists of purple *Thiospirillum* sp. and *Chromatium* sp. forms on the spring bed. Green bacteria are also abundant in this association. Of interest is the presence of the non-sulphur purple *Rhodobacter* sp. in high sulphide systems. Mechanical rigidity of the microbial mat in Talgi spring is provided by the filamentous cyanobacterium *Oscillatoria* sp. At lower sulphide concentrations (50 mg l^{-1}) downstream, *Phormidium* sp. and the green filamentous bacterium *Oscillochloris trichoides* provide rigidity. The total mat accumulates by anoxic photosynthesis (see Section 8.1). Sulphide inhibits photosystem II in cyanobacteria. Therefore, high sulphide concentrations result in the formation of completely anaerobic mats in sulphur springs.

8.2.4.3 pH

It appears that pH variations from 5.7 to 10 do not significantly affect the species composition of photosynthetic bacteria in springs (Brook, 1978; Gorlenko et al., 1985). Mildly acidic conditions and high sulphide concentrations are unfavourable for purple bacteria (Pfennig, 1967), which cannot compete with the more sulphide-tolerant green sulphur bacteria. The composition of cyanobacteria in association with *Chloroflexus aurantiacus* changes depending on pH and sulphide concentration. In alkaline low sulphide springs, it is *Synechococcus* sp.; in neutral springs lacking sulphide it is *Mastigocladus laminosus* (Nikitina and Gerasimenko, 1983); in mildly acidic and neutral springs with moderate sulphide concentrations—*Phormidium* sp. and *Oscillatoria* sp.

8.2.4.4 Combined effect of salinity and temperature

Salinity is a very important ecological factor. In mesophilic saline benthic sulfureta, the phototrophic species compositon differs greatly from those of freshwater basins (Gorlenko et al., 1984). Some hot sulphur springs with highly mineralized water are interesting for studying the combined effect of salinity and temperature. An example is the microbial mat in springs at Berikei, Dagestan (authors, unpublished data). These are NaCl springs with mineralization of 60–97 g l^{-1}; pH, 5.7–6.7 and sulphide, 30–40 mg l^{-1}.

Microbial mats begin to form at 54°C and lower whereas at higher temperatures, there is no visible development of microorganisms. Therefore, the upper temperature limit for cyanobacteria and anoxic phototrophic bacteria is significantly lower in saline springs than in others (authors, unpublished data).

The microbial mat of Berikei springs consists of species unique to saline water systems. The number of cyanobacterial species is limited, and filamentous *Phormidium* sp. predominate. There are also small quantities of unicellular *Synechocystis* sp. During the day, the mat is covered by a sulphur film indicating microbial sulphide oxidation. It was shown that cyanobacteria carry out O_2 photosynthesis in the presence of sulphide (>20 mg l^{-1}). These cyanobacteria can be assigned to the second or third group by their relation to sulphide (Table 8.1).

The microbial mat of Berikei saline springs is multilayered. The purple bacteria *Ectothiohodospira* sp. and *Rhodobacter* sp. and green bacterium *Prostecochlosis* sp. coexist in the upper layer. These purple and green bacteria were found for the first time in hot sulphur springs (50–54°C). Thus, the combined action of high salinity and high temperature promotes specific halophilic forms of cyanobacteria, and purple and green bacteria which are unable to develop in fresh water. In spite of a favourable temperature regime, *Chloroflexus aurantiacus* is absent from saline sulphur springs.

8.2.5 Horizontal and vertical zonality in microbial associations of hot sulphur springs

A zonal distribution and variation of microbial associations downstream in springs were observed by many investigators (Streczewski, 1913; Brock, 1967, 1969, 1978; Castenholz, 1984a, 1984b; Gorlenko et al., 1985). This results from gradual changes in physicochemical parameters of the medium. The temperature, pH and sulphide concentrations decrease, and the O_2 concentration increases, i.e. the conditions change gradually from anaerobic to aerobic in the horizontal direction. In CO_2 springs, the bicarbonate content decreases downstream. An example is the change in microbial associations observed in the Thermophilny spring (Gorlenko et al., 1987a). At 78°C, sulphide concentration of 16–19 mg l^{-1}, pH of 6.3, in the absence of O_2, the dominant processes are anaerobic sulphate reduction and methanogenesis. Downstream, at 67–76°C and O_2 concentration of 0.1–1.0 mg l^{-1}, one can observe an intensive development of the chemolithoautotrophic microorganisms *Thermothrix thiopara, Macromonas* sp., and *Thiobacterium bovista* (Figure 8.2). This zone is characterized by abundant S^0. In lower temperature zones (54–62°C) and sulphide concentrations of 7–10 mg l^{-1}, photosynthetic microorganisms form olive or orange mats (Figure 8.7). The latter are found at higher O_2 concentrations (1–3 mg l^{-1}). The basic components of the olive and orange zones are the green bacterium *Chloroflexus aurantiacus*, cyanobacteria *Phormidium* sp., and *Synechococcus* sp. Further downstream, the temperature drops below 50°C, the sulphide concentration decreases to 1 mg l^{-1}, and the O_2 content increases to 5–7 mg l^{-1}. Over the temperature range 28–50°C, a large multi-component green mat forms consisting of several species of cyanobacteria (Figure 8.7), *C. aurantiacus*, and sulphur and non-sulphur bacteria. The different biological regimes are presented on Eh versus pH and rH_2 versus temperature diagrams in Figures 8.2 and 8.3. The systematic change in microbial associations in response to a gradual decrease of temperature and an increase in redox potential is very evident in the diagrams.

For the development of photosynthetic and chemotrophic sulphur bacteria, the availability of sulphide as H-donor (in discharge water and generated during microbial destruction) is important.

Sulphur bacteria can develop in low-sulphide springs if the sulphate concentration is high.

The thickness of microbial mats in different springs and different zones of the same spring vary from 1 to more than 100 mm (Castenholz, 1984a). Photosynthesis proceeds actively in the upper 1–3 mm. Below the

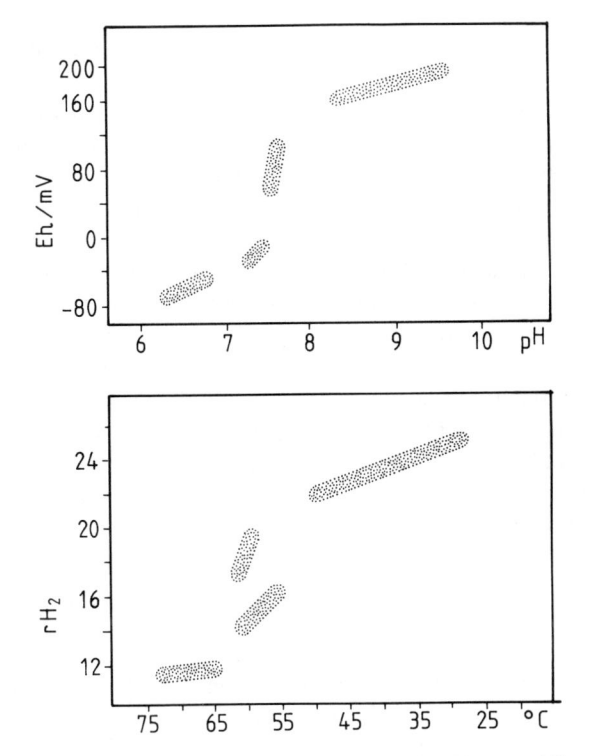

Figure 8.2 Location of different biological zones in Thermophilny spring vresus physicochemical conditions: Eh–pH; rH_2–$t°C$. 1, 2, 3, 4 are white, olive, orange and green mats, respectively

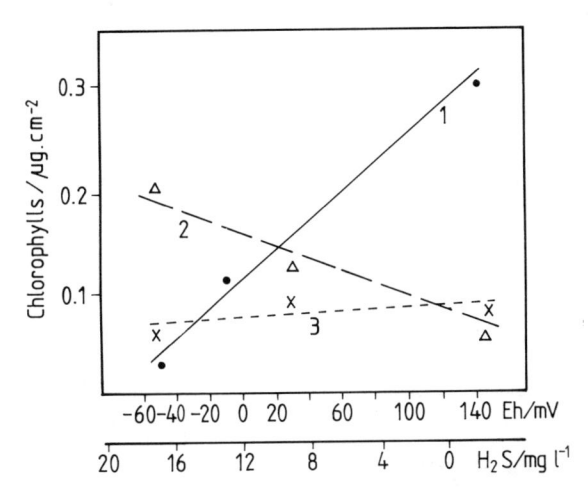

Figure 8.3 Redistribution of chlorophylls in different types of photosynthetic communities in Thermophilny spring versus physicochemical conditions. 1—chlorophyll-a; 2—bacteriochlorophyll-c; 3—bacteriochlorophyll-a

photosynthetic zone, anaerobic destruction proceeds with the participation of methanogenic, sulphate-reducing and sulphur-reducing microorganisms (Doemel and Brock, 1976). This causes an abrupt change in the redox potential in the uppermost layer of the mat. The Eh distribution in mats of the Thermophilny spring is shown in Figure 8.5. The slope of the Eh gradient determines the spatial distribution of microbial groups within the benthic association. A green low-temperature (40–50°C) loose mat, 1–2 cm thick, has the greatest potential for the development of physiologically different groups. Figure 8.4 shows the vertical distribution in green mats. Five species of cyanobacteria can develop in different microzones. *Chloroflexus aurantiacus* is found in close association with these producers (Figure 8.4). The highest number of filamentous green bacteria is observed during the day, 3–5 mm below the surface. The optimal sulphate reduction and methanogenesis are found at 15 and 20 mm depth respectively (Figure 8.6).

8.2.6 Productivity of microbial associations in different types of springs

The productivity of phototrophic associations in different springs correlates with the total chlorophyll pigment content. However, when making calculations, one should take into account that a significant part of the pigments is in the form of pheophytin (Gorlenko *et al.*, 1987a).

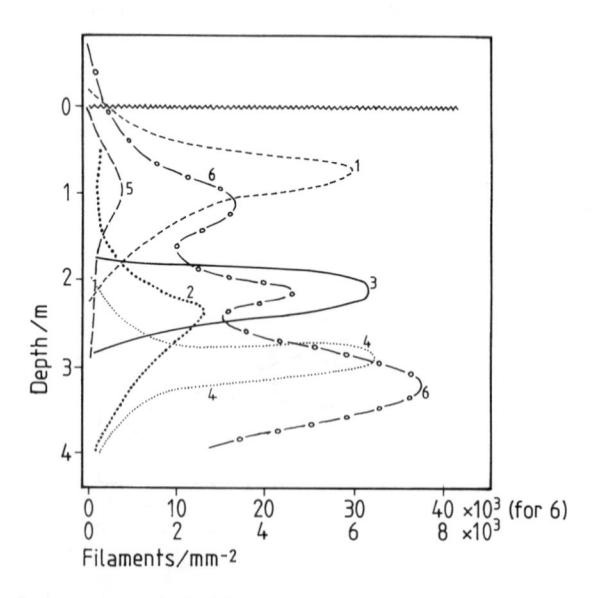

Figure 8.4 Vertical distribution of different morphological forms of cyanobacteria (1–5) and *Chloroflexus aurantiacus* (6) in the body of a green mat in Thermophilny spring

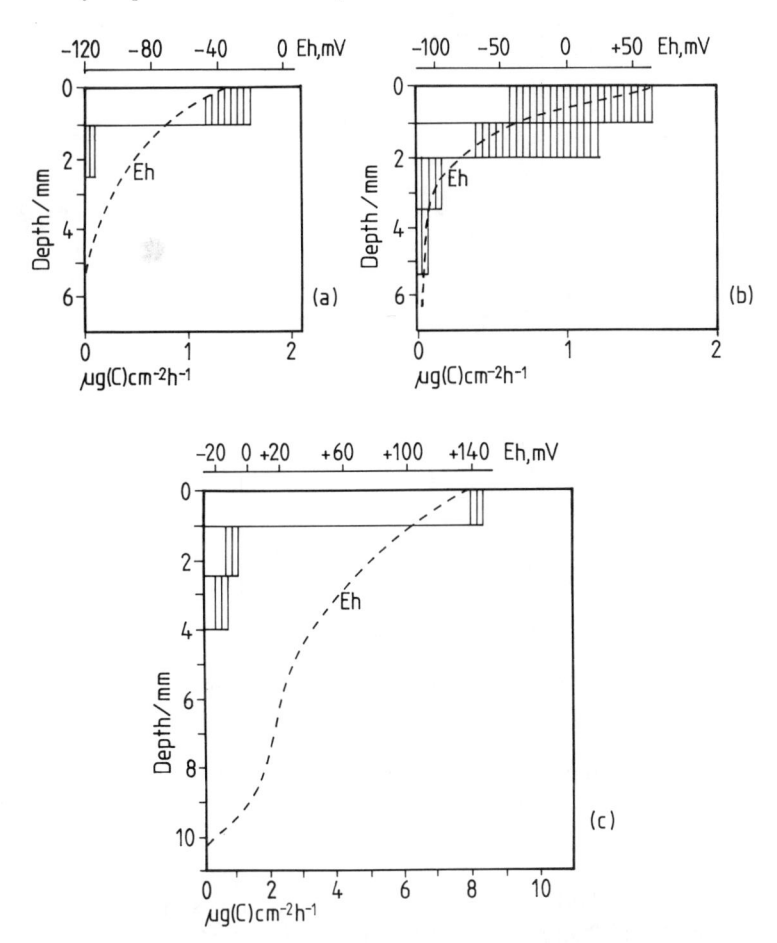

Interaction of Sulphur and Carbon Cycles in Microbial Mats

Figure 8.5 Rate of photosynthesis and redox potential in microbial mats of Thermophilny spring. □—oxygenic photosynthesis; ▥—anoxic photosynthesis; (a) olive mat; (b) orange mat; (c) green mat

Primary production changes along the stream gradient (Table 8.3), while the level of photosynthesis is generally high in all types of springs. For instance, in the hot saline Berikei springs, the photosynthetic production ranges from 4.7 to 53.8 g (C) cm^{-2}h^{-1} (which means about 0.37–4.3 g (C) m^{-2} for a typical 8 hour daylight period) at different sites. The maximum value of photosynthesis approaches the highest productivity of the cyanobacterial mat in Solar Lake (Cohen et al., 1977b).

Anoxic photosynthesis varies depending on the type of spring (Table 8.3), and is mainly performed by purple and green bacteria and to a lesser extent by unicellular and filamentous cyanobacteria.

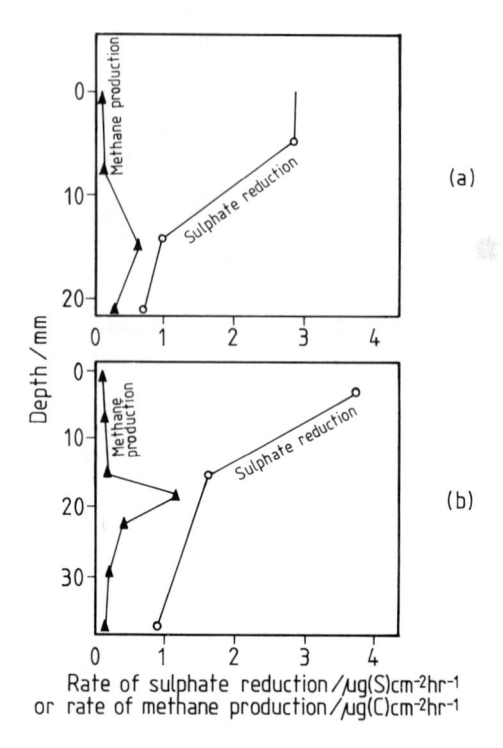

Figure 8.6 Rates of sulphate reduction and methanogenesis in orange (a) and green (b) mats in Thermophilny spring

In the springs studied, extracellular production makes up 53% of the total fixed carbon (Bauld and Brock, 1974; Gorlenko *et al.*, 1987b). Therefore excretions from photosynthetic microorganisms play an important role in trophic chains of ecosystems. Specifically, they create favourable conditions for the development of photoheterotrophic microorganisms such as *Chloroflexus* and non-sulphur purple bacteria.

The biomass accumulated during photosynthesis in microbial mats undergoes partial or complete destruction. Research into the degradation of mat components should indicate the pathways of burial and lithification of organic matter as well as the formation of some economic minerals.

8.2.7 Anaerobic destruction of organic matter in thermophilic mats

The bulk of the biomass of primary producers undergoes anaerobic destruction since the lower parts of mats are usually under reduced conditions (Table 8.5).

Table 8.3 Photosynthetic and dark production in microbial mats of Thermophilny spring

Type of mat, depth, (mm)	P-photo-synthesis (μg(C) cm^{-2} h^{-1})	P-extra cellular (%)	P-anoxic (%)	P-dark (μg(C) cm^{-2} h^{-1})	Chl-a BChl-a
White	—	20.0	—	0.074	—
Olive					
(0–1)	1.57	18.8	35.3	0.23	0.07
(1–2.5)	0.04	46.5	40.0		0.63
Orange					
(0–1)	1.62	2.9	59.3		1.1
(0–2)	1.40	7.4	62.0	0.68	1.2
(2–3.5)	0.13	24.0	76.0		—
(3.5–5.5)	0.07	—	—		—
Green					
(0–1)	7.63	22.2	1.0		6.0
(1–2.5)	0.77	8.5	10.1	1.21	4.35
(2.5–4)	0.54	3.8	2.9		—

Not much is known about the microbiota which hydrolyse biopolymers in thermophilic microbial mats. To date, the only hydrolytic organism isolated from a cyanobacterial mat is *Clostridium thermosulfurogenes* which can degrade pectin (Schink and Zeikus, 1983). It has the unique ability to form S^0 from thiosulphate. However, it is quite probable that this organism is not important in the destruction of cyanobacterial biomass because of its low population (10^3 cells cm^{-3}). Sugar-consuming bacteria are more abundant (10^8–10^9 cells cm^{-3}) (Zeikus *et al.*, 1980). Among the saccharolytic organisms is the earlier known species *Cl. thermohydrosulfuricum* (Zeikus *et al.*, 1980). In addition, three saccharolytic anaerobic organisms were isolated and assigned to new genera: *Thermoanaerobium brockii* (Zeikus, *et al.*, 1979), *Thermobacterioides acetoethylicus* (Ben Bassat and Zeikus, 1981) and *Thermoanaerobacter ethanolius* (Wiegel and Ljungdahl, 1981). They ferment hydrolysis products of algal polysaccharides into acetate, lactate, ethanol, H$_2$ and CO$_2$. However, the abundance of proteolytic bacteria in the thermophilic mat (45°C) of Thermophilny spring is 10^9 cells cm^{-3}, i.e. an order higher than the number of saccharolytic bacteria (Bonch-Osmolovskaya *et al.*, 1988). Consequently, one can infer that protein is the basic substrate for destruction processes whereas polysaccharide is decomposed more slowly. Intermediate products (acids, alcohols, hydrogen) from hydrolysis and fermentation serve as energy substrates for bacteria carrying out the final step of anaerobic destruction. Depending on the type of the spring temperature zone, and vertical position within the mat, the terminal processes may be sulphate reduction, sulphur reduction or methanogenesis.

The intensity of sulphate reduction in cyanobacterial mats is determined primarily by the sulphate content (Sandbeck and Ward, 1981). For example, at a concentration of 718 mg l^{-1}, sulphate reduction is the main terminal process of destruction, and H_2S is the principal reduced product (Ward and Olson, 1980). These authors found that sulphate reduction was the most intensive in the uppermost 5 mm of the mat. In Thermophilny spring, sulphate concentration was much lower (36 mg l^{-1}). The rate of sulphate reduction, determined by means of $^{35}SO_4^{2-}$ *in situ*, reached a maximum of 6 mg (S) $cm^{-2} h^{-1}$ in the most productive green mat (Bonch-Osmolovskaya *et al.*, 1988; Tables 8.4 and 8.5). In the dark, sulphate reduction is most intensive in the uppermost (0–1 mm) layer of the mat, below which it decreases rapidly (Figure 8.6). The residence time of sulphate in cyanobacterial mats is very short: 2 to 3.7 hours (Table 8.4). In the depth interval, 1.5–2 cm, methanogenesis is most intensive and sulphate reduction is probably limited by a slow sulphate diffusion.

At lower SO_4^{2-} concentrations (16.6 mg l^{-1}, Octopus spring in Yellowstone National Park), sulphate reduction is evidently a minor process and does not inhibit methanogenesis even in the upper layer (Ward, 1978).

In model systems, sulphate reduction usually inhibits methanogenesis (Bonch-Osmolovskaya *et al.*, 1978). This is usually explained by the high energy yield of sulphate reduction:

$$SO_4^{2-} + 4H_2 + H^+ \rightarrow HS^- + 4H_2O \quad \Delta G^0 = -152.5 \text{ kJ mol}^{-1}$$

compared to the reaction of methanogenesis:

$$CO_2 + 4H_2 \rightarrow CH_4 + 2H_2O \quad \Delta G^0 = -131.1 \text{ kJ mol}^{-1}$$

where ΔG^0 is the free energy change associated with these reactions (Thauer *et al.*, 1977).

However, in nature, methanogenesis and sulphate reduction often occur simultaneously (Oremland and Polcin, 1982). This is explained by the difference in affinities of methanogens and sulphate reducers to the substrate, which allows them to prevail in different ecological situations (Lovley *et al.*, 1982). A number of substrates (methanol, methylamines) are non-competitive since they are utilized only by methanogens. Thus methanogens can develop even in biotopes rich in sulphate (Oremland and Polcin, 1982).

In thermophilic cyanobacterial mats, methanogenesis does not occur at high SO_4^{2-} concentrations (Ward and Olson, 1980). During incubation of

Interaction of Sulphur and Carbon Cycles in Microbial Mats

Table 8.4 Rates of terminal reduction processes in different mats of Thermophilny spring

Type of mat	H$_2$S (mg cm^{-3})	S^0 (mg cm^{-3})	Sulphate reduction rate (μg cm^{-2} h^{-1})	Sulphur reduction rate (μg cm^{-1})	CH$_4$ production rate (μg cm^{-2} h^{-1})	Sulphate residence time (h)
White	1.3	3.1	0.16	0.4	trace	75
Olive	0.39	8.0	3.3	—	0.12	3.6
Orange	0.41	9.3	3.2	—	0.58	3.7
Green	0.47	9.6	6.0	—	1.77	2.0

Table 8.5 Sulphate reduction and methanogenesis in destruction processes in microbial mats of Thermophilny spring

Type of mat	Production (μg (C) cm^{-2} l^{-1})	Destruction, μg (C) cm^{-2} h^{-1}, due to: sulphate reduction	sulphur reduction	methanogenesis	$\dfrac{D(SO_4^{2-})}{D(CH_4)}$	P−D*
White	0.074	0.09	0.28	trace	900	−0.31
Olive	1.87	2.48	—	0.24	10.30	−0.85
Orange	3.90	2.40	—	1.16	2.00	+0.34
Green	10.80	4.50	—	3.54	1.27	+2.18

*P—production, D—destruction

samples from such a spring, methanogenesis started only after 200 hours, and according to these authors, it is exclusively a laboratory process which does not occur in nature. As mentioned above, at low concentrations (16.6 mg l^{-1}), sulphate reduction does not occur, and methanogenesis is the main terminal anaerobic process (Ward, 1978).

Both destructive processes occur at varying rates in Thermophilny spring, having a sulphate concentration of 36 mg l^{-1} (Table 8.4). The contribution of methanogenesis to the total destruction increases gradually, as the temperature decreases, and reaches a maximum in the green mat zone. Methanogenesis increases with depth and is a maximum immediately under the sulphate reduction zone (Figure 8.6).

A study of methanogenesis with different substrates showed that the main process in thermophilic mats is CO$_2$ reduction by hydrogen. This pathway in Thermophilny spring accounts for 89% of the methane produced (Bonch-Osomolovskaya et al., 1988). This is consistent with the high number (10^8

cells cm^{-3}) of the methanogen consuming H$_2$, which was identified as *methanobacterium thermoautotrophicum*.

Acetate did not appear to be consumed in hot spring mats of Yellowstone National Park during methanogenesis (Ward, 1978) and does not stimulate sulphate reduction (Ward and Olson, 1980). It has been shown that the bulk of the acetate in mats of Octopus spring, Yellowstone National Park, is assimilated by *Chloroflexus* (Ward et al., 1984). In contrast to this a thermophilic organism which produces CH$_4$ from acetate was isolated from Thermophilny spring (Nozhevnikova and Yagodina, 1982). Radioisotopic study *in situ* showed that up to 11% of methane in Thermophilny spring is formed from acetate (Bonch-Osomolovskaya et al., 1987).

Usually methane can be formed intensively from C$_1$-compounds by representatives of the genus *Methanosarcina*. However, microorganisms of this genus are not found in thermophilic cyanobacterial mats. A rod-like methanogenic microflora similar to *Methanobacillus kuzneceovii* (Pantshava and Pchyolkina, 1969) was isolated using a methanol medium (Zhilina et al., 1983). It was shown that such a culture is in a symbiotic association with *Clostridium thermoautotrophicum* producing H$_2$ from methanol, and *Methanobacterium thermoautotrophicum* consuming this hydrogen to reduce CO$_2$ to CH$_4$ (Ilarionov and Bonch-Osmolovskaya, 1986). Thus in anaerobic destruction of thermophilic cyanobacterial mats, hydrogen is the main substrate for competition between methanogenic and sulphate-reducing bacteria.

Sulphur reduction in microbial mats has been studied less. However, this process may also play an important role in the anaerobic destruction of thermophilic systems. Sulphur may be an electron acceptor during anaerobic destruction of organic matter, H$_2$S being the reduction product. Elemental sulphur in mats may have different origins.

For instance, in the white mat of Thermophilny spring, S^0 is the product of H$_2$S oxidation by *Thermothrix thiopara* (Figure 8.7a). Organic matter synthesized by this reaction undergoes decomposition in anaerobic conditions, and the terminal electron acceptor is again S^0. It was shown that *Thermothrix thiopara* exists in association with facultative anaerobic thermophilic bacteria including those able to reduce S^0 (Balashova, 1986).

We have established that under 1 cm^2 of the white mat, 0.4 µg (S^0) is reduced per hour. This is 2.3 times higher than the rate of sulphate reduction in the same zone. Therefore sulphur reduction dominates in the destruction of organic matter synthesized by *Thermothrix thiopara*.

Cyanobacterial mats of Thermophilny spring contain significant quantities of S^0 (8–9.6 mg cm^{-2}). The largest amount accumulates in the upper layers of green and orange mats, which correspond to the photosynthetic zone and the transition between aerobic and anaerobic zones (Figure 8.7b; Jørgensen, 1982). The source of S^0 may be anaerobic oxidation of H$_2$S by phototrophic

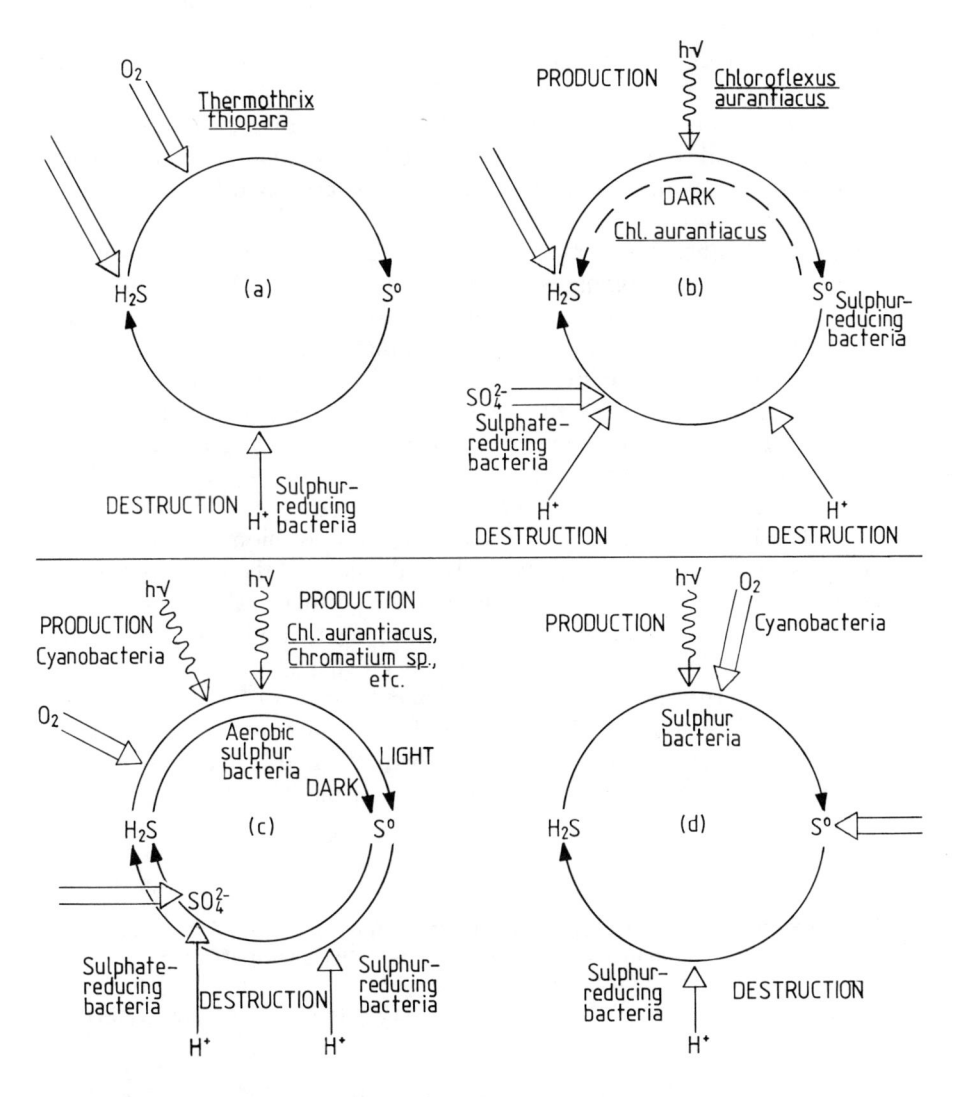

Figure 8.7 Hypothetical schemes of the sulphur cycle in thermophilic microbial mats of different types.
(a) Microbial mat associated with *Thermotrix thiopara* (72°C)
(b) Microbial mat with *Chloroflexus aurantiacus* as main producent (65°C)
(c) Multi-component cyanobacterial mat (45–55°C)
(d) Cyanobacterial mat with exogenic source of sulphur (50–55°C)
⇒ influx of compounds from outside;
→ the sulphur cycle inside the mat;
$H^+ \to$ formation of a reducer during destruction

bacteria, aerobic oxidation of H$_2$S by sulphur bacteria, or the activity of organisms similar to *Clostridium thermosulfurogenes* which can produce S^0 from thiosulphate (Schink and Zeikus, 1983). It would seem that in these zones, sulphur reduction contributes significantly to the anaerobic destruction of organic matter. When S^0 was added to samples of cyanobacterial mats from Thermophilny spring, H$_2$S was evolved and methanogenesis inhibited (Table 8.6). This points to the ability of cyanobacterial associations to reduce S^0 during destruction of organic matter.

When a cyanobacterial mat develops in a hot spring with suspended colloidal sulphur, the exogenous sulphur is found between algal layers (Figure 8.7a). In such a mat (found in the Uzon caldera at 50–57°C) sulphur reduction dominates during anaerobic destruction and completely inhibits methanogenesis. The addition of different organic substrates to samples from such a mat increases the yield of H$_2$S but methanogenesis is still inhibited (Table 8.6). The abundance of sulphur-reducing bacteria (up to 10^{10} cells cm^{-3}) also points to the dominance of sulphur reduction.

Therefore, sulphate- and sulphur-reducing bacteria of thermophilic cyanobacterial associations participate in the destruction of biomass. The balance equations of Krumbein (1983) were used to calculate their contribution to destruction of organic matter at Thermophilny spring (Table 8.5).

The participation of methanogenesis in the destruction of organic matter increases gradually in going from white to green mats. In the green mat, the contributions of SO$_4^{2-}$ reducers and methanogens to the decomposition of microbial biomass are almost equal. In the olive mat, destruction by sulphate reduction is 10 times higher than by methanogenesis. Sulphur reduction dominates in destruction occurring in the white mat.

Table 8.6 Reduction products in anaerobic destruction of different organic substrates with samples of sulphur-containing microbial mats

Substrate*	CH$_4$ (mM)	HS$^-$ (mM)
Pectin	0	12.5
Yeast extract	0	7.1
Casein hydrolysate	0	12.8
Lactate	0.97	17.6
Formate	0.87	19.2
Acetate	0	23.5
Methanol	0	22.3
Ethanol	0	21.5
Glycerol	0	21.3

*Samples of mat (1 cm^3) were incubated anaerobically for 14 days in 50 ml vials with mineral medium and organic substrate (5 mg ml^{-1}) (Pfennig, 1965).

Visual examination showed that organic remnants do not accumulate and are not buried in white and olive mats. This means that production is balanced by destruction. Calculations confirm this conclusion. Organic matter accumulates in the orange and especially green mats (3 and 15 cm thick, respectively). The calculated production is higher than the anaerobic destruction by methanogenesis and sulphate reduction.

8.2.8 Conclusion

Microbial associations of hot sulphur springs have a local character of development and do not appreciably affect terrestrial geochemical processes which occur nowadays. Nonetheless their study gives insight into ideas about the evolutionary pattern of the atmosphere and lithosphere, and indicate which microorganisms could play a leading part in biogeochemical processes at raised temperatures and O_2 deficiency. The species composition of microbial mats depends upon a set of physicochemical parameters and is different in various sources.

In thermophilic microbial mats the sulphur cycle is coupled with the cycle of carbon, i.e. the primary production and destruction of organic matter. In all cases studied, the production leads, directly or indirectly, to elemental sulphur formation.

In the *Thermothrix* community, the aerobic oxidation of H_2S yields elemental sulphur as the main catabolic product (Figure 8.7a). In a mat formed by *Chloroflexus aurantiacus*, sulphur is produced in the course of photosynthesis as a result of H_2S oxidation in the light (Figure 8.7b). In multi-component mats consisting mainly of cyanobacteria, elemental sulphur is the product of chemical or biological H_2S oxidation with O_2 brought about by phototrophs (Figure 8.7c).

Reduced compounds are formed during biomass destruction; H_2S is produced with elemental sulphur as terminal electron acceptor.

Thus we see that in thermophilic microbial mats of sulphur springs a short sulphur cycle $H_2S \rightarrow S^0$ is taking place. This supports the idea of an analogy between 'oxygenic' and 'sulphur' life (Wolfe and Pfennig, 1977). But, unlike the closed cycles in symbiotic associations, described by Pfennig *et al.* (Wolfe and Pfennig, 1977; Biebl and Pfennig, 1978), open sulphur cycles take place in thermal springs. In the *Thermothrix* community, exhalations of juvenile gases contribute to the H_2S pool in the system (Figure 8.7a). Elemental sulphur, when reaching high concentrations in the spring, also enters the cycle (Figure 8.7d). Sulphate can be of exogenous origin, or produced endogenously in the course of aerobic sulphur bacteria activity.

Thus, microbial communities in thermal sulphur springs can be regarded as natural model systems. Further investigations will contribute to a better understanding of processes that took place in the geological past of the Earth.

REFERENCES

Aizenshtat, Z., Lipiner, G., and Cohen, Y. (1984). Biogeochemistry of carbon and sulphur cycle in the microbial mats of Solar Lake (Sinai). In: Cohen, Y., Castenholz, R.W., and Halvorson, H.O. (Eds.) *Microbial Mats: Stromatolites*. Alan R. Liss, New York, pp., 281–312.

Awramik, S.M. (1984). Ancient stromatolites and microbial mats. In: Cohen, Y., Castenholz, R.W., and Halvorson, H.O. (Eds.) *Microbial Mats: Stromatolites*, Alan R. Liss, New York, pp. 1–22.

Baas-Becking, L.G.M., and Wood, E.J.F. (1955). Biological processes in the estuarine environment: 1–11. Ecology of the sulphur cycle. *Proc. K. Med. Akad. Wet. Serct.*, **B58**, 160–81.

Balashova, V.V. (1985). Use of molecular sulphur as H_2 oxidiser by facultative anaerobic pseudomonade. *Mikrobiologia*, **54**, 324–6 (in Russian).

Balashova, V.V. (1986). Thermophilic sulphureta modelling. In: Imshenetsky, A.A. (Ed.) *Biology of Thermophilic Microorganisms*, Nauka, Moscow, pp. 113–16 (in Russian).

Bauld, J. (1981a). Geobiological role of cyanobacterial mats in sedimentary environments: production and preservation of organic matter. *BMR J. Aust. Geol. Geophys.*, **6**, 307–17.

Bauld, J. (1981b). Occurrence of benthic microbial mats in saline lakes. *Hydrobiologia*, **81**, 87–111.

Bauld, J. (1984). Microbial mat in marginal marine environments: Shark Bay, Western Australia and Spencer Gulf, South Australia. In: Cohen, Y., Castenholz, R.W., and Halvorson, H.O. (Eds.) *Microbial Mats: Stromatolites*, Alan R. Liss, New York, pp. 39–58.

Bauld, J., and Brock, T.D. (1973). Ecological studies of *Chloroflexus*, a gliding photosynthetic bacterium. *Arch. Microbiol.*, **92**, 267–84.

Bauld, J., and Brock, T.D. (1974). Algal excretion and bacterial assimilation in hot spring algal mats. *J. Physol.*, **10**, 101–6.

Bauld, J., Chambers, L.A., and Skyring, G.W. (1979). Primary productivity, sulphate reduction and sulphur isotope fractionation in algal mats and sediments of Hamelin Pool, Chark Bay, W.A. *Aust. J. Mar. Freshwater Res.*, **30**, 753–64.

Belkin, S. (1983). Hydrogen metabolism in the facultative anoxygenic cyanobacteria. Ph.D. Thesis, Hebrew University, Jerusalem.

Belkin, S., and Padan, E. (1979). Hydrogen metabolism in the facultative anoxygenic cyanobacteria (blue green algae) *Oscillatoria limnetica* and *Aphanotheca halophilica*. *Arch. Microbiol.*, **116**, 109–11.

Ben-Bassat, A., and Zeikus, J.G. (1981). Thermobacteroides acetoethylicus *Gen.nov.* and *spec.nov.*, a new chemoorganotrophic, anaerobic thermophilic bacterium. *Arch. Microbiol.*, **128**, 365–70.

Biebl, H., and Pfennig, N. (1978). Growth yields of green sulphur bacteria in mixed cultures with sulphur and sulphate reducing bacteria. *Arch. Microbiol.*, **117**, 9–16.

Bildushkinov, S.S., and Gerasimenko, L.M. (1985). Thermophilic cyanobacteria from cyanobacterial community. *Mikrobiologia*, **54**, 490–3 (in Russian).

Bonch-Osmolovskaya, E.A. (1986). Reduction of elemental sulphur in thermophilic algal bacterial communities of Uzon caldera (Kamchatka). In: Imshenetsky, A.A., (Ed.) *Biology of Thermophilic Microorganisms*, Nauka, Moscow, pp. 112–13 (in Russian).

Bonch-Osmolovskaya, E.A., and Svetlichny, V.A. (1989). Extremely thermophilic sulphur-reducing archaebacteria. *Archaebacteria*, Puschino, USSR (in Russian).

Bonch-Osmolovskaya, E.A., Vedenina, I.Ya., and Balashova, V.V. (1978). The influence of inorganic electron acceptors on methane formation during anaerobic cellulose destruction. *Mikrobiologia*, **47**, 611–16 (in Russian).

Bonch-Osmolovskaya, E.A., Gorlenko, V.M., Karpov, G.A., and Starynin, D.A. (1987). Anaerobic destruction of organic matter in cyanobacterial mats of Thermophilny spring (Uzon, Kamchatka). *Mikrobiologia*, **56**, 1022–8 (in Russian).

Bonch-Osmolovskaya, E.A., Slesarev, A.I., Miroshnichenko, M.L., Svetlichnaya, T.P., and Alekseev, V.A. (1988). *Desulfurococcus amylolytuus* n.sp.—a new extremely thermophilic archaebacterium from Kamchatka and Kunashire hot springs. *Mikrobiologia*, **57**, 94–101 (in Russian).

Bogorov, L.V. (1974). On capacities of *Thiocapsa roseopersicina* strain BBS isolated from the White Sea estuaries. *Mikrobiologia*, **43**, 326–31 (in Russian).

Bothe, H., Tennigkeit, Y., and Eisbrinner, G. (1977). The utilisation of molecular hydrogen by the blue green alga *Anabaena cylindrica*. *Arch. Microbiol.*, **114**, 43–9.

Brannan, D.K., and Caldwell, D.E. (1980). *Thermotrix thiopara*: growth and metabolism of a newly isolated thermophile capable of oxidising sulphur and sulphur compounds. *Appl. Environ. Microbiol.*, **40**, 211–16.

Brock, T.D. (1967). Relationship between standing crop and primary productivity along a hot spring thermal gradient. *Ecology*, **48**, 566–71.

Brock, T.D. (1969). Vertical zonation in hot spring algal mats. *Phycologia*, **8**, 201–5.

Brock, T.D. (1978). *Thermophilic Microorganisms and life at High Temperatures*, Springer-Verlag, New York.

Cairns-Smith, A.G. (1978). Precambrian solution photochemistry, inverse segregation and banded iron-formations. *Nature*, **276**, 807–8.

Caldwell, D.E., Caldwell, S.J., and Laycock, J.P. (1976). Thermothrix thioparus gen. et sp.n., a facultatively anaerobic facultative chemolithotroph living at neutral pH and high temperature. *Can. J. Microbiol.*, **22**, 1509–17.

Campbell, R. (1983). Microbial ecol. In: Wilkinson, J.F. (ed.) 2 edn. Blackwell Scientific Publications, Oxford.

Castenholz, R.W. (1969). Thermophilic blue-green algae and the thermal environment. *Bact. Rev.*, **33**, 476–504.

Castenholz, R.W. (1973). The possible photosynthetic use of sulphide by the filamentous phototrophic bacteria of hot springs. *Limnol. Oceanogr.*, **18**, 863–76.

Castenholz, R.W. (1976). The effect of sulphide on blue green algae of hot springs. I. New Zealand and Iceland. *J. Phycol.*, **12**, 57–68.

Castenholz, R.W. (1977). The effect of sulphide on blue green algae of hot springs. II. Yellowstone National Park. *Microb. Ecol.*, **3**, 79–68.

Castenholz, R.W. (1984a). Composition of hot spring microbial mats: A summary. In: Cohen, Y., Castenholz, R.W., and Halvorson, H.O. (Eds.) *Microbial Mats: Stromatolites*, Alan R. Liss, New York, pp. 101–20.

Castenholz, R.W. (1984b). Habitats of *Chloroflexus* and related organisms. In: King, M.J., and Reddy, C.A. (Eds.) *Current Perspectives of Microbial Ecology*, ASM, Washington, DC.

Castenholz, R.W., and Utkilen, H.C. (1984). Physiology of sulphide tolerance in a thermophilic *Oscillatoria*. *Arch. Microbiol.*, **138**, 299–305.

Cloud, P. and Gibor, A. (1970). The oxygen cycle. *Scientific American* (Sept.).

Cohen, Y. (1985a). Microtechnique for *in situ* sulphate reduction measurement in proximity to oxygen. *Arch. Microbiol.* (in press).

Cohen, Y. (1985b). Sulphate reduction under oxygen in cyanobacterial mats and its coupling to primary production. *Limnol. Oceanogr.* (in press).

Cohen, Y. and Gack, E. (1985). Fe^{++}-dependent photosynthesis in cyanobacterial. *Nature* (in press).
Cohen, Y., Castenholz, R.W., and Halvorson, H.O. (Eds.) (1984). *Microbial Mats: Stromatolites*, MBL Lectures in Biology, Vol. 3. Alan R. Liss, New York, pp. 1–498.
Cohen, Y., Padan, E., and Shilo, M. (1975a). Facultative anoxygenic photosynthesis in the cyanobacterium *Oscillatoria limnetica*. *J. Bacteriol.*, **12**, 855–61.
Cohen, Y., Jørgensen, B.B., Padan, E., and Shilo, M. (1975b). Sulphide-dependent anoxygenic photosynthesis in the cyanobacterium *Oscillatoria limnetica*. *Nature*, **257**, 489–91.
Cohen, Y., Goldberg, M., Krumbein, W.E., and Shilo, M. (1977a). Solar Lake (Sinai). I. Physical and chemical limnology. *Limnol. Oceanogr.*, **22**, 597–607.
Cohen, Y., Krumbein, W.E., and Shilo, M. (1977b). Solar Lake (Sinai). 2. Distribution of photosynthetic microorganisms and primary production. *Limnol. Oceanogr.*, **22**, 609–10.
Cohen, Y., Aisenshtat, Z., Stoler, A., and Jørgensen, B.B. (1980). Microbial geochemistry of Solar Lake Sinai. In: Trudinger, P.A., and Walter, M.R. (Eds.) *Biogeochemistry of Ancient and Modern Environments*, Australian Academy of Science, Canberra, pp. 167–77.
Doemel, W.N., and Brock, T.D. (1976). Vertical distribution of sulphur species in benthic algal mats. *Limnol. Oceanogr.*, **21**, 237–44.
Doemel, W.N., and Brock, T.D. (1977). Structure, growth and decomposition of laminated algal-bacterial mats in alkaline hot springs. *Appl. Environ. Microbiol.*, **34**, 433–52.
Fisher, F., Zillig, W., Stetter, K.O., and Schreiber, G. (1983). Chemolithoautotrophic metabolism of anaerobic extremely thermophilic archaebacteria. *Nature*, **301**, 511–13.
Ganf, G.G., and Viner, A.B. (1973). Ecological stability in a shallow equatorial lake (Lake George, Uganda). *Proc. R. Soc. Lond. (Biol.)*, **184**, 321–46.
Garlick, S., Oren, A., and Padan, E. (1977). Occurrence of facultative anoxygenic photosynthesis among filamentous and unicellular cyanobacteria. *J. Bacteriol.*, **129**, 623–9.
Gerasimenko, L.M., Karpov, G.A., Orleansky, V.K., and Zavarzin, G.A. (1983). The role of cyanobacterial filter in gas component of hydrotherms transformation with caldera Uzon in Kamchatka as an example. *Zhurnal obshchei biologii*, **44**, 842–51 (in Russian).
Gerdes, G., and Krumbein, W.E. (1984). Animal communities in recent potential stromatolites of hypersaline origin. In: Cohen, Y., Castenholz, R.W., and Halvorson, H.O. (Eds.) *Microbial Mats: Stromatolites*. Alan R. Liss, New York, pp. 59–84.
Golovacheva, R.S. (1984). Aerobic thermophile chemolithotrophic bacteria participating in the sulphur cycle. *Uspehi Mikrobiologii*, **19**, 166–202 (in Russian).
Gorlenko, V.M. (1974). Dark thiosulphate oxidation by *Amoebobacter roseus* in microaerophilic conditions. *Mikrobiologia*, **43**, 729–31 (in Russian).
Gorlenko, V.M. (1981). Purple and green bacteria and their role in the carbon and sulphur cycles. Thesis, INMI Acad, Sci. USSR, p. 51 (in Russian).
Gorlenko, V.M., Dubinina, G.A., and Kuznetsov, S.I. (1983). The ecology of aquatic microorganisms. *Die Binnengewässer*, Volume XVIII, Stuttgart, E. Schweizerbartsche Verlagsbuchhandlung (Nägele und Obermiller), 252 pp.
Gorlenko, V.M., Kompantseva, E.I., and Puchkova, N.N. (1985). Influence of temperature on phototrophic bacteria distribution in thermal springs. *Mikrobiologia*, **54**, 848–53 (in Russian).

Gorlenko, V.M., and Korotkov, S.A. (1979). Morphological and physiological features of the new filamentous gliding green bacteria *Oscillochloris trichoides* nov.comb. *Izv. Akad. Nauk SSSR. Ser. Biol.*, **5**, 848–57 (in Russian).

Gorlenko, V.M., Kompantseva, E.I., Korotkov, S.A., Puchkova, N.N., and Savvichev, A.S. (1984). Growth conditions and diversity of phototrophic bacteria species in saline shallow waters of Crimea. *Izv. Akad. Nauk SSSR, Ser. Biol.*, **3**, 362–73 (in Russian).

Gorlenko, V.M., Bonch-Osmolovskaya, E.A., Kompantseva, E.I., and Starynin, D.A. (1987). Differentiation of microbial communities in connection with changes in physico-chemical parameters in Thermophilny spring. *Mikrobiologia*, **56**, 314–22 (in Russian).

Gorlenko, V.M., Starynin, D.A., Bonch-Osmolovskaya, E.A., and Kachalkin, V.J. (1987b). Primary production processes in microbial mats of Thermophilny spring. *Mikrobiologia* (in Russian).

Govindjee, R., Rabinovitch, E., and Govindjee, (1968). Maximum quantum yield and action spectrum of photosynthesis and fluorescence in *Chlorella*. *Biochem. Biophys. Acta.*, **162**, 539–44.

Gromet-Elhanan, Z. (1977). Electron transport and photophosphorylation in photosynthetic bacteria. In: Trebt, A., and Avron, M. (Eds.) *Encyclopedia of Plant Physiology*, Springer-Verlag, Berlin, pp.637–62.

Hansen, T.A., and Van Gemerden, H. (1972). Sulphide utilization by purple bacteria. *Arch. Microbiol.*, **86**, 49–56.

Hartman, H. (1983). The evolution of photosynthesis and microbial mats: A speculation on the Banded Iron Formation. In: Cohen, Y., Castenholz, R.W., and Halvorson, H.O. (Eds.) *Microbial Mats: Stromatolites*, Alan R. Liss, New York, pp. 441–54.

Howsley, R., and Pearson, H.W. (1979). pH dependent sulphide toxicity to oxygenic photosynthesis in cyanobacteria. *FEMS Lett.*, **6**, 287–92.

Ilarionov, S.A., and Bonch-Osmolovskaya, E.A., (1986). Methane production from methanol by bacterial associations. *Mikrobiologia*, **55**, 282–8 (in Russian).

Imhoff, J.F., Hashwa, F., and Truper, H.G. (1978). Isolation of extremely halatrophic bacteria from the alkaline Wadi Natrun, Egypt. *Arch. Hydrobiol.*, **84**, 381–8.

Javor, B.J., and Castenholz, R.W. (1984a). Invertebrate grazers of microbial mats, Lagoona Guerrera Negro, Mexico. In: Cohen, Y., Castenholz, R.W., and Halvorson, H.O. (Eds.) *Microbial Mats: Stromatolites*, Alan R. Liss, New York, pp. 85–94.

Javor, B.J., and Castenholz, R.W. (1984b). Productivity studies of microbial mats, Lagoona Guerrero Negro, Mexico. In: Cohen, Y., Castenholz, R.W., and Halvorson, H.O. (Eds.) *Microbial Mats: Stromatolites*, Alan R. Liss, New York, pp. 149–70.

Jørgensen, B.B. (1982). Ecology of the bacteria of the sulphur cycle with special reference to anoxic–oxic interface environments. *Phil. Trans. R. Soc. Lond. B.*, **298**, 543–61.

Jørgensen, B.B., and Cohen, Y. (1977). Solar Lake (Sinai). 5. The Sulphur cycle of benthic cyanobacterial mats. *Limnol. Oceanogr.*, **22**, 657–66.

Jørgensen, B.B., Revsbech, N.P., and Cohen, Y. (1983). Photosynthesis and structure of benthic microbial mats: microelectrode and SEM studies of four cyanobacterial communities. *Limnol. Oceanogr.*, **28**, 1075–93.

Jørgensen, B.B., Revsbech, N.P., Blackburn, T.H., and Cohen, Y. (1979). Diurnal cycle of oxygen and sulphide microgradients and microbial photosynthesis in a cyanobacterial mat sediment. *Appl. Environ. Microbiol.*, **38**, 46–58.

Kämpf, C., and Pfennig, N. (1980). Capacity of Chromatiaceae for chemotrophic growth. Specific respiration rates of *Thiocytis violaceae* and *Chromatium vinosum*. *Arch. Microbiol.*, **127**, 125–35.

Kaplan, I.R. (1956). Evidence of microbiological activity in some of the geothermal regions of New Zealand. *N.Z. J. Sci. Technol.*, **37**, 639–48.

Knobloch, K. (1969). Sulphide oxidation via photosynthesis in green alga. In: Muntzer, H. (Ed.) *Progress in Photosynthesis Research*, Vol. II, International Union of Biological Sciences, Tubingen, pp. 1032–4.

Knoll, A.H. (1979). Archaean photoautotrophy: some alternatives and limits. *Origins of Life*, **9**, 313–27.

Kompantseva, E.I. (1985). New halophilic purple bacteria *Rhodobacter euryhalinus* sp.nov. *Mikrobiologia*, **55**, 974–82 (in Russian).

Krumbein, W.E. (Ed.) (1983). *Microbial Geochemistry*, Blackwell Scientific Publications, Oxford, 330 pp.

Krumbein, W.E., and Cohen, Y. (1974). Klastische und evaporitische Sedimentation in einem mesothermen monomiktischen ufernahen See (Golf von Aqaba, Sinai). *Geol. Rundschau*, **63**, 1035–65.

Krumbein, W.E., and Cohen, Y. (1977). Primary production, mat formation and lithification contribution of oxygenic and facultative anoxygenic cyanobacteria. In: Flugel, E. (Ed.) *Fossil Algae*, Springer-Verlag, New York, pp. 37–56.

Krumbein, W.E., Cohen, Y., and Shilo, M. (1977). Solar Lake (Sinai). 4. Stromatolitic cyanobacterial mats. *Limnol. Oceanogr.*, **22**, 635–56.

Krumbein, W.E., Buchholz, H., Franke, P., Giani, D., Giele, C., and Wonneberger, C. (1979). O_2 and H_2S coexistence in stromatolites. A model for the origin of mineralogical lamination in stromatolites and banded iron formations. *Naturwissenschaften*, **66**, 381–9.

Kuznetsov, S.I. (1955). Microorganisms of Kamchatka thermal springs. *Trudy Inst. Mikrobiologii*, **4**, 130–154 (in Russian).

Kuznetsov, S.I. (1970). *The Microflora of Lakes and its Geochemical Activity*. University of Texas Press, Austin, 424 pp.

Lauterborn, R. (1915). Sapropelische Lebewelt. *Verh. Dtsch. Naturhist. Med. Ver. Heidelberg, N.F.*, **13**, 395–481. (in German).

LeGall, J., and Postgate, J.R. (1973). The physiology of sulphate reducing bacteria. In: Rose, H.A., and Tempest, D.W. (Eds.) *Methods in Microbiology*, Vol. 3A, Academic Press, London, New York, pp. 81–133.

Lemasson, C., Tandeau de Marsac, N., and Cohen-Basier, G. (1973). Role of allophycocyanin as a light harvesting pigment in cyanobacteria. *Proc. Natl Acad. Sci. USA* **70**, 3130–3.

Logan, B.W., Davies, G.R., Read, J.F., and Cebulski, D.E. (1970). Carbonate sedimentation and environments, Shark Bay, Western Australia. *Tulsa Am. Assoc. Petrol. Geol. Mem.*, **13**, 223.

Lovley, D.R., Dwyer, D.F., and Klug, M.J. (1982). Kinetic analysis of competition between sulphate reducers and methanogens for hydrogen in sediments. *Appl. Environ. Microbiol.*, **43**, 1373–9.

Madigan, M.T. (1984). A novel photosynthetic bacterium isolated from a Yellowstone hot spring. *Science*, **225**, 313–15.

Madigan, M.T., and Brock, T.D. (1973). CO_2 fixation in photosynthetically growing *Chloroflexus aurantiacus*. *FEMS Microbiol. Lett.*, **1**, 301–4.

Madigan, M.T., and Brock, T.D. (1975). Photosynthetic sulphide oxidation by *Chloroflexus aurantiacus*, a filamentous photosynthetic gliding bacterium. *J. Bacteriol.*, **122**, 782–4.

Madigan, M.T., Petersen, S.R., and Brock, T.D. (1974). Nutritional studies on

Chloroflexus, a filamentous photosynthetic, gliding bacterium. *Arch. Microbiol.*, **100**, 97–103.

Margulis, L., Ashendorf, S., Banerjee, S., Francis, S., Giovannoni, S., Stolz, J., Barghoorn, E.S., and Chase, O. (1980). The microbial community in the layered sediments at Laguna Figueroa, Baja California, Mexico: does it have Precambrian analogues? *Precambrian Res.*, **11**, 93–123.

Marshall, K.C. (1980). Reactions of microorganisms, ions and macromolecules at interfaces. In: Ellwood, D.C., Hedger, J.N., Latham, M.J., Lineh, J.M., and Slater, J.H. (Eds.) *Contemporary Microbial Ecology*, Academic Press, London, New York, pp. 93–106.

Meeks, J.C., and Castenholz, R.W. (1971). Growth and photosynthesis in an extreme thermophile *Synechococcus lividus* (Cyanophyta). *Arch. Microbiol.*, **78**, 25–41.

Miyake, Ya. (1965). *Elements of Geochemistry*, Marazen Company Ltd, Tokyo, 326 pp.

Miyoshi, M. (1897). Studien über die Schwefelbakterien der Thermen von Yumoto bei Nikko. J. College Sci. Imp, Univ. Tokyo, X. Pt. 2. 143, Referat: *Zbl. Bakteriol. Parasitenkunde, Infektionskrankh. und. Hyg.*, **11**, Abt., B.3, 526.

van Niel, C.N. (1931). On the morphology and physiology of the purple and green sulphur bacteria. *Arch. Microbiol.*, **3**, 1–112.

Nikitina, V.N., and Gerasimenko, L.M. (1983). An unusual thermophilic form of *Mastigocladus laminosus*. *Mikrobiologia*, **52**, 477–82 (in Russian).

Nozhevnikova, A.N., and Yagodina, T.G. (1982). A thermophilic acetate-consuming methane-producing bacteria. *Mikrobiologia*, **51**, 642–7 (in Russian).

Odum, E. (1983). *Basic ecology*, CBS College Publ., Philadelphia, USA.

Oremland, R.S., and Polcin, S. (1982). Methanogenesis and sulphate reduction: competitive and noncompetitive substrates in estuarine sediments. *Appl. Environ. Microbiol.*, **44**, 1270–6.

Oren, A. and Padan, E. (1978). Induction of anaerobic photoautotrophic growth in the cyanobacteriaum *Oscillatoria limnetica*. *J. Bacteriol.*, **133**, 558–63.

Oren, A., Padan, E., and Avron, M. (1977). Quantum yields for oxygenic and anoxygenic photosynthesis in the cyanobacterium *Oscillatoria limnetica*. *Proc. Natl Acad. Sci. USA*, **73**, 2152–6.

Oren, A., Paden, E., and Malkin, S. (1979). Sulphide inhibition of photosystem 2 in cyanobacteria (blue green algae) and tobacco chloroplast. *Biophys. Acta*, **546**, 270–9.

Oren, A., and Shilo, M. (1979). Anaerobic heterotrophic dark metabolism in the cyanobacterium *Oscillatoria limnetica*: sulphur respiration and lactate fermentation. *Arch. Microbiol.*, **122**, 77–84.

Padan, E., and Cohen, Y. (1982). Anoxygenic photosynthesis. In: Carr, N.C., and Whitton, B.A. (Eds.) *The Biology of Cyanobacteria*, Blackwell Scientific, Oxford, pp. 215–35.

Pantshava, E.S., and Pchyolkina, V.V. (1969). Methane fermentation of methyl alcohol by the culture of Methanobacillus kuzneceovii. *Prikl. biochim. Mikrobiologia*, **5**, 416–20 (in Russian).

Pfennig, N. (1965). Anreicherungskulturen für rote und grüne schwefelbacterium. *Zbl. Bakter. I Abt, Suppl.*, (**1**), p.179 (in German).

Pfennig, N. (1967). Photosynthetic bacteria. *Ann. Rev. Microbiol.*, **21**, 285–324.

Pfennig, N. (1975). Phototrophic bacteria and their role in the sulphur cycle. *Plant Soil*, **43**, 1–16.

Pfennig, N. (1977). Phototrophic green and purple bacteria: a comparative systematic survey. *Ann. Rev. Microbiol.*, **31**, 275–90.

Pfennig, N. (1978). General physiology and ecology of photosynthetic bacteria. In: Clayton, R.K., and Sistron, W.R. (Eds.) *The Photosynthetic Bacteria*, Plenum Press, New York, London, pp. 3–18.

Pfennig, N., and Biebl, H. (1976). *Desulfuromonas acetooxidans* gen.nov. and sp.nov, a new anaerobic sulphur-reducing, acetate-oxidising bacterium. *Arch. Microbiol.*, **110**, 3–12.

Pfennig, N., Widdel, F., and Truper, H.G. (1981). The dissimilatory sulphate-reducing bacteria. In: Starr, M.P., Stop, H., Truper, H.G., Balows, A., and Schlegel, H.C. (Eds.) *The Prokaryotes*, Springer, Berlin, pp. 926–40.

Pierson, B.K., and Castenholz, R.W. (1974). A phototrophic gliding filamentous bacterium of hot springs. *Chloroflexus aurantiacus* gen. and sp.nov. *Arch. Microbiol.*, **100**, 5–24.

Playford, P.E., and Cockbain, A.E. (1969). Algal stromatolites: deep water forms in the Devonian of Western Australia. *Science*, **1965**, 1008–10.

Raymond, J.C., and Sistrom, W.R. (1969). *Ectothiophodospira halophila*. A new species of the genus *Ectothiorhodospira*. *Arch. Microbiol.*, **69**, 121–6.

Revsbech, N.P., and Ward, D.M. (1984). Microelectrode studies of interstitial water chemistry and photosynthetic activity in a hot spring microbial mat. *Appl. Environ. Microbiol.*, **48**, 270–5.

Revsbech, N.P., Jørgensen, B.B., Blackborn, T.H., and Cohen, Y. (1983). Microelectrode studies of the photosynthetic and O_2, and H_2S and pH profiles of a microbial mat. *Limnol. Oceanogr.*, **28**, 1062–74.

Sandbeck, K.A., and Ward, D.M. (1981). Fate of immediate methane precursors in low-sulphate hot spring algal-bacterial mats. *Appl. Environ. Microbiol.*, **41**, 775–82.

Schink, B. and Zeikus, J.G. (1983). *Clostridium thermosulfurogenes* sp.nov., a new thermophile that produces elemental sulphur from thiosulphate. *J. Gen. Microbiol.*, **129**, 1149–58.

Schopf, W.H. (Ed.) (1983). *Earth's Earliest Atmosphere, its Origin and Evolution*, Princeton University Press, Princeton, NJ., 543 pp.

Schwabe, G.H. (1960). Uber den thermobioten Kosmopoliten *Mastigocladus laminosus*. Cohn. Blaugen und lebensraum. V. Schweeig. *J. Hidrol.*, **22**, 757–92.

Skyring, F.W. (1984). Sulphate reduction in marine sediments associated with cyanobacterial mats, Australia. In: Cohen, Y., Castenholz, R.W., and Halvorson, H.O. (Eds.) *Microbial Mats: Stromatolites*, Alan R. Liss, New York, pp. 265–76.

Stanier, R.Y. (1974). The origin of photosynthesis in eukaryotes. *Soc. Gen. Microbiol. Symp.*, **24**, 219–40.

Stanier, R.Y. (1977). The position of the cyanobacteria in the world of phototrophs. *Carlsberg Res. Commun.*, **42**, 77–98.

Stetter, K.O. (1985). Extremthermophile Bacterien. *Naturwissensch.*, **72**, 281–301.

Steward, W.D.P., and Pearson, W.H. (1970). Effects of aerobic and anaerobic conditions on growth and metabolism of blue-green algae. *Proc. R. Soc. London, Ser. B*, **175**, 293–311.

Stolz, J.F. (1984). Fine structure of the stratified microbial community at Laguna Figueroa, Baja California, Mexico. II. Transmission electron microscopy as a diagnostic tool in studying microbial communities *in situ*. In: Cohen, Y., Castenholz, R.W., and Halvorson, H.O. (Eds.) *Microbial Mats: Stromatolites*, Alan R. Liss, New York, pp. 23–38.

Streczwski, B. (1913). Beitrag zur Kenntnis der Schwefelflora in der Umgebung von Krakau. *Bull. Acad. Sci. Cracovie, Ser. B*, pp. 309–12.

Tel-Or, E., Luijk, C.W., and Packer, L. (1977). As inductible hydrogenase in cyanobacteria enhances NH_2 fixation. *PEBS Lett.*, **78**, 49–52.

Thauer, R.K., Jungermann, K., and Decker, K. (1977). Energy conservation in chemotrophic anaerobic bacteria. *Bact. Rev.*, **41**, 100–80.
Trüper, H.G. 1973. The present state of knowledge of sulphur metabolism in phototrophic bacteria. *Abstr. Symp. Prokyotic. Photosyn. Organisms, Freiburg*, pp. 160–6.
Utkilen, H.C. (1976). Thiosulphate as electron donor in the blue-green alga *Anacystis nidulance*. *J. Gen. Microbiol.*, **95**, 177–80.
Utkilen, H.C., and Castenholz, R.W. (1979). Physiological aspects of adaptation to sulphide in a thermophilic *Oscillatoria*. In: Nichols, J.M. (Ed.) *Abstract of Third International Symposium on Photosynthetic Prokaryotes*, Oxford, Liverpool, England, p. 12.
Van Gemerden, H. (1968). Utilisation of reducing power in growing cultures of Chromatium. *Arch. Microbiol.*, **64**, 111–17.
Walter, M.R. (Ed.) (1976). *Developments in Sedimentology 20, Stromatolites*, Elsevier, Amsterdam.
Ward, D.M. (1978). Thermophilic methanogenesis in a hot spring algal-bacterial mat (71–30°C). *Appl. Environ. Microbiol.*, **35**, 1019–26.
Ward, D.M., and Olson, G.I. (1980). Terminal processes in the anaerobic degradation of an algal-bacterial mat in a high-sulphate hot spring. *Appl. Environ. Microbiol.*, **40**, 67–74.
Ward, D.M., Beck, E., Revsbech, N.R., Sandbeck, K.A., and Winfrey, M.R. (1984). Decomposition of hot spring microbial mats. In: Cohen, Y., Castenholz, R.W., and Halvorson, H.O. (Eds.) *Microbial Mats: Stromatolites*, Alan R. Liss, New York, pp.191–214.
Weisman, J.C., and Benemann, J.R. (1977). Hydrogenase production by nitrogen starved culture of *Anabena cylindrica*. *Appl. Environ. Microbiol.*, **35**, 123–31.
Weller, D., Doemel, W., and Brock, T.D. (1975). Requirement of low oxidation–reduction potential for photosynthesis in blue green alga (*Phormidium* sp.). *Arch. Microbiol.*, **104**, 7–13.
Wiegel, J., and Ljungdahl, L.G. (1981). Thermoanaerobacter ethanolicus gen. nov., sp.nov., a new extreme thermophilic anaerobic bacterium. *Arch. Microbiol.*, **128**, 343–8.
Wolfe, R.S., and Pfennig, N. (1977). Reduction of sulphur by *Spirillium 5175* and syntrophism with Chlorobium. *Appl. Environ. Microbiol.*, **33**, 427–33.
Zavarzin, G.A. (1984). *Bacteria and the Atmosphere Composition*, Nauka, Moscow, 192pp. (in Russian).
Zeikus, J.G., Hegge, P.W. and Anderson, P.W. (1979). *Thermoanaerobium brockii* gen.nov. and sp.nov., a new chemoorganotrophic caldoactive anaerobic bacterium. *Arch. Microbiol.*, **122**, 41–8.
Zeikus, J.G., Ben-Bassat, A., and Hegge, P.W. (1980). Microbiology of methanogenesis in thermal volcanic environments. *J. Bacteriol.*, **143**, 432–40.
Zeikus, J.G., Dawson, M.A., Thompson, T.E., Ingoorsen, K., and Hatchikian, E. (1983). Microbial ecology of volcanic sulfidogenesis: isolation and characterisation of *Thermodesulfobacterium commune* gen.nov. and sp.nov. *J. Gen. Microbiol.*, **129**, 1159–69.
Zhilina, T.N., Chudina, V.I., Ilarionov, S.A., and Bonch-Osmolovskaya, E.A. (1983). Thermophilic methane-producing bacteria from *Methanobacillus kuzneceovii* methylotrophic associations. *Mikrobiologia*, **52**, 328–34 (in Russian).
Zillig, W., Stetter, K.O., Prangishvilli, D., Schäfer, W., Wunderl, S., Janeovic, D., Holz, I., and Palm, P. (1982). Desulfurococcaceae, the second family of the extremely thermophilic anaerobic sulphur-respiring Thermoproteales. *Zbl. Bakt. Hyg.*, Abt I Orig. C3, pp. 304–17.

Zillig, W., Gierl, A., Schreiber, G., Wunderl, S., Jankovic, D., Stetter, K., and Klenk, H.P. (1983). The archaebacterium *Thermophilum pendensreprenus*, a novel genus of the thermophilic anaerobic sulphur-respiring Thermoproteales. *System. Appl. Microbiol.*, **4**, 79–87.

Index

acetate oxidation 158
alkaline lakes 192
anaerobic diagenesis 145–149, 162
anhydrite 5, 57, 59
Antarctic ice cores 65, 94–105
Aphanotheca halophilica 195–196
Archaean–Proterozoic boundary 11
Archaean atmosphere 9
Archaean oceans 8–9, 13
Atlantic Ocean 161, 165

bacteria
  green sulphur 215
  purple non-sulphur 215
  purple sulphur 210, 215
  sulphur oxidizing 9
Baltic Sea 112, 160
Bathymodiolus thermophilus 188
Bering Sea 160, 168
Black Sea 23, 112, 160, 168
box model 38–40

California, Gulf of 160–161
Calyptogena magnifica 182
carbon
  dissolved organic (DOC) 129
  particulate organic (POC) 129
carbon dioxide, volcanic 67–68
carbon dioxide assimilation 198
carbon disulphide 78, 80, 82–85, 108, 112, 115
carbon in sediments 145–149
carbon isotopes 33–34, 69–72, 159–161, 166, 203
carbon reservoir 42
carbon-14 128–129, 144
carbonyl sulphide 78, 80, 82–85, 108, 112, 115
charge balance in ice cores 95
Chlorobiaceae 9
Chlorobium sp. 193, 211

Chloroflexis sp. 193, 197, 207–209, 211, 216–218, 220, 226–229
chlorophylls, bacterio- 192
Chlostridium sp. 223, 226, 228
Chromatiaceae 9
Chromatium sp. 193–194, 211, 216
Chromatius sp. 210
cyanobacterial mats 127–129, 136–137, 139–142, 191–204

Desulfomonas sp. 3
Desulfotomaculum sp. 3
Desulfovibrio sp. 3
Desulfuroccus sp. 214
dimethylsulphide 78, 80, 82–85, 108, 115
  anthropogenic effects 91–2
  photo-oxidation 91
dimethylsulphoxide 115

Ectothiorhodospira sp. 211, 217
Enteromorpha sp. 128
erosion constant 43, 53
evaporite basins 57
evaporites 5, 57

fatty acids 125, 159

Greenland ice cores 65, 94–105, 110
gypsum 5, 42, 57, 59, 184

historical analyses
  precipitation 93–94
  soils and sediments 105–107, 116
  surface water 94, 116
Holocene sediments 23
hot springs 192, 205–229
  composition 205–207
hydrogen sulphide 78, 80, 82–85, 181, 184

239

hydrothermal systems 15–17, 26, 35, 43, 65–66, 181–188
hydroxyl radical 9–10, 91
hypersaline environments 192

Indian Ocean 165
iron dependent photosynthesis 202
iron hydroxides 182
iron sulphur interactions 204
iron (III) reduction 145, 158

kinetic isotope effect 4

Lamprocystis sp. 194
lead-210 106
leaks in sulphur cycle 25
Lynogbia 7104 195

Macromonas sp. 213, 218
manganese hydroxides 182
manganese (IV) reduction 145, 158
mass extinctions 73
Mastigocladus luminosus 207–208, 217
methane
  dissolved 167
  formation 159, 163, 168, 224, 228
  oxidation 127, 149–158, 168
methane sulphonic acid 115
Methanobacillus kuzneceovii 226
Methanobacterium thermoautotrophicum 226
Methanosarcina sp. 226
Microcoleus sp. 128, 136, 197, 200, 203

nitrogen oxides 91

Oscillatoria amphigranulata 197, 200, 208
Oscillatoria limnetica 129, 193–196, 199–200, 207, 216
Oscillatoria okenii 207
Oscillochloris trichoides 209, 211, 216
oxygen consumption 144
oxygen in ancient atmosphere 9, 41
oxygen in mats 197–198
oxygen isotopes 33
oxygen toxicity 202

Pacific Ocean 161, 165
pH of ice cores 95–98

Phodopseudomonas sp. 211
Phormidium sp. 207–208, 217–218
photosynthesis 9–10, 192–197
  iron dependent 202–203
  pigments 192
photosystem II inhibition 199
planktonic ecosystems 131, 138
primary productivity 127–130, 132–133, 143–145
Prosthecochloris sp. 193, 211–212
Pseudomonas mendocina 213
pyrite 4–5, 11, 23, 135

Rhodobacter sp. 211, 216–217
Rhodocyclus sp. 211, 216
Rhodopseudomonas viridis 216
Rhodospirilium mediosalinum 211, 216
Riftia pachyptila 182
Röt event 37, 58–59

salt marsh ecosystems 130
sea-level 35, 69–72
sedimentary rocks, mass of 67
shales 23, 26
smog 111
soil erosion 90
Spartina sp. 130, 135, 137
Spirulina sp. 207
steady state 25
stromatolites 139, 191–192, 204
strontium isotopes 34–35, 69
sulphate
  aeolian 78–79, 82–85, 89–90
  ice cores 94–105
  marine flux 78, 80, 82–85
  oceanic reservoir 8, 15, 23, 26, 51
  present inventory 29
  reduction 3, 5, 11, 15, 23, 125–127, 132–143, 201–204
  sedimentary 7, 26, 46, 49, 51
Sulphate in Archaen oceans 8, 13
sulphate in evaporites 24, 57–62
sulphide
  Precambrian 11
  sedimentary 7, 26, 47, 51
  toxicity 193
sulphite, disproportionation 9
sulphur
  anthropogenic 77–78, 80, 82–83, 110

Index

current global budget 80, 82–83
cycle box model 38–40
cycle events 36–38, 72–73
cycle leaks 25
deposition in sediments 24
exogenic cycle 22, 27, 32–34
fossil fuels 85–86
from fertilizers 79
in rivers 25, 87–89, 112
industrial emissions 83–87
isotope age record 31–32, 52, 62–64
isotope fractionation 12, 16–17, 27
isotope geochemistry 29–31
isotopes 6–7, 11, 14, 49–50, 57–64, 69–72, 184
lake sediments 92–93
metabolism, prokaryotic 14
metallurgical source 85
oxidizing bacteria 9
primordial 9
Quaternary 93
reservoir 48
volcanic 65, 73, 78–79, 82–85
sulphur/carbon
  interactions 41, 69, 143–158
  ratio 23, 32–34
sulphur-35 135–136

sulphuric acid production 86
Synechococcus sp. 200, 207–208, 217–218

Thermoanaerobacter ethanolius 223
Thermoanaerobium brockii 223
Thermobacterioides acetoethylicus 223
Thermodesulfobacterium commune 213
Thermofilum sp. 214
Thermoproteus sp. 214
Thermothrix sp. 212–214, 218, 226–227, 229
Thiobacillus thermophila 212
Thiobacterium bovista 213, 218
Thiocapsa sp. 193, 211
Thiocystis sp. 193
Thiospirillum sp. 212, 216
transport scales 108–109

volcanic eruptions, evidence in ice cores 98–105
volcanic rocks, mass of 67–68

Zechstein 61–62
Zostera sp. 131, 137